TRENDS IN STEM CELL RESEARCH

TRENDS IN STEM CELL RESEARCH

ERIK V. GREER
EDITOR

Nova Biomedical Books
New York

For permission to use material from this book please contact us:
Telephone 631-231-7269; Fax 631-231-8175
Web Site: http://www.novapublishers.com

Library of Congress Cataloging-in-Publication Data

Trends in stem cell research / Erik V. Greer, editor.
p. ; cm.
Includes bibliographical references and index.
ISBN 1-59454-315-1 (hardcover)
1. Stem cells.
[DNLM: 1. Stem Cells. 2. Biomedical Research--trends. QH 588.S3 T794 2005] I. Greer, Erik V.
QH588.S83T745 2005
616'.02774--dc22 2005005609

Published by Nova Science Publishers, Inc. ✦ New York

Contents

Preface

Among the many applications of stem cell research are nervous system diseases, diabetes, heart disease, autoimmune diseases as well as Parkinson's disease, end-stage kidney disease, liver failure, cancer, spinal cord injury, multiple sclerosis, Parkinson's disease, and Alzheimer's disease. Stem cells are self-renewing, unspecialized cells that can give rise to multiple types all of specialized cells of the body. Stem cell research also involves complex ethical and legal considerations since they involve adult, fetal tissue and embryonic sources. This new book brings together leading research from throughout the world in this frontier field.

In Chapter I, the author describes that the hematopoietic system serves as a useful model for studies of stem cell biology and cell fate determination. Most mature blood cells have a limited life span and therefore must be continuously replaced. This is achieved by pluripotential hematopoietic stem cells, which have the capacity to both self-renew and undergo sequential commitment to all of the distinct lineages of blood. Previously, it had been assumed that progressive restriction of lineage potential represented irreversible steps. In this chapter, the author will discuss recent studies that provide evidence for potential lineage plasticity during early stages of hematopoietic development. In addition, I will review experiments demonstrating functional cross antagonism of opposing lineage-specific transcription factors. These latter studies suggest that alternate lineage gene programs remain held in an active state of repression, providing a possible mechanistic understanding of the observed lineage plasticity. Taken together, these findings indicate that a window of lineage instability exists during early hematopoietic development in which the final cell identity rests on the balance of forces dictated by alternate lineage-specific transcription factors and their cofactors.

As discussed in chapter II retrovirus vectors suffer severe gene silencing in stem cells that hampers their use in stem cell gene therapy. Transduced retrovirus vector can be completely silenced, or dynamically silenced in a subset of cells resulting in variegation. Expressing retrovirus also can be extinguished during stem cell differentiation. Multiple epigenetic modifications including histone modifications and DNA methylation are functionally involved in establishing and maintaining retrovirus silencing via the formation of silent chromatin. A model is proposed that the balance of epigenetic effects at the integration site and retroviral silencer elements on the vectors determines the decision to be silent, variegated or expressing in stem cells. Silencing resistant retrovirus vectors have been developed by removal of these silencer elements, and silencing in stem cells may ultimately be blocked using appropriate insulator elements.

Stem cell research has raised hopes of addressing the difficulties inherent in whole organ transplantation: the limited availability of donor tissue and the side effects of long-term immunosuppression, as discussed in chapter III. Exciting work has been performed within the last several years, providing evidence that tissue-specific stem cells capable of self-renewal and differentiation are present in some adult organs, enabling life-long organ regeneration. Moreover, plasticity of adult stem cells has been described. This, in conjunction with described cell fusion phenomena challenges traditional concepts of cell biology. The underlying mechanisms of adult stem cell growth and differentiation are still unclear. However, there is increasing evidence that these processes are greatly influenced by the surrounding milieu. The development of culture systems offering a complex, tissue-specific microenvironment for the cells could greatly enhance further *in vitro* studies. In chapter III, the authors review recent findings on adult stem cell research with a focus on cell biological and technological strategies for the possible clinical use of such cells. Three-dimensional four-compartment capillary membrane bioreactors with integral oxygenation and decentralized mass exchange support high-density cell co-culture and tissue reorganization under *in vivo*-like conditions. Initial results on liver progenitor cell culture in these bioreactors illustrate the potential of innovative culture technology in the field of stem cell-based regenerative medicine.

Chapter IV reviews that connective tissue growth factor/hypertrophic chondrocyte-specific gene product 24 (CTGF/Hcs24/CCN2) is the second member of the CCN family of proteins and a multifunctional growth factor for fibroblasts, chondrocytes, osteoblasts and vascular endothelial cells. Depending on the type of cell with which it interacts, it promotes chomotaxis, migration, adhesion, proliferation, differentiation and/or extracellular matrix formation. Because the gene expression of CCN2/CTGF is very high in hypertrophic chondrocytes in the physiological state, a major physiological role for this factor has been suggested to be the promotion of endochondral ossification. However, its expression was also recently detected in various types of chondrocytes, osteoblasts and endothelial cells during embryonic development. The expression was also found to be up regulated in undifferentiated mesenchymal cells, osteoblasts, chondrocytes and endothelial cells during wound healing of bone, cartilage or skin, indicating the involvement of this factor in tissue regeneration. In fact, since CCN2/CTGF enhanced the proliferation and phenotypic maintenance of articular chondrocytes *in vitro*, without causing them to enter the hypertrophic state, recombinant CCN2/CTGF was administered *in vivo* to both rats of an experimental osteoarthritis model and to those with full-thickness defects in their articular cartilage. The results showed the factor to be therapeutically useful for the regeneration and reconstruction of damaged cartilaginous tissues. Also, CCN2/CTGF enhanced the phenotypic expression of cartilage by mouse bone marrow-derived stromal cells, and induced chondrogenesis *in vitro*, indicating that CCN2/CTGF promotes at least chondrogenic differentiation of mesenchymal stem cells. Furthermore, CCN2/CTGF is known to be involved in fibrotic disorders, angiogenic diseases, and malignancies, and these pathological states may be due to uncontrolled overexpression of this growth factor. This review focuses on the role of CCN2/CTGF as a regeneration factor (regenerin) acting on various types of cells including mesenchymal stem cells.

Chapter V describes peripheral blood progenitor cells (PBPC) increasingly used as a source of stem cells for either autologous or allogeneic hematopoietic transplantation. Technical problems are derived from the size of the patient or donor in the pediatric setting. Thereby, vascular access, anticoagulation and extracorporeal volume need special consideration in children. Several groups have published over the last years their experience in mobilization and transplantation in small children. In this paper our own experience and that of other groups is reviewed. Mobilization with cytokines alone (in donors and patients) or in combination with chemotherapy (patients) has been reported to be efficient and well tolerated by these children. Several variables are related to the collection in donors and patients. This issue is discussed in this paper based on published studies. PBPC collection is a safe procedure as reported in the literature, when performed by experienced apheresis teams.

As discussed in chapter VI the neural crest (NC) is a multipotent cell population generated from the tip of the neural fold. NC cells differentiate into neurons, glial cells, osteoblasts, odontoblasts and melanocytes, and participate in organogenesis of the cranio-face, parathyroid, heart and thymus. The thymus is a vertebrate-specific organ related to the adaptive immune system. Since the failure of NC cells results in immunodeficiency, NC cells should be concerned with thymic organogenesis. The differentiation of NC cells has been thought to be regulated by environmental factors during the migration of these cells into peripheral tissues, and in accord with this, multipotent NC-derived cells can rarely be detected in the peripheral tissues. Recently, we found that NC-derived cells with the potential to differentiate into neurons, glial cells and melanocytes are present in the developing thymus, and that these cells are rarely detected at late embryonic stages. Multipotent NC-derived cells were also found to be present in the developing teeth. These findings suggest that multipotent NC-derived cells are present in peripheral tissues and may participate in organogenesis. We discuss here whether NC cells may contribute to the organogenesis of lymphoid organs other than thymus, and of the central nervous system.

The spermatogenic lineage starts with the formation of primordial germ cells (PGCs) that migrate to the genital ridges, as reviewed in chapter VII. There they become enclosed by Sertoli cell precursors in seminiferous cords and differentiate into gonocytes. After birth the gonocytes give rise to adult type spermatogonial stem cells and various types of spermatogonia. Spermatogonial stem cells are at the beginning of the adult spermatogenic process. In non-primate mammals this are single cells (A_s spermatogonia). Their daughter cells either migrate away from each other and become two new stem cells or stay together, connected by an intercellular bridge. The latter designates these cells to differentiate and ultimately become spermatozoa.

Spermatogonial stem cells can be counted by morphologically distinguishing these cells in whole mounts of seminiferous tubules or by way of a functional assay, the spermatogonial stem cell transplantation technique. To work with spermatogonial stem cells it would be nice if one could purify and/or culture these cells. Fortunately, the search for specific markers that would facilitate the purification of these cells is progressing rapidly. Furthermore, pure spermatogonial stem cells proved difficult to culture, better results have been obtained with co-cultures of Sertoli cells with mouse and bovine spermatogonia. Nevertheless no culture protocol through which one can study the regulation of proliferation and differentiation of these cells is yet available. Infertile mutant and genetically modified mice have indicated the presence of regulatory mechanisms governing the proliferation, differentiation, migration and survival of PGCs, gonocytes and adult spermatogonial stem cells. For example, a major factor that plays a role in regulating stem cell fate is glial cell line derived neurotrophic factor (GDNF) produced by Sertoli cells which inhibits stem cell differentiation. Furthermore, the *Plzf* gene encoded protein, expressed in spermatogonial stem cells, also inhibits their differentiation. Finally, spermatogonial stem cells are mainly localized to those tubule areas that border on the interstitial tissue. Apparently, also factor(s) from the interstitium, possibly testosterone or FSH, inhibit stem cell differentiation. Recent progress in understanding the regulation of spermatogonial stem cell behavior is discussed. Finally, spermatogonial multiplication and stem cell renewal in primates is discussed which differs from that in other mammals.

The author of chapter VIII summarizes the available evidence suggesting that humoral or cellular immune mechanisms may damage the bone marrow hematopoietic progenitor cell compartment and may also affect the bone marrow stromal cell hematopoiesis supporting capacity in patients with systemic lupus erythematosus and rheumatoid arthritis. Since there is currently much interest in exploring the role of hematopoietic stem cell transplantation as an alternative treatment approach in patients with systemic autoimmune diseases refractory to conventional medication including systemic lupus erythematosus and rheumatoid arthritis, data on patients' bone marrow function are of particular importance. An injured bone marrow may potentially impact on harvesting of stem cells and their engraftment potential.

The aim of chapter IX is to review the role of political ideology in health care decisions which has become increasingly important in the past decade. As science pushes the boundaries of pain relief and life replication (cloning and stem cells) issues it confronts political, ideological and ethical issues that may have detrimental impacts upon patient care, the relief of pain and cures for diseases. For this study, we examine that portion of political ideology which effects one's beliefs about the origins and preservation of life and its effect upon health care policies, specifically patient pain care and basic research regarding stem cells. We hypothesize that ideological belief systems (based partly upon religious beliefs) can have a detrimental effect upon future patient care. While we are aware that it is difficult to distinguish between political and ideological policy entrepreneurship, we also agree with Colin Campbell (2004) that ideological entrepreneurship involves a depth of feeling that makes compromise difficult. It also involves interests tied to a president's or congressperson's political base which are particularly persistent in their policy preferences. This study examines a number of variables associated with these two areas. First we discuss the new emerging model of patient health care from the perspective of pain management practitioners. Second, we apply the model to two case studies involving political decisions. One decision is the congressionally based Hyde – Nichols bill with implications for patient pain control and the other is President Bush's decision concerning stem cells and its effect upon basic research and ultimately patient care. Third, we derive some lessons from both experiences that might help manage some of the problems concerning patient pain care and basic research in the stem cell area.

In: Trends in Stem Cell Research
Editor: Erik V. Greer, pp. 1-26
ISBN 1-59454-315-1

Chapter I

Hematopoietic Stem Cell Differentiation: A Question of Commitment

*Alan B. Cantor**
Division of Hematology/Oncology, Department of Pediatrics, Children's Hospital Boston and Dana Farber Cancer Institute, Harvard Medical School,
Boston MA, USA

Abstract

The hematopoietic system serves as a useful model for studies of stem cell biology and cell fate determination. Most mature blood cells have a limited life span and therefore must be continuously replaced. This is achieved by pluripotential hematopoietic stem cells, which have the capacity to both self-renew and undergo sequential commitment to all of the distinct lineages of blood. Previously, it had been assumed that progressive restriction of lineage potential represented irreversible steps. In this chapter, I will discuss recent studies that provide evidence for potential lineage plasticity during early stages of hematopoietic development. In addition, I will review experiments demonstrating functional cross antagonism of opposing lineage-specific transcription factors. These latter studies suggest that alternate lineage gene programs remain held in an active state of repression, providing a possible mechanistic understanding of the observed lineage plasticity. Taken together, these findings indicate that a window of lineage instability exists during early hematopoietic development in which the final cell identity rests on the balance of forces dictated by alternate lineage-specific transcription factors and their cofactors.

* For correspondences: Tel: 617-919-2015; Fax 617-730-0222; e-mail: alan.cantor@childrens.harvard.edu

Introduction

A universal theme in developmental biology is the derivation of cells of two or more distinct phenotypes from a common multipotential precursor cell. How these cell fates are determined at the molecular level remains incompletely understood. In addition, the degree to which these decisions are irreversible has been the subject of considerable controversy recently. In this chapter, I will discuss experiments within the hematopoietic system that indicate potential plasticity of cell fate determination during early stages of commitment, and potential molecular mechanisms involving lineage-specific transcription factors (for additional reviews on this topic, see [1,2]).

The hematopoietic system has long proven to be a useful model for examining questions of cell fate determination. Most mature blood cells have a limited lifespan and therefore must be continuously replaced. This process relies on the presence of pluripotential hematopoietic stem cells, which have the capacity to both self-renew and undergo sequential commitment to all of the distinct elements of mature blood. In humans, this represents at least nine different cell types. Use of *in vitro* colony assays, animal transplantation studies, and immunophenotype marker analysis has made it possible to identify characteristic intermediate progenitor populations that under normal conditions exhibit restricted cell fate potential [3,4] This hierarchy is illustrated in Figure 1.

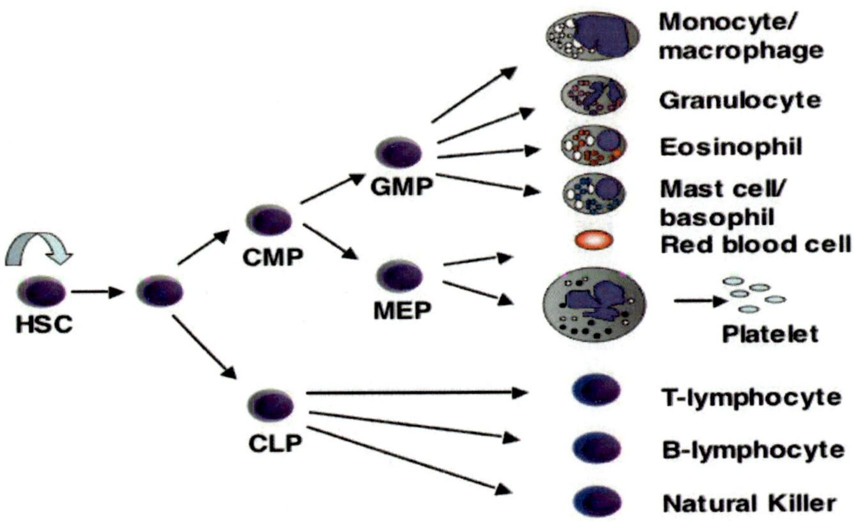

Figure 1. Hierarchy of blood cell development. Gradual restriction of differentiation potential occurs as the cells progress from the left to the right side of the figure. Immunophenotypically defined multipotent progenitor intermediate cells are indicated, as per[3]. HSC, hematopoietic stem cell; CMP, common myeloid progenitor; CLP, common lymphoid progenitor; GMP, granulocyte/macrophage progenitor; MEP, megakaryocyte/erythroid progenitor. The curved arrow above HSCs indicates self-renewal function.

It is also worth noting that hematopoiesis in vertebrates is an extremely active process. It has been estimated that an adult human produces almost one trillion new blood cells a day (5). This high flux is evident from the rapid onset of peripheral blood cytopenias following bone marrow insults.

Lineage Commitment

The classical view of hematopoietic development holds that once multipotent progenitor cells commit to a particular lineage, the fate of that cell is irreversibly fixed. However, experiments over the past fifteen years have challenged this model, and instead suggest that there may be considerable plasticity of "committed" multipotent progenitor cells during at least early stages of hematopoietic differentiation. Initial studies made use of chemical treatment, viral expression of oncogenes, or manipulation of established transformed cell lines. As provocative as these studies were, the possibility existed that the transformed (or chemically-altered) state of the cell unmasked an artificial plasticity potential. More recent experiments using primary, immunopurified cell populations, minimal cell culturing, and/or whole animal systems, have confirmed these earlier observations, suggesting that cell fate instability during early lineage commitment may be physiologically relevant (Table 1).

Lineage Conversion Using Chemically-Treated or Transformed Cell Lines

Some of the earliest evidence for developmental plasticity during hematopoietic differentiation comes from the work of Boyd and Schrader who showed in 1982 that treatment of a pre-B cell lymphoma cell line with the DNA demethylating agent 5-azacytidine resulted in their conversion to macrophage-like cells [6]. Similar findings were reported by Adams and colleagues using oncogene transformation of pre-B cells [7].

In 1995, Graf and his colleagues showed that enforced expression of GATA-1, a zinc finger transcription factor normally restricted to erythroid, megakaryocytic, eosinophil and mast cell lineages, converts myb-ets transformed avian myelomonocytic cells (MEPs) to either eosinophils, erythroblasts or thromboblasts (equivalent of mammalian megakaryoblasts) [8]. The different outputs depend on the expression level of GATA-1, and/or the presence of the ets transcription factor. Similar erythroid/megakaryocytic lineage conversion has been reported utilizing enforced GATA-1 expression in several murine myeloid transformed cell lines including 416B, M1, and FDCW2 [9-11]. Not only did these studies demonstrate phenotypic plasticity of hematopoieitc cells, they provided evidence for the instructive role of key transcription factors in hematopoietic cell fate determination.

Table 1. Examples of lineage conversion of hematopoietic cells

Starting Lineage Phenotype	Ending Lineage Phenotype	Manipulation	Reference
Transformed Cell Systems			
Pre-B cell lymphoma Cell line	Macrophages	5-azacytidine treatment	6
Pre-B cells from Emu-myc transgenic mice	Macrophages	Retroviral expression of v-raf	7
Myb-ets transformed avian myelo-monocytic cells (MEPs)	Eosinophils, thromboblasts,* erythroblasts	GATA-1 enforced expresssion	8
416B (early myeloid) cell line	Megakaryocytic	GATA-1 or GATA-2 enforced expression	9
M1 (myeloid) cell line	Erythroid, megakaryocytic	GATA-1 enforced expression	10
FDCW2 (myeloid) cell line	Erythroid	GATA-1 enforced expression	11
Primary Cell Systems			
Lymphoid committed progenitors (CLPs), pro-T cells	Macrophage/granulocyte	IL-2 receptor signaling	12
Lymphoid committed progenitors (CLPs), pro-T cells	Macrophage/granulocyte	GM-CSF receptor signaling	13
Granulocyte/monocyte progenitors	Erythroid, eosinophilic, basophil-like	Enforced expression of GATA-1/ER	14
Lymphoid (CLPs) and myelomonocytic progenitors (GMPs)	Erythroid, megakaryocytic	Enforced expression of GATA-1	15
Committed mast cell progenitors	Erythroid, megakaryocytic	Enforced expression of FOG-1	(Cantor AB, Orkin SH, unpublished observation)

* Equivalent of megakaryoblasts in mammals.

Lineage Conversion of Primary Hematopoietic Progenitor Cells

Over the past three years, several groups have re-examined the question of hematopoietic lineage plasticity using primary committed progenitor cells. Weissman and his colleagues reconstituted an active IL-2 receptor in prospectively isolated common lymphoid progenitors (CLPs), which normally lack IL-2 signaling [12]. Unmanipulated CLPS give rise exclusively to B and T lymphocytes, and natural killer cells. However, when the transduced cells were cultured in the presence of IL-2, they instead generated macrophages and granulocytes. This was true of more committed pro-T cells as well. Clonal analysis by limited dilution excluded the possibility of selection of contaminating multipotent precursor cells. Furthermore, about 5% of Pro-T cell derived granulocyte-macrophage colonies exhibited genomic rearrangements of their T-cell receptor locus, suggesting that the cells had progressed to at least the DN2 stage of T cell maturation before converting to a granulocyte-macrophage phenotype. Ectopic GM-CSF receptor signaling has also been reported to convert CLPs and pro-T cells into myelomonocytic cells [13].

Enver and his colleagues ectopically expressed GATA-1 in immunopurified bipotent neutrophil/monocytes precursor cells using a tamoxifen-activatable GATA-1:estrogen receptor domain fusion molecule [14]. Activation of GATA-1 resulted in lineage conversion to cells with features of erythrocytes, eosinophils, and/or basophils. Importantly, clone-marking and daughter cell experiments ruled out the possibility that this was due to selection of a minor contaminating pluripotent precursor cell.

Akashi and his colleagues have extended these observations this past year. They showed that ectopic expression of GATA-1 not only reprograms committed granulocyte/monocytes progenitors, but also more distantly related CLPs, into erythroid and megakaryocytic cells [15]. Competitive repopulation studies in mice using marked cells demonstrated lineage conversion of common lymphoid progenitors to erythroid cells in the context of whole animals.

A final example comes from studies of Friend of GATA (FOG) proteins, a family of large multitype zinc finger containing polypeptides that act as essential cofactors for GATA transcription factors in certain tissues [16-19]. Within the hematopoietic system, FOG-1 is required for the function of GATA-1 in erythroid and megakaryocytic development, and GATA-2 in early megakaryopoiesis [19-21]. However, FOG proteins are not expressed in eosinophils or mast cells, where GATA-1 and GATA-2 also play functional roles [22, 23 24] and (Cantor AB, Iwasaki H, Orkin SH, and Akashi A, unpublished observation). This implies that GATA factors operate in a FOG-independent manner in these lineages. In fact, FOG proteins appear to antagonize GATA factors in eosinophil and mast cell development since enforced expression of FOG-1 in myeloid multipotent progenitor cells blocks their ability to generate eosinophils or mast cells [22] and (Cantor AB, Iwasaki H, Orkin SH, and Akashi K, unpublished observation). Ectopic expression of FOG-1 in maturing mast cells results in down regulation of the high affinity IgE receptor, a marker of mature mast cells, and acquisition of erythroid and megakaryocyte features in a subpopulation of cells (Cantor AB, Orkin SH, unpublished observation). Similar observations using clonal primary mast cells indicate direct lineage conversion. Together, these studies reveal the potential for early hematopoietic progenitor cells, which have seemingly initiated a selective differentiation program, to be redirected to an alternate fate.

Lineage Conversion Pathways

If the finding of lineage plasticity during early hematopoiesis is true, it raises several important questions. First, what pathways do the cells take during this developmental switch? One possibility is that the cells undergo "transdifferentiation", in which activation of one lineage-specific gene program is directly replaced by an alternate program (Figure 2A). An alternate possibility is that the cells first "dedifferentiate" into a more primitive (multipotent) state, and then "redifferentiate" along an alternate pathway (Figure 2B). The work of Nerlov and his colleagues provides some evidence for the latter [22]. In their experiments, it was shown that down regulation of FOG-1 expression is a pre-requisite for multipotent myb-ets transformed myelomonocytic avian cells to differentiate along the eosinophilic pathway. When FOG-1 was ectopically expressed in the eosinophilic cell line 1A1, the cells lost expression of the terminal maturation marker EOS47, and acquired expression of MEP21, a marker of primitive multipotent cells. Morphologic changes were also consistent with FOG-1 induced acquisition of a more primitive state. Although the cells were not reported to take on characteristics of erythroid or megakaryocytic cells in this study, the initial changes indicated reversal to a more primitive multipotent state. Careful kinetic analysis using colony and transplantation assays will be required to fully distinguish between "transdifferentiation" and "dedifferentiation-rediifferentiation" pathways.

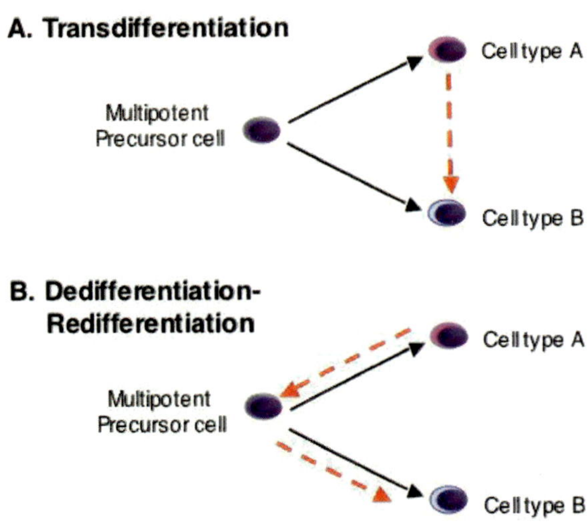

Figure 2. Transdifferentiation versus Dedifferentiation/Redifferentiation Pathways in Hematopoietic Cell Lineage Plasticity. (A) Transdifferentiation Pathway. Cell type A is converted directly to cell type B, without traversing a more primitive state. (B). Dedifferentation-Redifferentiation Pathway. Cell type A first changes to a more primitive, multipotent state. This primitive cell then redifferentiates into cell type B.

Lineage Conversion Signaling

How long must an "alternate lineage" signal be present before the cells fully re-commit to a new lineage? Kondo *et al.* found that lineage conversion of committed CLPs to myeloid cells by exogenous IL-2 receptor signaling was complete after two days of IL-2 stimulation [12]. That is, after two days, IL-2 could be withdrawn and the CLPs would still convert to myeloid cells when grown under permissive conditions. Conversely, Heyworth *et al.* studying GATA-1 mediated lineage conversion of committed myeloid/monocytes progenitors to erythroid cells found that sustained GATA-1 signaling for longer than two days was necessary for lineage conversion [14]. It seems likely that the duration of signal necessary may vary depending on the maturation state of the target cells, the particular lineage conversion, and perhaps expression levels of the relevant molecules. It will be of interest to examine this question further as new systems of lineage conversion are uncovered.

A Point of No Return?

The preceding discussion provides examples of how the developmental journey of hematopoietic stem cells can be altered by manipulation of lineage-specific transcription factors, cofactors, or cytokine receptor expression. Is there a point of no return? That is, is there a stage at which a differentiating hematopoietic cell is refractory to lineage conversion manipulations? The work of Kondo *et al.* suggests that this may be the case, at least in one example of lymphoid lineage conversion [12]. As discussed earlier, these workers found that exogenous IL-2 receptor signaling in CLPs and pro-T cells induces lineage conversion to myeloid lineage cells. In this system, the CLPS are cultured on OP9 stromal cells in the presence of IL-7, conditions that induce T-cell maturation. When the investigators first allowed the CLPs to incubate on the stromal cell system for two days, and then added IL-2, no lineage conversion was observed. This suggests that the cells had reached a point of irreversible T cell lineage commitment.

Work from the author's laboratory examining the ability of ectopically expressed FOG-1 to convert mature primary mast cells into erythroid/megakaryocytic cells may provide further evidence for this. While a small proportion of cells undergo lineage conversion, a much larger population of cells die upon this manipulation (Cantor AB and Orkin SH, unpublished observation). Since the starting cell population is somewhat heterogeneous, in terms of mast cell maturation, it is possible that more immature mast cell progenitors undergo lineage conversion, whereas more terminally differentiated cells succumb to apoptotic crisis. Further analysis is in progress to directly test this possibility.

If these examples hold true, it suggests that there is a limited "window" during differentiation in which lineage conversion is possible. Beyond this point, the cells are resistant to alternate lineage transcription factor signaling or react by activating apoptotic pathways. What mechanism(s) underlies this transition to a truly "committed" cell fate is unknown. Possibilities include epigenetic modifications, and/or establishment of essential cytokine regulatory networks.

Lineage Exclusion

The preceding discussion involves cases in which ectopic expression of transcription factors, or cytokine signals, results in conversion of one hematopoietic lineage to another. There are also several examples in which removal of a lineage-specific transcription factor results in the development of alternate cell types. The transcription factor Pax5 is expressed exclusively in the B-lymphoid lineage in hematopoietic tissues and is required for progression of pro-B cells to more mature B-lymphoid cells *in vivo* [25-28]. Loss of Pax5, by gene targeting in mice, enables pro-B cells to differentiate into multiple other lineages such as macrophages, osteoclasts, dendritic cells, and natural killer cells when cultured in the presence of appropriate cytokines [29]. However, restoration of Pax5 expression once again restricts their differentiation potential to the B lymphoid lineage. This implies that Pax5 actively suppresses alternate lineage choices. Similarly, FOG-1$^{-/-}$ embryos have markedly elevated numbers of mast cell-like progenitors compared to wild type, suggesting that it may normally repress mast cell development [17].

PU.1 is a member of the ets family of transcription factors. It plays an essential role in the development of B lymphoid cells, granulocytes, macrophages, and mast cells [30-32]. PU.1 is also expressed in the earliest thymic T cell precursor cells, but its levels markedly diminish as the cells mature to the pro-T cell stage [31,33]. Enforced expression of PU.1 in T-cell progenitors results in a complete block in T cell maturation, implying that sustained PU.1 expression normally excludes T cell lineage commitment [33].

Further evidence for alternate lineage exclusion comes from studies utilizing forced expression of lineage-specific transcription factors in cell populations that normally lack these factors. Purified B220^{+}CD43^{+}IgM^{-} proB cells, which lack GATA-1 expression, can be induced to mature along the B-lineage when cultured under appropriate conditions. When GATA-1 is ectopically expressed in these cells by retroviral transduction, maturation is blocked, whereas cells transduced with control retroviruses mature readily [15]. Careful examination of the GATA-1 expressing cells shows that they undergo one or two cell divisions and then die through an apoptotic pathway. Thus, promiscuous expression of GATA-1 excludes a B-lymphoid outcome.

Lessons from Drosophila

Similar examples of lineage conversion and exclusion by transcription factors have been reported in studies of Drosophila blood cell development. Drosophila contain a primitive hematopoietic system consisting of three distinct lineages; plasmatocytes, crystal cells and lamellocytes. Plasmatocytes comprise about 90-95% of circulating hemocytes. They disperse throughout the organism and serve a macrophage-like function. Crystal cells neutralize pathogenic material through a process known as “melanization”. Lamellocytes are rare cells under normal circumstances, but expand to high numbers under certain conditions of immune challenge. All three cell types are derived from a common hemocyte precursor cell (Figure 3A). A remarkable finding is that orthologues of many of the transcription factors essential to vertebrate hematopoiesis also play critical roles in drosophila blood development (for review, see [34]). *Serpent (srp),* a GATA factor orthologue, is expressed in hemocyte precursor cells and required for the development of both plasmatocytes and crystal cells. *U-shaped (Ush)*, an ortholgue of the Friend of GATA family of cofactors, is expressed in hemocyte precursor cells and plasmatocytes, but not crystal cells [35]. *Lozenge (Lz)*, an ortholgue of mammalian Runx1 (AML1), is expressed only in the crystal cell lineage. *Glial cells missing (Gcm)* is selectively expressed in plasmatocytes.

Ush loss of function mutants exhibit enhanced crystal cell formation, whereas ectopic expression of *Ush* in crystal cell precursors represses their development [35]. Similarly, ectopic expression of *Gcm* in crystal cell precursors blocks their ability to form mature crystal cells and converts them into plasmatocyte-like cells [36]. Thus, the plasmatocyte-specfic factors *Ush* and *Gcm* appear to normally exclude a crystal cell fate of developing hemocyte precursors.

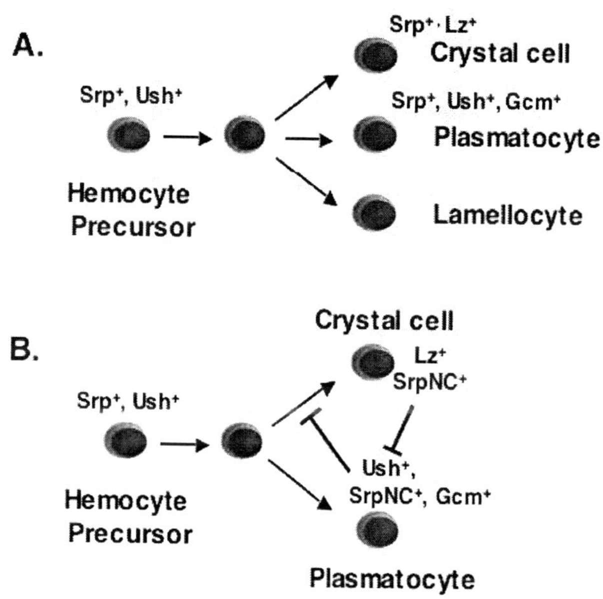

Figure 3. Schematic of Drosophila Hematopoiesis. (A) Derivation of crystal cells, plasmatocytes, and lamellocytes from a common hemocyte precursor cell. Expression of the transcription factors/cofactors *Srp*, *Lz*, *Gcm* and *Ush*. (B) Negative regulatory transcriptional network. "T" lines indicate inhibitory actions. *SrpNC* refers to *Srp* isoform containing two zinc fingers. *Ush* in combination with *SrpNC* inhibits crystal cell formation. *SrpNC* and *Lz* transcriptionally repress *Ush*, and thus plasmatocyte development.

Given that *Ush* is also expressed in multipotent hemocyte precursors, what mediates its down regulation in developing crystal cells to allow for their formation? Recent work by Schultz and his colleagues provides evidence for a combinatorial regulatory network involving *Srp*, *Ush* and *Lz* (Figure 3B). Mammalian GATA factors contain two zinc fingers.

The carboxyl zinc finger mediates high affinity DNA binding activity, whereas the amino terminal zinc finger interacts with FOG proteins. The predominant form of *Srp* in drosophila contains only a single zinc finger, equivalent to the carboxyl DNA binding zinc finger of mammalian GATA factors (*SrpC*), and is therefore not expected to interact with *Ush* (37). However, alternative splicing of *Srp* mRNA leads to the generation of an isoform containing both the amino and carboxyl zinc fingers (*SrpNC*), which is capable of interacting with *Ush* [37 38]. Enforced expression of *Ush* and *SrpNC*, but not *SrpC*, blocks crystal cell formation, implying that a physical association between *Ush* and *SrpNC* is required for *Ush's* negative influence on crystal cell formation. *SrpNC* also physically interacts with *Lz*. This combination represses *Ush* expression, thereby allowing crystal cell formation [37 39]. Thus, combinatorial control of different lineage-specific transcription factors, and cofactors, promotes one lineage choice, while actively excluding an alternate fate.

Multilineage Priming

These examples of lineage conversion and lineage exclusion suggest that there may be a balance of forces that ultimately determine the lineage fate of a multipotent cell. If this were the case, then it might be possible to detect low levels of promiscuous gene expression during early stages of development, before one factor is able to dominate. Indeed, Enver and his colleagues found evidence for this in an elegant study using single-cell polymerase chain reaction [40]. These researchers found that a single multipotent progenitor cell co-expressed low levels of marker genes of two distinct lineages, namely the erythroid maker β-globin, and the myeloid marker myeloperoxidase, as well as cytokine receptors from multiple lineages. They termed this phenomenon "multilineage priming." Similar findings have been reported using DNA microarray analysis of immunopurified populations of hematopoietic progenitor cells [41].

In these earlier experiments, only a subset of cells was observed to promiscuously express lineage-specific marker genes. This led to the question of whether the subset of "multilineage primed" cells is functionally active, or rather represents an inert developmental "error". In a recent study, Graf and his colleagues provide strong evidence for the former using a genetic cell marking approach in hematopoietic stem cells [42]. In these experiments, mice were generated in which the gene for enhanced yellow fluorescence protein (EYFP) was knocked into the ubiquitously expressed ROSA26 locus. However, upstream of the EYFP gene, they inserted a transcriptional stop signal flanked by two loxP sites. In this configuration, expression of EYFP is blocked. However, breeding these mice to ones that express Cre recombinase results in excision of the STOP sequence, and expression of EYFP. In their experiments they used mice in which the Cre recombinase gene had been knocked into the locus for lysozyme M, a gene whose expression had previously been shown to be restricted to maturing myelomoncytic cells. This places Cre under the endogenous lysozyme M transcriptional regulatory elements and is expected to mimic its expression pattern. When the two mice are crossbred, progeny receiving both engineered alleles exhibit stable activation of EYFP in cells (and their daughter cells) that had at one time expressed lysozyme M, permanently "marking" them. When this was done, EYFP+ cells were observed not only in myelomonocytic cells as expected, but also in a fraction of HSCs and B and T lymphocytes, consistent with promiscuous expression of lysozyme M in HSCs. Importantly, when they purified EYFP+ HSCs by fluorescence activated cell sorting and injected them into lethally irradiated mice, the cells were able to fully reconstitute the hematopoietic system of the recipient mice. Nearly 100% of B and T lymphocytes in the host animals expressed EYFP after 8 weeks following transplantation. This indicates that the "primed" HSCs retain the capacity for multilineage engraftment and long-term repopulation, hallmarks of functional stem cells. Moreover, when they transplanted the EYFP- fraction of HSCs, the recipient mice developed gradually increasing percentages of EYFP+ B and T cells, indicating continued oscillating promiscuous expression of lysozyme M in the transplanted HSCs.

Emergence of Unilineage Gene Regulation

How then does unilineage gene regulation emerge from an early hematopoietic cell undergoing multilineage priming? Here, I describe a model in which a stochastic event, or possibly an environmental cue, allows one master transcriptional regulator to gain an advantage over alternate lineage regulators. These initial "choices" are then further reinforced through a series of positive and negative transcriptional feedback loops.

Autoregulation of Lineage-Restricted Transcription Factors

It has long been speculated that lineage-selective transcription factors activate their own expression. Such a positive feedback loop would serve to reinforce expression of the "selected" factor. Direct evidence for this is only beginning to emerge, mostly through the use of chromatin immunoprecipitation assays. This technique allows for *in vivo* detection of specific transcription factor occupancy at selected promoter and/or enhancer DNA elements.

Positive Auto- Regulation of the GATA-1 Gene

Two known promoters regulate expression of the GATA-1 gene. An upstream promoter, termed IT, predominantly directs expression in Sertoli cells of the testis. A more proximal promoter, termed IE, is involved in hematopoietic expression. At least two DNAseI hypersensitive(HS) sites have been detected within the vicinity of IE in erythroid and megakaryocytic cells. The more distal region (HSI), located about 3700 base pair upstream of IE, contains a core 317 base pair region that is able to direct expression in erythroid and megakaryocytic lineages in transgenic mice [43]. This core contains a GATA-binding site separated by 10 base pairs from an E-box binding motif. Together, this composite element is thought to bind a protein complex consisting of GATA-1, SCL/tal-1, E2A, Lmo2 and Ldb-1 based on electromobility shift assays [43,44]. Point mutations in the GATA site, which disrupt GATA factor binding, abolish the ability of this element to direct erythroid and megakaryocytic expression in transgenic mouse experiments. This suggests that GATA factor binding to this region is involved in GATA-1 gene regulation. However, these early experiments could not distinguish whether it was GATA-1 binding to this region, or a different GATA family member. Recent analyses using chromatin immunoprecipitation assays in erythroid cells confirm predominant occupancy of HSI by GATA-1 *in vivo*, at least during later stages of erythroid maturation [45]. Taken together, these data suggest that GATA-1 binding to this site serves to activate and/or maintain its own expression in GATA-1 containing cells.

The more proximal Dnase HS I site, located ~500 bp upstream of the heterogeneous hematopoietic transcriptional start sites, contains a high-affinity palindromic ("double") GATA-1 binding site. This site is required for full promoter activity in transient transfection reporter assays, and its removal by targeted deletion in mice results in complete ablation of eosinophil development [23]. *In vivo* DNA footprinting experiments in erythroid cells show that this region is occupied by protein, and chromatin immunoprecipitation assays confirm the presence of GATA-1 (France S, Katz SG, and Orkin SH, personal comm.).

Yamamoto and his colleagues have provided further evidence for positive autoregulation of GATA-1 gene expression using a transgenic zebrafish system [46]. In these studies, transgenic reporter fish were generated in which the reporter gene green fluorescent protein (GFP) was placed under the control of 8.1 kilobases of upstream zebrafish GATA-1 genomic DNA sequence. These fish express GFP in a pattern similar to that of endogenous GATA-1 as determined by *in situ* hybridization. However, when GATA-1 mRNA was ectopically injected into the transgenic reporter fish embryos, GFP expression was observed at sites corresponding to the location of exogenous GATA-1 expression. This ectopic activation required the presence of a high affinity double GATA binding site located in a distal enhancer element present in the transgene.

Autoregulation of PU.1

During hematopoietic development, PU.1 is initially expressed in multipotent progenitor cells, including erythroblasts. Thereafter, its expression is down regulated during erythroid development, and up regulated in myeloid development. In mature cells, PU.1 expression is restricted to myeloid and B lymphoid cells. DnaseI HS site mapping has identified at least three clusters of sensitive sites upstream of the PU.1 gene. A 3.5 kilobase enhancer fragment located about 14 kilobases upstream of the transcriptional start site, and containing one of these HS sites, confers myeloid-specific expression in stably transfected cell lines [47]. Targeted deletion of this region in mice leads to a decrease in PU.1 expression in bone marrow to about 20% of normal levels, and a profound decrease in B cells and macrophages [48,49]. This region contains active binding sites for PU.1 itself, and chromatin immunoprecipitation assays indicate occupancy of this site by PU.1 *in vivo* [49]. PU.1 binding sites are also located in a proximal promoter region located +20 base pairs downstream of the transcriptional start site [50]. A reporter construct under the control of this minimal promoter element could be activated by transfecting PU.1 into non-PU.1 containing cells. These findings suggest that, like GATA-1, PU.1 probably acts in a positive autoregulatory fashion to reinforce its own expression.

Autoregulation of the GATA-2 Gene

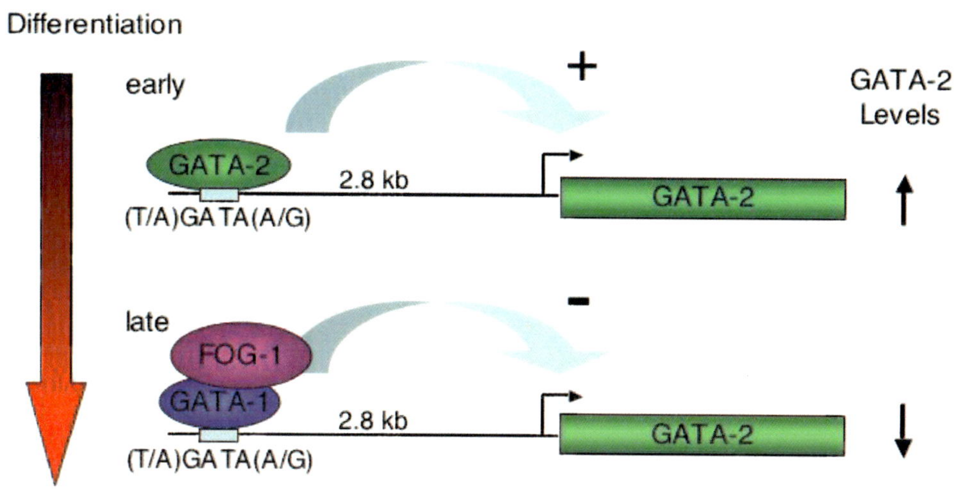

Figure 4. GATA Factor Switching at an Enhancer in the GATA-2 Locus. An enhancer element, containing a functional GATA binding site, located approximately 2.8 kilobases upstream of the transcription start site of the GATA-2 gene is shown schematically. Hematopoietic development is depicted progressing from the top of the figure to the bottom. During early stages of hematopoietic cell development, the enhancer element is predominantly occupied by GATA-2. The chromatin is in an "open" configuration, and the GATA-2 gene is actively transcribed. As hematopoietic cell maturation progresses, GATA-1 replaces GATA-2 at this enhancer element in a FOG-1 dependent manner. The chromatin throughout the locus changes to a "closed" configuration, and transcription of the GATA-2 gene diminishes. Adapted from [45,59].

GATA-1 and GATA-2 play predominantly reciprocal roles during hematopoietic development. GATA-2 is required for the expansion of multipotent progenitors, and must be down regulated for terminal maturation of certain hematopoietic lineages [51-54]. In contrast, GATA-1 activates terminal maturation of erythroid, megakaryocytic, mast and eosinophilic cells, and inhibits cell proliferation [23,24,55-58]. Using chromatin immunoprecipitation assays, Bresnick and colleagues scanned a large region of DNA sequence upstream of the GATA-2 gene for occupancy by GATA factors [59]. They identified a region located ~ 2.8 kilobases upstream of the transcriptional start site that is avidly bound by GATA-2 in early erythroid cell lines. Analysis of histone modifications around this region, and extending across the entire GATA-2 locus, showed enrichment in acetylated histones H3 and H4 and methylation of histone H3 lysine 4, consistent with an “open”, transcriptionally active chromatin state. This correlates with high expression levels of GATA-2 in these cells. Induction of erythroid maturation in these cell systems correlates with switching of GATA-1 for GATA-2 at this site, an activity that requires the presence of FOG-1 [45]. This is associated with down regulation of GATA-2 expression, and a global reduction in histone acetylation across the GATA-2 locus. Based on these findings, the investigators have proposed a model in which GATA-2 autoregulates its own expression in early progenitor cells to maintain high GATA-2 levels (Figure 4). As GATA-1 levels rise during cellular maturation, GATA-1 competes with GATA-2 for binding at the GATA-2 enhancer element leading to transcriptional silencing of the GATA-2 locus. How GATA-1 represses transcription in this case is not known, but recent work using DNA microarray techniques shows that GATA-1 acts to repress transcription of many target genes, in addition to activating others [58].

Autoregulation of other Lineage-Selective Transcription Factors

The proto-oncogene c-jun is a component of the AP-1 transcription factor complex and a coactivator of PU.1. c-jun expression levels are thought to play a role in the differentiation outcome of myelomonocytic progenitor cells [60]. Increased expression of c-jun promotes monocyte development, whereas decreased expression is necessary for granulocytic differentiation. The c-jun gene contains two AP-1 binding sites in its promoter, which have been proposed to mediate its autoregulation [61].

Cross-Antagonism of Alternate Lineage-Specific Transcription Factors

It was previously believed that lineage-specific transcription factors played principally positive roles in directing cell fate determination. However, studies performed over the past several years indicate that they can simultaneously repress alternate lineage outcomes by antagonizing opposing lineage-specific transcription factors. This may occur via direct physical interaction and/or by inhibiting transcription of the opposing transcription factor (Table 2).

Table 2. Functional antagonism of alternate lineage specific transcription factors

Transcription Factor	Antagonistic Transcription Factor	Proposed Mechanism(s) of Inhibitiion	References
GATA-1	PU.1	Prevents DNA binding; Tethers corepressor pRB to GATA-1; decreases expression	65, 67
PU.1	GATA-1	Blocks binding of essential co-factor c-jun	64, 72
FOG	C/EBP-β	Down regulates transcription	22
C/EBP-β	FOG	Transcriptionally corepresses C/EBP-β target genes	22
PU.1	C/EBPα	Blocks binding of essential co-factor c-jun	71
c-jun	C/EBPα	Prevents DNA binding, and autoregulation	60
GATA-2	PU.1	Transcriptional repression	64, 73
PU.1	GATA-2	Blocks binding of essential cofactor c-jun	64

Cross-Antagonism between GATA-1 and PU.1

The first demonstration of direct interaction and cross-antagonism of two alternate lineage transcription factors involved GATA-1 and PU.1, master regulators of the erythroid/megakaryocytic and myeloid lineages, respectively. Previously, it had been known that misexpression of PU.1 in erythroid lineage cells is associated with the development of certain murine erythroleukemias [62]. Based on this, Skoultchi and his colleagues hypothesized that PU.1 antagonizes GATA-1 [63]. Indeed, they demonstrated direct physical association between these two factors using co-immunoprecipitation experiments. Transcriptional reporter assays revealed potent inhibition of GATA-1 transactivation by PU.1. Last, they showed that ectopic expression of PU.1 in *Xenopus* embryos blocks erythroid development, but this can be overcome by overexpressing GATA-1. Thus, PU.1 acts to antagonize GATA-1 in erythroid development through direct physical interaction.

Several other groups have also reported a direct interaction and functional cross-antagonism between GATA-1 and PU.1 [64-66]. All of the studies have identified the interaction domains to be the respective DNA binding regions of each molecule, the carboxyl zinc finger of GATA-1 and the Ets domain of PU.1. While the interaction and functional cross antagonism has been well established, the mechanism(s) of inhibition remain somewhat controversial. Rehktman *et al.* have shown that both the amino terminal transactivation domain and the Ets domain are required for PU.1 mediated antagonism of GATA-1. They have proposed that PU.1 inhibits GATA-1 function by tethering the transcriptional co-repressor retinoblastoma protein (pRb) to GATA-1 bound to its target DNA sequences [67]. Evidence has also been presented indicating that PU.1 directly blocks GATA-1 DNA binding

[65]. GATA-1 has been proposed inhibit PU.1 by blocking the interaction with its essential cofactor c-jun [64].

Cross-Antagonism between CEBPα and PU.1

Since these earlier studies, other examples of direct cross antagonism between alternate lineage-specific transcription factors have been reported. Monocytes and granulocytes are derived from a common myelomonocytic bipotent precursor cell. CCAAT enhancer binding protein α (C/EBPα), a basic-region leucine zipper transcription factor, is required for granulocyte differentiation, but not monocyte differentiation [68]. Overexpression of C/EBPα in myelomonocytic bipotent precursor cells selectively drives granulocytic differentiation, and blocks phorbol ester induced monocyte differentiation [69]. In contrast, PU.1 is required for monocyte/macrophage development, but not granulocyte development [31]. Thus, C/EBPα and PU.1 appear to play reciprocal roles in granulocyte and monocyte/macrophage differentiation. Reciprocal roles for C/EBP family members and PU.1 have also been reported for Langerhans cell commitment [70].

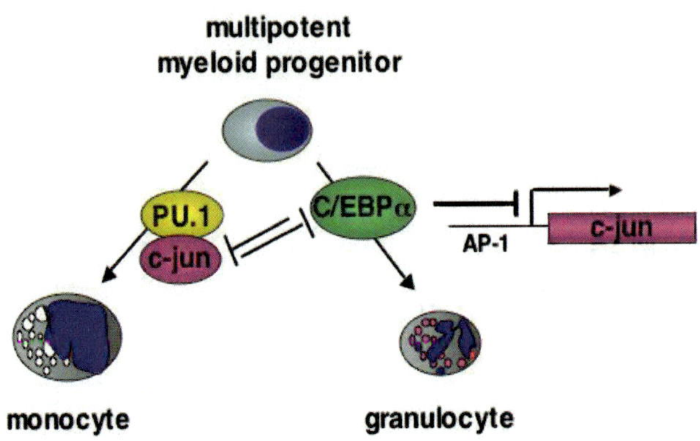

Figure 5. Cross antagonism of C/EBPα and c-jun in Monocyte and Granulocyte Differentiation. The differentiation of monocytes/macrophages and granulocytes from a common myeloid progenitor is illustrated. The transcription factor PU.1, along with its essential cofactor c-jun, is required for monocyte/macrophage differentiation. The transcription factor C/EBPα is required for granulocyte differentiation. C/EBPα inhibits PU.1 activity by displacing c-jun via a direct physical interaction. In addition, C/EBPα represses c-jun expression by interfering with its autoregulation. Conversely, c-jun inhibits granulocyte development. "T" lines represent inhibitory activity. Adapted from [60].

Behre and his colleagues have recently shown that C/EBPα functionally antagonizes PU.1 in monoctye differentiation [60,71]. Two mechanisms have been proposed. One involves the direct physical interaction between CEBPα and PU.1 via their DNA binding domains. This displaces PU.1's essential cofactor c-jun, and functionally inactivates PU.1. The other involves direct physical association of C/EBPα with c-jun itself, via their leucine zipper domains. This blocks c-jun DNA binding activity, and results in down regulation of c-jun expression by impairing its ability to autoregulate itself. Based on these findings, they propose a model in which C/EBPα simultaneously inhibits monocytes/macrophage differentiation, via antagonism of PU.1/c-jun, while promoting granulocyte maturation (Figure 5). A reciprocal direct inhibition of C/EBPα seems likely since overexpression of c-jun blocks granulocyte development.

Antagonism Via Negative Transcriptional Cross-Talk

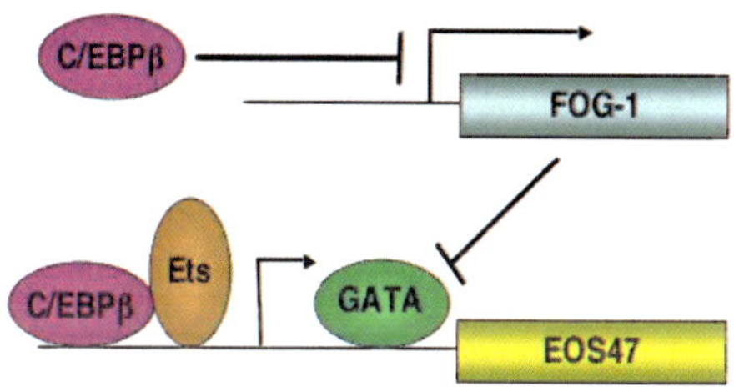

Figure 6. Negative Regulatory Network Involving FOG-1 and C/EBPβ in Eosinophil Development. C/EBPβ, a positive regulator of the eosinophil gene program, is shown activating the avian eosinophil gene EOS47 in combination with the Ets and GATA transcription factors. C/EBPβ simultaneously represses transcription of FOG-1. FOG-1 represses EOS47 transcription through its interaction with GATA factors. "T" lines represent inhibitory action. Adapted from [22].

In addition to cross antagonism via direct physical interaction, examples have been found in which a transcription factor active in one lineage directly inhibits expression of an alternate lineage-specific factor. As discussed earlier, FOG-1 is highly expressed in avian myb-ets transformed myelomonocyte progenitors (MEPs) and must be down regulated in order for eosinophil maturation to proceed. In contrast, the transcription factor C/EBPβ is required for eosinophil maturation. Nerlov and his colleagues have provided evidence for antagonistic transcriptional cross talk between C/EBPβ and FOG-1 [22] (Figure 6). They showed that FOG-1 behaves as a transcriptional corepressor of eosinophil genes, presumably via its interaction with GATA-factors and interference with C/EBPβ/GATA factor cooperativity. Conversely, C/EBPβ represses transcription of FOG-1. Thus, a regulatory network exists in which C/EBPβ promotes eosinophil development by directly transactivating eosinophil specific genes and simultaneously repressing transcription of the negative regulator FOG-1.

As described above, PU.1 and GATA-1 cross antagonize one another via direct physical interaction. There is also evidence that PU.1 represses transcription of the GATA-1 gene, although it is unknown if this is a direct target [72].

Cooperative and Antagonistic Interplay between GATA-2 and PU.1

Singh and his colleagues have recently reported a case in which two lineage-restricted transcription factors either cooperate or antagonize one another depending on their expression levels [73]. Mast cells and macrophages are derived from common myeloid progenitor cells. Mast cell development requires both PU.1 and GATA-2, whereas macrophage development only requires PU.1. As expected, PU.1$^{-/-}$ myeloid progenitor cells generate both macrophages and mast cells upon restoration of PU.1 activity when cultured under appropriate conditions. However, high level PU.1 expression generated by a tamoxifen-inducible fusion molecule leads to marked down regulation of GATA-2 expression and the exclusive development of macrophages. Enforced expression of GATA-2 in these progenitor cells restores mast cell development potential. This antagonistic activity of high level PU.1 toward GATA-2 does not appear to require DNA binding activity of PU.1. Whether it involves direct physical association between GATA-2 is unknown, but seems possible since these factors have previously been shown to interact [64]. A reciprocal inhibition of PU.1 activity by GATA-2 has not been reported.

A Balance of Forces

The findings described above provide significant new insight into the manner by which lineage-specific transcription factors direct unilineage differentiation. In addition to their positive influences on a particular lineage choice, they simultaneously repress alternate lineage outcomes by directly antagonizing opposing lineage-specific transcription factors (Figure 7). This, in effect, “turns up the gain” on cell fate decisions once they have been initiated.

One can envision an early progenitor cell emerging from a state of low-level multilineage gene expression. Initially, no lineage-specific transcription factor dominates. Then, through stochastic or possibly environmental cues, one lineage-specific transcription factor gains a small advantage over others. Autoregulation of the factor quickly amplifies its levels, while antagonism of opposing factors neutralizes their influence. This allows for further domination of the selected factor and further suppression of alternate choices. Gradually, unilineage gene expression patterns emerge and gain increasing stability as the balance of forces get tipped further in one direction.

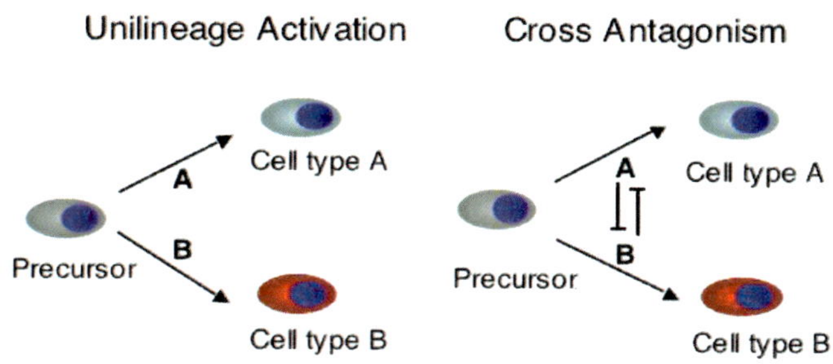

Figure 7. Cross-antagonistic Model of Hematopoietic Lineage Commitment (A) A “unilineage” model is depicted in which lineage-specific transcription factors play primarily positive roles in activating genes of a particular lineage. (B) A cross-antagonistic model is shown in which lineage-specific transcription factors, in addition to activating one lineage gene program, simultaneously repress alternate lineage choices by antagonizing opposing lineage-specific transcription factors. This may occur through direct physical association (as shown), or via negative transcriptional regulation of the opposing transcription factor.

Lineage Instability During Early Hematopoiesis

The finding that alternate lineage gene programs are held in an active state of repression by opposing lineage-specific transcription factors fits well with the cases of lineage plasticity and lineage exclusion described at the beginning of this chapter. Manipulations, such as enforced or ectopic high-level expression of a lineage-specific transcription factor, might break emerging positive and negative feedback loops and establish new ones. This would re-direct the cell to an alternate fate. In this model, a window of lineage instability exists during early hematopoietic differentiation in which the term “commitment” should be viewed as relative, not absolute (Figure 8). Left unperturbed, the cell begins expressing certain unilineage features and appears “committed” to a given pathway. However, alterations in the balance of forces between lineage-specific transcription factors may cause the cell to “re-commit” to a different pathway. Whether such back and forth “lineage indecisions” occur during normal hematopoiesis is not known, but the experiments described in this chapter provide evidence for its potential.

A subset of leukemias express markers of multiple hematopoietic lineages, and have been termed “mixed lineage” or “biphenotypic” leukemias. It is tempting to speculate that this may be the result of perturbations of the cross antagonistic transcription factor networks that normally reinforce lineage identity, as described above. Thus, these cells may represent the equivalent of early progenitors that exist in a state of lineage instability.

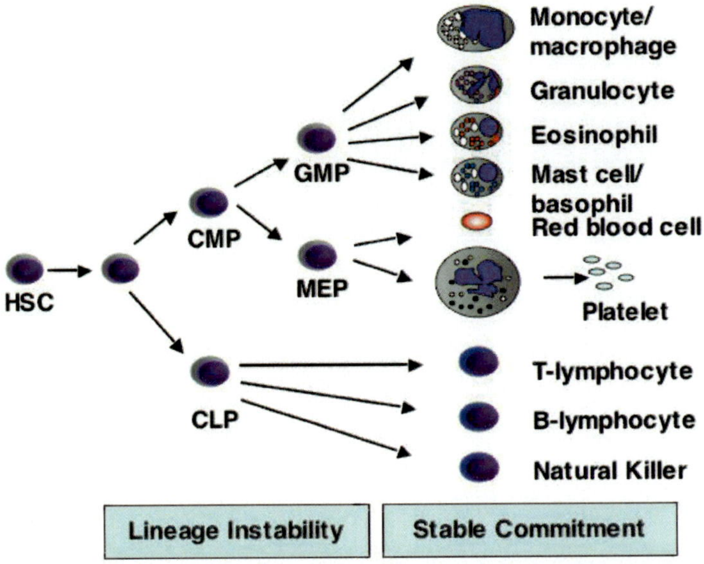

Figure 8. Lineage Instability During Hematopoietic Differentiation. In this model, a window of potential lineage instability exists during early stages of hematopoietic differentiation. During this period, cells beginning to commit to a specific lineage can be redirected to an alternate lineage under appropriate conditions. Later, commitment decisions become irreversible. The illustrated boundaries of lineage instability are drawn arbitrarily.

Lineage Plasticity in Non-Hematopoietic Systems

Is functional cross antagonism of alternate lineage specific transcription factors confined to the hematopoietic system, or does it play roles in other developmental settings? Recently, Gruss and colleagues reported a functional balance between the lineage-specific transcription factors Arx and Pax4 in endocrine pancreas development, quite reminiscent of findings of hematopoietic transcription factors described above [74]. In pancreas development, β-cells (which produce insulin), δ-cells (which produce somatostatin) and α-cells (which produce glucagon) are generated from a common endocrine precursor cell (Figure 9). The homeodomain-containing transcription factor Arx is expressed early in the developing pancreatic epithelium, and is later restricted to maturing α and δ cells. Pax4, a paired-box containing transcription factor, is co-expressed with Arx in endocrine progenitor cells, but then becomes restricted to β-cells as differentiation progresses. Arx $^{-/-}$ mice die shortly after birth from severe hypoglycemia and early-onset loss of mature α cells. However, there is a concomitant increase in the number of β– and δ-cells. Conversely, Pax4 deficient mice have drastically reduced β– and δ-cell numbers, and an expanded α-cell population [75]. Multiplex RT-PCR analysis demonstrated a 3.3-fold increase in Arx transcript levels in Pax4-deficient mice, and a 4-fold increase of Pax4 mRNA transcripts in Arx-depleted mice, suggesting that these two lineage-specific factors may mutually inhibit one another. Based on these finding, Sosa-Pineda *et al.* propose that during early endocrine progenitor development both factors co-exist, but that one ultimately predominates and antagonizes the other. If it is Pax4, the cells take on a β or δ-cell fate; if it is Arx, the cells become α-cells.

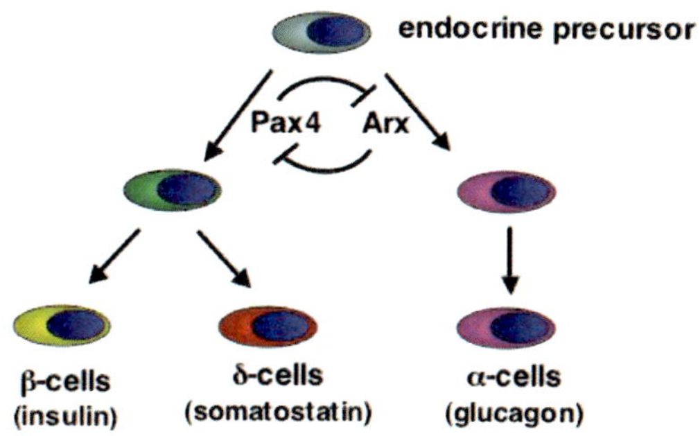

Figure 9. Cross antagonism of Pax4 and Arx in Pancreatic Endocrine Cell Differentiation. Derivation of β-cells (which produce insulin), δ-cells (which produce somatostatin) and α-cells (which produce glucagon) from a common endocrine precursor cell is shown. Pax4 is required for β- and δ-cell differentiation, whereas Arx is required for α-cell development. Functional cross-antagonism of Pax4 and Arx (shown by the curved "T" bars) inhibits gene activation (or repression) specific to one lineage, in cells committing to an alternate lineage. Adapted from [74].

Limitations of Lineage Plasticity

Over the past several years, claims have been made that certain organ-specific adult stem cells have the capacity to transdifferentiate into alternate tissues (for recent review, see)[76]. Among others, hematopoietic stem cells have been reported to change into muscle, liver, and neural tissue; and neural, liver and muscle stem cells have been reported to become blood. As exciting as these initial reports were, subsequent studies indicate that most of these cases were due to cell fusion or other artifacts, or could not be reproduced [77-79]. The lineage plasticity described in this chapter involves a single organ system, the hematopoietic system, and appears confined to early stages of development. It is therefore a distinct scenario from that described for adult stem cell transdifferentiaton.

The finding of lineage conversion of early hematopoietic cells nevertheless illustrates the potential for phenotypic plasticity of cells. Why this appears to be limited to early stages of differentiation and within a single organ system, is not evident. Clearly, under extraordinary circumstances, such as the transfer of a nucleus to an oocyte environment, broader plasticity is possible as exemplified by the cloning of whole animals from somatic cells. Elucidation of the molecular mechanisms that normally limit plasticity should be of significant scientific and clinical interest in the future. Perhaps these emerging experimental systems will provide some important clues.

References

[1] Graf T. Differentiation plasticity of hematopoietic cells. *Blood.* 2002;99:3089-3101.

[2] Rothenberg EV, Dionne CJ. Lineage plasticity and commitment in T-cell development. *Immunol Rev.* 2002;187:96-115.

[3] Akashi K, Traver D, Miyamoto T, Weissman IL. A clonogenic common myeloid progenitor that gives rise to all myeloid lineages. *Nature*. 2000;404:193-197.

[4] Kondo M, Weissman IL, Akashi K. Identification of clonogenic common lymphoid progenitors in mouse bone marrow. *Cell.* 1997;91:661-672.

[5] Emerson S. The stem cell model of hematopoiesis. In: Hoffman R, Benz EJ, Shattil SJ, Furie B, Cohen HJ, eds. *Hematology: Basic Principles and Practice*. New York: Churchill Livingstone; 1991:72-81.

[6] Boyd AW, Schrader JW. Derivation of macrophage-like lines from the pre-B lymphoma ABLS 8.1 using 5-azacytidine. *Nature*. 1982;297:691-693.

[7] Klinken SP, Alexander WS, Adams JM. Hemopoietic lineage switch: v-raf oncogene converts Emu-myc transgenic B cells into macrophages. *Cell.* 1988;53:857-867.

[8] Kulessa H, Frampton J, Graf T. GATA-1 reprograms avian myelomonocytic cell lines into eosinophils, thromboblasts, and erythroblasts. *Genes Dev.* 1995;9:1250-1262.

[9] Visvader JE, Elefanty AG, Strasser A, Adams JM. GATA-1 but not SCL induces megakaryocytic differentiation in an early myeloid line. *Embo J.* 1992;11:4557-4564.

[10] Yamaguchi Y, Zon LI, Ackerman SJ, Yamamoto M, Suda T. Forced GATA-1 expression in the murine myeloid cell line M1: induction of c-Mpl expression and megakaryocytic/erythroid differentiation. *Blood.* 1998;91:450-457.

[11] Seshasayee D, Gaines P, Wojchowski DM. GATA-1 dominantly activates a program of erythroid gene expression in factor-dependent myeloid FDCW2 cells. *Mol Cell Biol.* 1998;18:3278-3288.

[12] Kondo M, Scherer DC, Miyamoto T, et al. Cell-fate conversion of lymphoid-committed progenitors by instructive actions of cytokines. *Nature*. 2000;407:383-386.

[13] Iwasaki-Arai J, Iwasaki H, Miyamoto T, Watanabe S, Akashi K. Enforced granulocyte/macrophage colony-stimulating factor signals do not support lymphopoiesis, but instruct lymphoid to myelomonocytic lineage conversion. *J Exp Med.* 2003;197:1311-1322.

[14] Heyworth C, Pearson S, May G, Enver T. Transcription factor-mediated lineage switching reveals plasticity in primary committed progenitor cells. *Embo J.* 2002;21:3770-3781.

[15] Iwasaki H, Mizuno S, Wells RA, Cantor AB, Watanabe S, Akashi K. GATA-1 converts lymphoid and myelomonocytic progenitors into the megakaryocyte/erythrocyte lineages. *Immunity*. 2003;19:451-462.

[16] Tsang AP, Visvader JE, Turner CA, et al. FOG, a multitype zinc finger protein, acts as a cofactor for transcription factor GATA-1 in erythroid and megakaryocytic differentiation. *Cell.* 1997;90:109-119.

[17] Tsang AP, Fujiwara Y, Hom DB, Orkin SH. Failure of megakaryopoiesis and arrested erythropoiesis in mice lacking the GATA-1 transcriptional cofactor FOG. *Genes Dev.* 1998;12:1176-1188.

[18] Tevosian SG, Deconinck AE, Tanaka M, et al. FOG-2, a cofactor for GATA transcription factors, is essential for heart morphogenesis and development of coronary vessels from epicardium. *Cell.* 2000;101:729-739.

[19] Chang AN, Cantor AB, Fujiwara Y, et al. GATA-factor dependence of the multitype zinc-finger protein FOG-1 for its essential role in megakaryopoiesis. *Proc Natl Acad Sci U S A.* 2002;99:9237-9242.

[20] Crispino JD, Orkin SH. The use of altered specificity mutants to probe specific protein-protein interactions involved in the activation of GATA-1 target genes. *Methods.* 2002;26:84-92.

[21] Nichols KE, Crispino JD, Poncz M, et al. Familial dyserythropoietic anaemia and thrombocytopenia due to an inherited mutation in GATA1. *Nat Genet.* 2000;24:266-270.

[22] Querfurth E, Schuster M, Kulessa H, et al. Antagonism between C/EBPbeta and FOG in eosinophil lineage commitment of multipotent hematopoietic progenitors. *Genes Dev.* 2000;14:2515-2525.

[23] Yu C, Cantor AB, Yang H, et al. Targeted deletion of a high-affinity GATA-binding site in the GATA-1 promoter leads to selective loss of the eosinophil lineage in vivo. *J Exp Med.* 2002;195:1387-1395.

[24] Migliaccio AR, Rana RA, Sanchez M, et al. GATA-1 as a regulator of mast cell differentiation revealed by the phenotype of the GATA-1low mouse mutant. *J Exp Med.* 2003;197:281-296.

[25] Adams B, Dorfler P, Aguzzi A, et al. Pax-5 encodes the transcription factor BSAP and is expressed in B lymphocytes, the developing CNS, and adult testis. *Genes Dev.* 1992;6:1589-1607.

[26] Li YS, Wasserman R, Hayakawa K, Hardy RR. Identification of the earliest B lineage stage in mouse bone marrow. *Immunity.* 1996;5:527-535.

[27] Urbanek P, Wang ZQ, Fetka I, Wagner EF, Busslinger M. Complete block of early B cell differentiation and altered patterning of the posterior midbrain in mice lacking Pax5/BSAP. *Cell.* 1994;79:901-912.

[28] Nutt SL, Urbanek P, Rolink A, Busslinger M. Essential functions of Pax5 (BSAP) in pro-B cell development: difference between fetal and adult B lymphopoiesis and reduced V-to-DJ recombination at the IgH locus. *Genes Dev.* 1997;11:476-491.

[29] Nutt SL, Heavey B, Rolink AG, Busslinger M. Commitment to the B-lymphoid lineage depends on the transcription factor Pax5. *Nature.* 1999;401:556-562.

[30] Scott EW, Simon MC, Anastasi J, Singh H. Requirement of transcription factor PU.1 in the development of multiple hematopoietic lineages. *Science.* 1994;265:1573-1577.

[31] McKercher SR, Torbett BE, Anderson KL, et al. Targeted disruption of the PU.1 gene results in multiple hematopoietic abnormalities. *Embo J.* 1996;15:5647-5658.

[32] Hromas R, Orazi A, Neiman RS, et al. Hematopoietic lineage- and stage-restricted expression of the ETS oncogene family member PU.1. *Blood.* 1993;82:2998-3004.

[33] Anderson MK, Weiss AH, Hernandez-Hoyos G, Dionne CJ, Rothenberg EV. Constitutive expression of PU.1 in fetal hematopoietic progenitors blocks T cell development at the pro-T cell stage. *Immunity.* 2002;16:285-296.

[34] Evans CJ, Hartenstein V, Banerjee U. Thicker than blood: conserved mechanisms in Drosophila and vertebrate hematopoiesis. *Dev Cell.* 2003;5:673-690.

[35] Fossett N, Tevosian SG, Gajewski K, Zhang Q, Orkin SH, Schulz RA. The Friend of GATA proteins U-shaped, FOG-1, and FOG-2 function as negative regulators of blood, heart, and eye development in Drosophila. *Proc Natl Acad Sci U S A.* 2001;98:7342-7347.

[36] Lebestky T, Chang T, Hartenstein V, Banerjee U. Specification of Drosophila hematopoietic lineage by conserved transcription factors. *Science.* 2000;288:146-149.

[37] Fossett N, Hyman K, Gajewski K, Orkin SH, Schulz RA. Combinatorial interactions of serpent, lozenge, and U-shaped regulate crystal cell lineage commitment during Drosophila hematopoiesis. *Proc Natl Acad Sci U S A.* 2003;100:11451-11456.

[38] Waltzer L, Bataille L, Peyrefitte S, Haenlin M. Two isoforms of Serpent containing either one or two GATA zinc fingers have different roles in Drosophila haematopoiesis. *Embo J.* 2002;21:5477-5486.

[39] Waltzer L, Ferjoux G, Bataille L, Haenlin M. Cooperation between the GATA and RUNX factors Serpent and Lozenge during Drosophila hematopoiesis. *Embo J.* 2003;22:6516-6525.

[40] Hu M, Krause D, Greaves M, et al. Multilineage gene expression precedes commitment in the hemopoietic system. *Genes Dev.* 1997;11:774-785.

[41] Akashi K, He X, Chen J, et al. Transcriptional accessibility for genes of multiple tissues and hematopoietic lineages is hierarchically controlled during early hematopoiesis. *Blood.* 2003;101:383-389.

[42] Ye M, Iwasaki H, Laiosa CV, et al. Hematopoietic stem cells expressing the myeloid lysozyme gene retain long-term, multilineage repopulation potential. *Immunity.* 2003;19:689-699.

[43] Vyas P, McDevitt MA, Cantor AB, Katz SG, Fujiwara Y, Orkin SH. Different sequence requirements for expression in erythroid and megakaryocytic cells within a regulatory element upstream of the GATA-1 gene. *Development.* 1999;126:2799-2811.

[44] Wadman IA, Osada H, Grutz GG, et al. The LIM-only protein Lmo2 is a bridging molecule assembling an erythroid, DNA-binding complex which includes the TAL1, E47, GATA-1 and Ldb1/NLI proteins. *Embo J.* 1997;16:3145-3157.

[45] Pal S, Cantor AB, Johnson KD, et al. Coregulator-dependent facilitation of chromatin occupancy by GATA-1. *Proc Natl Acad Sci U S A.* 2004;101:980-985.

[46] Kobayashi M, Nishikawa K, Yamamoto M. Hematopoietic regulatory domain of gata1 gene is positively regulated by GATA1 protein in zebrafish embryos. *Development.* 2001;128:2341-2350.

[47] Li Y, Okuno Y, Zhang P, et al. Regulation of the PU.1 gene by distal elements. *Blood.* 2001;98:2958-2965.

[48] Rosenbauer F, Wagner K, Kutok JL, et al. Myeloid and B-Lymphoid Development Require High PU.1 Expression by a Distal Element in Vivo. *American Society of Hematology, 45th Annual Meeting*. San Diego, CA: Blood; 2003:342a.

[49] Huang G, Y. O, Tenen DG. Regulation of PU.1 by AML1(RUNX1). *American Society of Hematology, 45th Annual Meeting*. San Diego, CA: Blood; 2003:345a.

[50] Chen H, Ray-Gallet D, Zhang P, et al. PU.1 (Spi-1) autoregulates its expression in myeloid cells. *Oncogene*. 1995;11:1549-1560.

[51] Tsai FY, Keller G, Kuo FC, et al. An early haematopoietic defect in mice lacking the transcription factor GATA-2. *Nature*. 1994;371:221-226.

[52] Tsai FY, Orkin SH. Transcription factor GATA-2 is required for proliferation/survival of early hematopoietic cells and mast cell formation, but not for erythroid and myeloid terminal differentiation. *Blood*. 1997;89:3636-3643.

[53] Persons DA, Allay JA, Allay ER, et al. Enforced expression of the GATA-2 transcription factor blocks normal hematopoiesis. *Blood*. 1999;93:488-499.

[54] Briegel K, Lim KC, Plank C, Beug H, Engel JD, Zenke M. Ectopic expression of a conditional GATA-2/estrogen receptor chimera arrests erythroid differentiation in a hormone-dependent manner. *Genes Dev*. 1993;7:1097-1109.

[55] Fujiwara Y, Browne CP, Cunniff K, Goff SC, Orkin SH. Arrested development of embryonic red cell precursors in mouse embryos lacking transcription factor GATA-1. *Proc Natl Acad Sci U S A*. 1996;93:12355-12358.

[56] Shivdasani RA, Fujiwara Y, McDevitt MA, Orkin SH. A lineage-selective knockout establishes the critical role of transcription factor GATA-1 in megakaryocyte growth and platelet development. *Embo J*. 1997;16:3965-3973.

[57] Vyas P, Ault K, Jackson CW, Orkin SH, Shivdasani RA. Consequences of GATA-1 deficiency in megakaryocytes and platelets. *Blood*. 1999;93:2867-2875.

[58] Rylski M, Welch JJ, Chen YY, et al. GATA-1-mediated proliferation arrest during erythroid maturation. *Mol Cell Biol*. 2003;23:5031-5042.

[59] Grass JA, Boyer ME, Pal S, Wu J, Weiss MJ, Bresnick EH. GATA-1-dependent transcriptional repression of GATA-2 via disruption of positive autoregulation and domain-wide chromatin remodeling. *Proc Natl Acad Sci U S A*. 2003;100:8811-8816.

[60] Rangatia J, Vangala RK, Treiber N, et al. Downregulation of c-Jun expression by transcription factor C/EBPalpha is critical for granulocytic lineage commitment. *Mol Cell Biol*. 2002;22:8681-8694.

[61] Angel P, Hattori K, Smeal T, Karin M. The jun proto-oncogene is positively autoregulated by its product, Jun/AP-1. *Cell*. 1988;55:875-885.

[62] Moreau-Gachelin F, Ray D, Mattei MG, Tambourin P, Tavitian A. The putative oncogene Spi-1: murine chromosomal localization and transcriptional activation in murine acute erythroleukemias. *Oncogene*. 1989;4:1449-1456.

[63] Rekhtman N, Radparvar F, Evans T, Skoultchi AI. Direct interaction of hematopoietic transcription factors PU.1 and GATA-1: functional antagonism in erythroid cells. *Genes Dev*. 1999;13:1398-1411.

[64] Zhang P, Behre G, Pan J, et al. Negative cross-talk between hematopoietic regulators: GATA proteins repress PU.1. *Proc Natl Acad Sci U S A*. 1999;96:8705-8710.

[65] Zhang P, Zhang X, Iwama A, et al. PU.1 inhibits GATA-1 function and erythroid differentiation by blocking GATA-1 DNA binding. *Blood.* 2000;96:2641-2648.

[66] Nerlov C, Querfurth E, Kulessa H, Graf T. GATA-1 interacts with the myeloid PU.1 transcription factor and represses PU.1-dependent transcription. *Blood.* 2000;95:2543-2551.

[67] Rekhtman N, Choe KS, Matushansky I, Murray S, Stopka T, Skoultchi AI. PU.1 and pRB interact and cooperate to repress GATA-1 and block erythroid differentiation. *Mol Cell Biol.* 2003;23:7460-7474.

[68] Zhang DE, Zhang P, Wang ND, Hetherington CJ, Darlington GJ, Tenen DG. Absence of granulocyte colony-stimulating factor signaling and neutrophil development in CCAAT enhancer binding protein alpha-deficient mice. *Proc Natl Acad Sci U S A.* 1997;94:569-574.

[69] Radomska HS, Huettner CS, Zhang P, Cheng T, Scadden DT, Tenen DG. CCAAT/enhancer binding protein alpha is a regulatory switch sufficient for induction of granulocytic development from bipotential myeloid progenitors. *Mol Cell Biol.* 1998;18:4301-4314.

[70] wama A, Osawa M, Hirasawa R, et al. Reciprocal roles for CCAAT/enhancer binding protein (C/EBP) and PU.1 transcription factors in Langerhans cell commitment. *J Exp Med.* 2002;195:547-558.

[71] Reddy VA, Iwama A, Iotzova G, et al. Granulocyte inducer C/EBPalpha inactivates the myeloid master regulator PU.1: possible role in lineage commitment decisions. *Blood.* 2002;100:483-490.

[72] Nerlov C, Graf T. PU.1 induces myeloid lineage commitment in multipotent hematopoietic progenitors. *Genes Dev.* 1998;12:2403-2412.

[73] Walsh JC, DeKoter RP, Lee HJ, et al. Cooperative and antagonistic interplay between PU.1 and GATA-2 in the specification of myeloid cell fates. *Immunity.* 2002;17:665-676.

[74] Collombat P, Mansouri A, Hecksher-Sorensen J, et al. Opposing actions of Arx and Pax4 in endocrine pancreas development. *Genes Dev.* 2003;17:2591 2603.

[75] Sosa-Pineda B, Chowdhury K, Torres M, Oliver G, Gruss P. The Pax4 gene is essential for differentiation of insulin-producing beta cells in the mammalian pancreas. *Nature.* 1997;386:399-402.

[76] Camargo FD, Chambers SM, Goodell MA. Stem cell plasticity: from transdifferentiation to macrophage fusion. *Cell Prolif.* 2004;37:55-65.

[77] Wagers AJ, Sherwood RI, Christensen JL, Weissman IL. Little evidence for developmental plasticity of adult hematopoietic stem cells. *Science.* 2002;297:2256-2259.

[78] Wang X, Willenbring H, Akkari Y, et al. Cell fusion is the principal source of bone-marrow-derived hepatocytes. *Nature.* 2003;422:897-901.

[79] Ying QL, Nichols J, Evans EP, Smith AG. Changing potency by spontaneous fusion. *Nature.* 2002;416:545-548.

In: Trends in Stem Cell Research
Editor: Erik V. Greer, pp. 27-66
ISBN 1-59454-315-1

Chapter II

Retrovirus Vector Silencing in Stem Cells

Shuyuan Yao and James Ellis*
Developmental Biology Program, Hospital for Sick Children, Toronto, Ontario, Canada; and Department of Molecular & Medical Genetics, University of Toronto

Abstract

Retrovirus vectors suffer severe gene silencing in stem cells that hampers their use in stem cell gene therapy. Transduced retrovirus vector can be completely silenced, or dynamically silenced in a subset of cells resulting in variegation. Expressing retrovirus also can be extinguished during stem cell differentiation. Multiple epigenetic modifications including histone modifications and DNA methylation are functionally involved in establishing and maintaining retrovirus silencing via the formation of silent chromatin. A model is proposed that the balance of epigenetic effects at the integration site and retroviral silencer elements on the vectors determines the decision to be silent, variegated or expressing in stem cells. Silencing resistant retrovirus vectors have been developed by removal of these silencer elements, and silencing in stem cells may ultimately be blocked using appropriate insulator elements.

* Tel: 416-813-7295; Fax: 416-813-8883; E-mail: jellis@sickkids.ca

Introduction

Gene therapy holds the potential to treat cancer, inherited or acquired monogenic disorders and other diseases. A broad arsenal of gene transfer systems is currently available and is still expanding. Among them, the retrovirus vector system has attracted a lot of interest for high efficiency transfer into a wide range of cell types and stable integration into the host genome. Because of their ability to repopulate an entire system, stem cells are often targets for therapeutic gene delivery. The use of retrovirus vectors has been particularly valuable for introducing and expressing foreign genes in stem cells, for genetic marking of stem cells to trace their differentiation and for mutating and tagging genes associated with specific phenotypes. However, the use of retrovirus vectors has been significantly hampered by gene silencing of retrovirus vector expression in primitive cell types such as embryonic stem (ES) cells, embryonal carcinoma (EC) cells, hematopoietic stem cells (HSC) and early embryos. Understanding retrovirus vector silencing and its mechanisms will benefit the development of retrovirus based stem cell gene transfer vectors and the design of stem cell based gene therapy strategies.

Background and Phenomena

Retrovirus Silencing

Retrovirus silencing was documented as early as 1975 in the study of Moloney Murine Leukemia Virus (Mo-MuLV) infection of 4-8 cell stage mouse preimplantation embryos [Jaenisch et al., 1975]. Although one of the animals raised from infected embryos developed leukemia, no virus was detected from the infected embryos by both co-cultivated viral infection plaque forming assay and electron microscopy. Preimplantation mouse embryos are susceptible to virus integration but cannot express viral functions. In contrast, midgestation stage embryos or newborn mice can be successfully infected by Mo-MuLV and produce viral proteins [Jaenisch, 1980]. Non-virus-producing cell lines derived from a mouse embryo carrying a genetically transmitted Mo-MuLV genome can be reactivated by 5-bromodeoxyuridine (BrdU) to express viral p30 protein in almost all the cells and produce infectious virus in some of the cells [Bacheler et al., 1979]. Clearly, there is a block in preimplantation embryos making them non-permissive to Mo-MuLV gene expression and viral production.

Murine embryonic carcinoma (EC) cells share with preimplantation embryos the potential to differentiate into all three germinal layer cells [Illmensee and Mintz, 1976]. EC cells have been used as an *in vitro* model to study retrovirus silencing in preimplantation embryos. Among the many publications describing retrovirus silencing in EC cells [Gautsch, 1980; Peries et al., 1977; Teich et al., 1977], Speers et al convincingly demonstrated it by picking single cell clones from Mo-MuLV infected EC cells [Speers et al., 1980]. While viral protein and viral production were never observed in undifferentiated EC cells, after differentiation and BrdU drug treatment 2 of 4 clones were reactivated to express viral p30 protein and produce virus particles. The reactivation indicates that retrovirus can be infected into EC cells and that the lack of viral production is due to a block in retrovirus expression. Comparison of the same retrovirus infection in EC cells and differentiated cells is a useful model to study retrovirus silencing.

The features of retrovirus silencing can be summarized as being most prevalent in primitive cells, heritable and transcriptional. Both preimplantation embryo cells and EC cells are primitive cells in their early differentiation stages. Other more differentiated cells are permissive for retrovirus expression indicating that retrovirus silencing is primitive cell type specific. Silent retrovirus is kept silenced during culture, during *in vitro* and *in vivo* differentiation, and even during passage from generation to generation. Therefore, the silenced retrovirus state is heritable. The silent provirus genome is intact and functional [Speers et al., 1980; Stewart et al., 1982]. Retrovirus silencing happens after viral integration and before mRNA translation. In fact, Gautsch and Wilson showed that retrovirus expressing EC cells produce 400-700 fold higher virus based nascent RNA than silent EC cells [Gautsch and Wilson, 1983]. Obviously, transcriptional gene silencing plays an important role in retrovirus silencing, although post-transcriptional silencing of retrovirus has been reported in differentiated cells [Gao et al., 2002].

Retrovirus Vector Silencing

Retrovirus vectors are derived from isolated retroviruses to transfer and integrate foreign genes into target cells. Retrovirus vectors are created by removing viral genes, leaving only the cis-acting sequences necessary for a single round of replication to avoid vector multiplication and spread to other cells [Buchschacher and Wong-Staal, 2000]. Retrovirus silencing was originally observed with infectious retrovirus. With the development of retrovirus vectors, more and more vectors with easily detectable reporter genes have been used in stem cell gene transfer. The use of retrovirus vectors extends retrovirus silencing studies into other types of primitive cells such as embryonic stem (ES) cells [Cherry et al., 2000; Laker et al., 1998], hematopoietic stem cells (HSC) [Challita and Kohn, 1994; Klug et al., 2000; Lange and Blankenstein, 1997] and hematopoietic progenitor cells [Zentilin et al., 2000]. Retrovirus vector silencing is also found in human SRCs (SCID repopulating cells, a type of HSC), confirming that retroviral silencing occurs in humans and thus will likely pose a significant problem to clinical application of retroviral mediated gene therapy [Guenechea et al., 2001]. The use of retrovirus vectors helped to identify several types of retrovirus vector silencing phenomena. In the literature, retrovirus vector silencing has referred to any phenomena related to non-expression of genes from an integrated retrovirus vector. In the early retrovirus silencing studies, Mo-MuLV was extensively used. Since Mo-MuLV induces severe silencing in EC cells, those studies were focused on silencing established directly after viral infection. Later studies using wildtype or modified retrovirus vectors found the existence of variegated expression and extinction during differentiation or long-term culture.

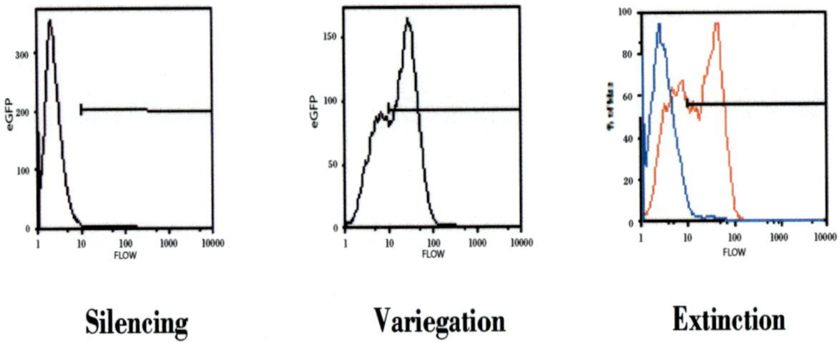

Figure 1. Retrovirus vector silencing, variegation and extinction in embryonic stem cells. Flow cytometry for eGFP expression shows retrovirus vector silencing, variegation and extinction. While silent retrovirus does not espress in any cells, variegated retrovirus expresses in some cells but is silent in others. In extinction, a variegated retrovirus in ES cells (red line) is silenced after differentiation (blue line) [Yao, 2004].

Retrovirus Vector Variegation

Retrovirus vectors can be silenced at different levels. The most extreme example is that the retrovirus vector is totally shut down with no transgene expression. Less well known is that it can be variegated in which retrovirus integrated at the same integration site is either silent or expressing (Figure 1). Single cell derived clones picked from MESV based vector transduced murine ES cells showed a varied percentage of positive cells among different clones and varied expression level among cells in the same clone [Laker et al., 1998]. Variegation of the MSCV retrovirus vector in transduced ES cells was also reported by Christopher Klug's laboratory [Swindle et al., 2004]. In their work about 35% (21 out of 55) of the integrated provirus showed variegated expression. An interesting feature of variegated retrovirus is instability. Sorted expressing and non-expressing cells from the variegated clones revert to the variegated expression pattern after brief expansion [Swindle et al., 2004]. Clones picked from a variegated clone by limit dilution are also variegated (Yao, et al., unpublished data). Unstable variegated clones were also isolated from retrovirus transduced de novo methylase null (dnmt3a-/-3b-/-) ES cells [Yao, 2004]. Therefore, variegation represents a dynamic silencing phenomenon occuring on some of the integrated retroviral proviruses in ES cells. Variegated expression was also observed in retrovirus vector transduced hematopoietic progenitor cell clones [Zentilin et al., 2000].

The expressing and non-expressing cells of the variegated clones contain the same retrovirus vector with the same integration site. Reactivation of the non-expressing cells with 5-AzaC in wildtype variegated ES cells further shows that the retrovirus vector in these cells is intact and functional [Yao, 2004]. Therefore, variegated retrovirus expression is due to epigenetic regulation.

Retrovirus Vector Extinction after Long-Term Culture

Retrovirus silencing can occur at different time points. Retrovirus expression can be shut down immediately post-integration which results in no transgene expression at any time. In silent preimplantation embryos no virus production was observed from both co-culture plaque forming assay and electron microscopy [Jaenisch et al., 1975]. With EC cells, the nascent RNA produced in silent EC cells is the same as uninfected EC cell control [Niwa et al., 1983]. This kind of silencing is also observed in ES cells [Laker et al., 1998] and long-term HSC cells [Klug et al., 2000]. In contrast, retrovirus that initially express may be silenced in a process known as extinction (Figure 1). Extinction is often observed after long term culture. For example, a GFP retrovirus vector is subject to extinction where both the percentage of cells that are GFP positive and the GFP expression level decrease in tranduced murine ES cells after 26 days of passage and culture [Cherry et al., 2000]. Retrovirus vector extinction was also observed in hematopoietic cell lines [McInerney et al., 2000]. Long-term extinction also occurs *in vivo*. Mo-MuLV based retrovirus vector transduced HSC can be assayed as colony forming units-spleen (CFU-S) generated after serial transplantation of infected born marrow cells. While high level gene expression was achieved in all of the primary CFU-S and tissues of the reconstituted marrow transplantation recipients, vector expression is inactive in >90% of the secondary CFU-S and 100% of the tertiary CFU-S [Challita and Kohn, 1994]. Since the primary CFU-S is derived from a committed progenitor cell restricted to the myeloid lineage, this result suggests retrovirus vector in HSC is extinguished during serial transplantation. Interestingly, stable transgene expression can be achieved after long-term culture. The factors that control the decision for long-term transgene extinction are not yet clear.

Retrovirus Vector Extinction after Differentiation

In vivo and *in vitro* differentiation of retrovirus vector infected primitive cells can also induce extinction of transgene expression. MESV based retrovirus vector transduced ES cell clones are subject to extinction after *in vitro* differentiation. This was observed with two different reporter genes (LacZ and IL3neoR) and in two different ES cell lines (W9.5 and R1). MESV vector expression was also extinguished after *in vivo* differentiation by making ES cell derived embryos [Laker et al., 1998]. Extinction of retrovirus vectors was also reported after differentiation of HSC. Retrovirus vector expression in transduced long-term HSC was extinguished when differentiated into lymphoid cell populations after transplantation [Klug et al., 2000]. When retrovirus transduced ES cell clones were subjected to 8 days *in vitro* differentiation, extinction occurred primarily on the variegated clones [Yao, 2004], suggesting retrovirus extinction during differentiation may be a consequence of variegation.

Implications of Retrovirus Vector Silencing

Retrovirus vector silencing is a heterogeneous phenomena. The existence of different kinds of silencing on retrovirus vectors implies that a constant monitoring system exists in cells to ensure genomic integrity. The identification of these different retrovirus silencing phenomena are beneficial in terms of evaluating new designs for retrovirus vectors and new gene therapy strategies. Because most of the *ex vivo* transplantation experiments use expressing cells obtained by drug selection or FACS sorting, variegated expressing cells are not excluded. The existence of the unstable variegated expressing cells and their extinction during differentiation may account for some of the unsuccessful gene therapy transplants. Therefore, methods should be developed to distinguish expressing cells and variegated expressing cells before therapeutic transplantation.

Lentivirus Vector Silencing

Lentivirus Vector

The newest and perhaps most promising vectors developed for the infection of non-dividing quiescent stem cells are Lentiviral vectors. Lentivirus vectors gained popularity since Naldini et al [Naldini et al., 1996] demonstrated in 1996 that human immunodeficiency virus type 1 (HIV-1) based vectors were capable of transducing nondividing cells. Conventional retrovirus vectors are limited by their inability to infect terminally differentiated and nondividing cell types. Lentivirus vectors do not rely on the passive mechanism of nuclear envelope breakdown during the cell cycle to gain entry to the nucleus. Instead, they utilize active transport mechanisms and achieve entry through the nucleopore [Chicurel, 2000]. This feature is particularly relevant when infecting slowly dividing stem cells, because a greater transduction efficiency is achieved in comparison to Mo-MuLV based retroviruses [Uchida et al., 1998]. Experiments with Lentiviruses have shown that they are able to efficiently transduce human SRCs [Schilz et al., 1998; Woods et al., 2000], which will be necessary for extending findings in mice to clinical trials.

Like Mo-MuLV, HIV-1 was selected for deriving gene transfer vectors for its well understood viral biology and extensively studied molecular biology. Other Lentiviruses also have been used to develop gene transfer vectors [Quinonez and Sutton, 2002]. Three refinements have been achieved to improve HIV-1 based Lentivirus vector usability and especially safety: Lentivirus vector particle pseudotyping, improvement of the packaging constructs and cell lines [Dull et al., 1998; Kim et al., 1998; Naldini et al., 1996; Naldini and Verma, 2000; Zufferey et al., 1997], and self-inactivating deletions in SIN-Lentivirus vector LTRs [Iwakuma et al., 1999; Miyoshi et al., 1998; Zufferey et al., 1998]. Other cis-acting elements such as the central polypurine tract (cppt) [Sirven et al., 2000; Sirven et al., 2001], Rev response element (RRE) [May et al., 2000] and woodchuck posttranscriptional element (WPRE) [Zufferey et al., 1999] were also included into Lentivirus vectors to improve viral titer or viral expression.

Lentivirus Vector Silencing

Lentivirus is subject to some gene silencing effects in primitive cells. Lentivirus vector silencing was documented in transduced murine hematopoietic progenitor cells. Primitive hematopoietic progenitor cells were transduced with HIV-1 based Lentivirus vector expressing the green fluorescent protein (GFP) gene and then plated for hematopoietic progenitor cell assays. The primary colonies formed are mosaic for GFP expression. When secondary colonies were cultured from the primary colonies 13% of them are GFP negative while PCR analysis showed the existence of Lentivirus provirus [Mikkola et al., 2000]. While this report showed that Lentivirus vector silencing occurs at least at the progenitor stage, we reported Lentivirus silencing in ES cells by identification of silent single copy vector transduced ES cell clones [Yao, 2004]. Lentivirus was also reported to extinguish during ES cell differentiation into CD34+ hematopoietic progenitor cells [Hamaguchi et al., 2000; Ma et al., 2003].

Some reports suggest there is no gene silencing in Lentivirus vector transduced primitive cells and during their differentiation [Lois et al., 2002; Pfeifer et al., 2002]. In Pfeifer's work, a SIN Lentivirus vector with *cppt* and WPRE was used to harbor either PGK or CAG (contains a modified chicken β-actin promoter and enhancer sequence from cytomegalovirus) promoter driven eGFP or LacZ genes. With multiplicity of infection (moi) of 5, about 50% of the cells are GFP positive with an integrated provirus copy of 1 to 3. An moi of 50 was needed to reach 100% positive cells in both murine and human ES cells. Transduced ES cell colonies stayed GFP positive when differentiated into EBs *in vitro* and into teratomas *in vivo*. Transduced ES cells can be readily used to make chimeras by blastocyst injection, and transgene expression was achieved in most of the organs and transmitted through the germ line. Infected eight-cell stage embryos showed transgene expression after 48 hours and expression continued during embryogenesis and after birth. Lois also reported that Lentivirus vectors escaped developmental silencing during embryogenesis [Lois et al., 2002].

Copy number differences may account for the conflicting observations on Lentivirus vector silencing. Multiple provirus copies can exist in one cell and an expressing provirus can mask silent provirus in the same cell. Single cell clones with a single functionally intact provirus are the ideal system for Lentivirus silencing studies.

Retrovirus Silencing is Epigenetic

Epigenetics by definition is "the study of mitotically and/or meiotically heritable changes in the function of a gene that cannot be explained by changes in its DNA sequences" [Russo, 1996]. Epigenetic regulation plays a key role in achievement of developmental differentiation and maintenance of cell lineage identities [McNairn and Gilbert, 2003].

Expression of a retrovirus is controlled by cellular epigenetic regulation mechanisms. Indeed, a silent retrovirus stays silent after cell division, embryonic and postnatal development indicating that retrovirus silencing is heritable. The lack of expression of a silent retrovirus is not due to DNA sequence mutations. A silent retrovirus can be reactivated at least partially by drugs such as 5-bromodeoxyuridine (BrdU) [Bacheler et al., 1979] and the methylation inhibitor 5-azacytidine (5-AzaC) [Stewart et al., 1982] demonstrating that the integrated retrovirus is functional. The existence of variegated retrovirus vectors is another excellent demonstration of epigenetic regulation in retrovirus silencing. Cell clones with a variegated vector show intraclonal mosaicism with vector expression. Since the integrated retrovirus vector and its flanking endogenous genes are identical in different cells from the same clone, the variation of vector expression cannot be explained by DNA sequence differences. According to the definition of epigenetics, retrovirus silencing is an epigenetic phenomenon.

Epigenetic Gene Silencing Background

It is well accepted that chromatin structure plays an important role in controlling and modulating eukaryotic gene expression [Wolffe and Guschin, 2000]. The basic unit of chromatin is the nucleosome composed of histone components with DNA wrapped around. In each nucleosome there is an octamer of histone H2A, H2B, H3 and H4. The histones forming the basic nucleosomal octamer are called core histones. About 145bp of DNA wraps 1.65 turns around the octamer. Linker DNA, the DNA between adjacent histone octamers, varies in length in different species and in metazoa is bound by linker histones such as histone H1. Arrays of nucleosomes form 30nm diameter chromatin fibers, and linker histone H1 is involved in this process [Graziano et al., 1994]. Chromatin can be bound by non-histone proteins which are referred to as chromatin factors. Like linker histones the binding of chromatin factors induces special chromatin conformations. Heterochromatin, which was originally a term to define cytologically condensed regions of the genome throughout the cell cycle, is now widely used to refer to any silent chromatin domain. Correspondingly, euchromatin refers to active chromatin domains. Chromatin is the platform of gene transcription in eukaryotic cells.

Epigenetic silencing is achieved by establishing and maintaining a relatively inaccessible repressive chromatin domain [Grewal and Moazed, 2003; Wolffe and Matzke, 1999]. DNA methylation and other chromatin modifications induce a silent chromatin conformation by either recruiting repressive chromatin proteins or avoiding positive chromatin protein binding. These epigenetic chromatin modifications serve as heritable silencing marks to pass silent chromatin conformation through cell division [Moazed, 2001]. The inaccessibility of silent chromatin can be studied by its insensitivity to DNase digestion. Chromatin modifications can be studied by chromatin imunoprecipitation (ChIP) assays using modification specific antibodies [Crane-Robinson et al., 1999].

DNA Methylation Induced Gene Silencing

About 70-80% of the CpG dinucleotides are methylated at cytosine in mammalian cells. Early stage embryos of mammals experience a dramatic demethylation and remethylation [Santos et al., 2002]. These processes are believed to be essential in determining somatic DNA methylation patterns and critical for proper development of the embryo. Treatment with the methylation inhibitor 5-azacytidine or by genetically reducing overall genomic methylation levels through gene targeting of the methyltransferases can reactivate some gene expression. In fact, knockout of the mouse DNA methyltransferase Dnmtl is embryonic lethal [Li et al., 1992]. These results show that DNA methylation plays a critical role in gene expression and development.

DNA Methyltransferases

Three DNA methyltransferases have been identified in mammals (Figure 2A). Dnmtl is the maintenance methylase with a preference for hemi-methylated DNA. It co-localizes with replication forks [Leonhardt et al., 1992]. The recent finding that Dnmtl interacts with the clamping protein PCNA in the replication complex is consistent with the maintenance function of Dnmtl [Iida et al., 2002]. Knocking down dnmtl expression by antisense oligonucleotides triggers the intra-S-phase arrest of DNA replication [Milutinovic et al., 2003]. Dnmt3a and 3b are the two known de novo methylases [Okano et al., 1999; Xie et al., 1999]. Recent publications also demonstrate that Dnmt3a and 3b function in DNA methylation maintenance by assisting Dnmtl [Liang et al., 2002].

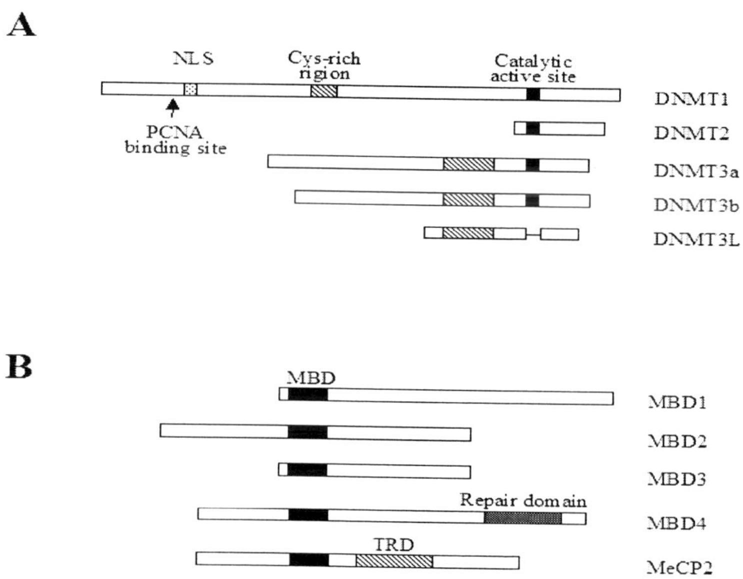

Figure 2. Mammalian DNA methyl-transferases and methylated DNA binding proteins. A). Mammalian DNA methyl-transferases. The methyl-transferase catalytic sites locate at the C-terminal. DNMT3L does not have a catalytic site. HDAC complex binds at the Cys-rich rigion. There is a NLS domain and a replication protein PCNA binding site at DNMT1 N-terminus. B). Mammalian methylated DNA binding proteins. MBD, methyl-CpG binding domain. The DNA repair domain of MBD4 locates at its C-terminus. MeCP2 has a transcription repression domain TRD [Bird and Wolffe, 1999].

Knocking out these two de novo methylases does not reduce the overall methylation levels in the genome. However, it does affect methylation on some genes such as endogenous and exogenous retrovirus, imprinting genes and satellite repeats. The dnmt3a knock out mice die 3 to 4 weeks after birth, while the dnmt3b knockout embryos do not develop further than 9.5 days after fertilization [Okano et al., 1999]. Dnmt3a and 3b have different isoforms. The existence of different isoforms in different cell types may determine the different methylation patterns on their genomic DNA. Dnmt2 has a domain homologous with other methylase catalyzing domains. No methylation activity by Dnmt2 has been observed [Okano et al., 1998], but *in vivo* antibody labeling does show Dnmt2 binding to genomic DNA [Liu et al., 2003]. Dnmt3L (dnmt3 like) shares homology with 3a and 3b, but lacks a functional catalytic domain. Dnmt3L assists other methylases in targeting maternal imprinted genes [Hata et al., 2002].

Methylated DNA Binding Proteins and Gene Repression

DNA methylation can result in two modes of repression which are likely biologically relevant. First, methylation of DNA may interfere with trans-acting factors binding to the cognate DNA sequence. The transcriptional activator CpG binding protein (CGBP) serves as an example. This protein exhibits a unique binding specificity for DNA elements that contain unmethylated CpG motifs [Voo et al., 2000]. Knocking out CGBP is embryonic lethal demonstrating that this protein is crucial for early embryogenesis [Carlone and Skalnik, 2001]. The second mode of repression involves recruiting proteins to methyl-CpG (Figure 2B). The first methyl-CpG binding protein identified was MeCP2 [Lewis et al., 1992]. Dissection of MeCP2 revealed that an 85 amino acid methyl-CpG binding domain (MBD) can bind nonspecifically to one or more symmetrically methylated CpGs [Nan et al., 1993]. By searching expressed sequence tag (EST) databases, four other methyl-CpG binding proteins were identified named MBD1, MBD2, MBD3 and MBD4. MBD4 participates in DNA damage repair [Ballestar and Wolffe, 2001]. The other 4 MBD containing proteins have been implicated in methylation-dependent repression of transcription [Bird and Wolffe, 1999].

Nan et al [Nan et al., 1997] identified the transcriptional repression domain (TRD) in MeCP2 responsible for long-range repression *in vivo*. The TRD domain was later found to be associated with a corepressor complex containing sin3 and a histone deacetylase (HDAC) [Jones et al., 1998; Nan et al., 1998]. HDAC and other repressors were also found to associate with MBD1, MBD2 and MBD3 [Ballestar and Wolffe, 2001]. These findings established a direct causal relationship between DNA methylation and gene repression via chromatin modification. Recent publications further show that MeCP2 complexed with Sin3a is targeted directly to some promoters and represses gene expression [Chen et al., 2003; Martinowich et al., 2003; Stancheva et al., 2003]. However, lines of evidence suggest that other pathways may exist in MeCP2 induced gene repression. First, binding of MeCP2 can form a specific secondary condensed structure *in vitro* [Georgel et al., 2003]. Second, recent publications demonstrate that MeCP2 associates with histone methyltransferase *in vivo* and directs repressive histone Lys9 methylation [Fuks et al., 2003a; Fuks et al., 2003b].

Chromatin Structure and Gene Silencing

Silent chromatin structure can be induced by histone covalent modification, or by linker histones or other accessory chromatin proteins. All histone proteins share similar structural characteristics. They have a globular domain essential for histone interaction with each other in nucleosomes [Wolffe and Guschin, 2000]. Flanking the globular domain, histones can have one or two non-structured tail domains that can be heavily covalently modified. Modification includes phosphorylation, acetylation, methylation, ADP ribosylation and ubiquitination [Berger, 2002]. Each of these modifications combined with their position on different histones has been reported to relate to different chromatin functions and chromatin structural alterations [Cheung et al., 2000]. In addition, findings of the interplay between these modifications on the same histone tail add another layer of complexity. Notably, histones are not the only proteins whose functions are regulated by post-translational modifications. Enzymes responsible for histone modification such as kinases are also involved in signaling transduction pathways. It was proposed that signaling pathways act directly on chromatin components to regulate different DNA-templated processes [Cheung et al., 2000]. Histone tail modifications and their combinatorial functions have been referred to as a Histone Code [Strahl and Allis, 2000].

Histone Modification and Gene Silencing

Core histone deacetylation and methylation have been found to strongly correlate with gene silencing. Studies of bulk acetylation levels in yeast have shown that up to 13 of the 30 tail lysine residues in a histone octamer are acetylated [Roth et al., 2001]. Histone acetylation levels are maintained by the opposing actions of histone acetyltransferase (HAT) and histone deacetylase (HDAC) complexes [Reid et al., 2000; Vogelauer et al., 2000]. Both HATs and HDACs act as part of large complexes. There are multiple mechanisms by which deacetylation of histone tails might facilitate transcription silencing. Deacetylation may stabilize the interaction between histones and DNA by alteration of the N-terminal charge; or may facilitate the internucleosomal interaction via the histone tail; or may recruit additional chromatin accessory proteins [Narlikar et al., 2002]. Also, deacetylation of histone tails may facilitate other modifications on the same histone tail which may induce further chromatin alteration.

Histone lysine methylation occurs on lysine residues 4, 9, 27 and 36 in H3, position 20 in H4 and also the H1 amino terminus. Suv39 is the first histone methyltransferase (HMTase) to be identified. Identification of the HMTase catalytic domain (*S*uv39-*E*(z)-*T*rithorax domain, SET domain) led to the identification of more than 70 gene sequences containing SET domains in mammals. Suv39h, G9a and Ezh2 are involved in histone H3 Lys9 and Lys27 methylation [Cao et al., 2002; Czermin et al., 2002; Rea et al., 2000; Tachibana et al., 2001]. Suv39h induced histone H3 Lys9 methylation can be bound by HP1 and induce heterochromatin [Peters et al., 2001]. Suv39h is also reported to be involved in repression of euchromatic genes [Nielsen et al., 2001]. Lys27 tri-methylation is observed in X-inactivation and bound by PcG proteins [Cao et al., 2002]. Histone H3 Lys4 methylation is induced by another SET protein MLL and associates with gene expression [Milne et al., 2002]. Interestingly, both Ezh2 and MLL are polycomb/trithrox group (PcG/trxG) proteins.

PcG/trxG are involved in epigenetic cellular memory in *Drosophila* [Pirrotta, 1998]. These findings suggest that histone H3 Lys9 and Lys4 tri-methylation may serve as signals for eukaryotic epigenetic regulation memory.

Accessory Chromatin Proteins and Gene Silencing

In addition to nucleosomes, the chromatin fiber contains a diversity of other accessory proteins including linker histone and other non-histone proteins. The binding of linker histone H1 is dynamic [Lever et al., 2000; Misteli et al., 2000]. The displacement of H1 is modulated by cdk2 mediated phosphorylation [Bhattacharjee et al., 2001; Horn et al., 2002]. The binding of linker histone H1 is believed to form a closed chromatin conformation [Karymov et al., 2001] and induce gene silencing on some genes but not all [Shen and Gorovsky, 1996].

Non-histone chromatin accessory proteins are the proteins present in some chromatin structures, which are not essential to form a functional chromatin structure. This includes methylated DNA binding proteins (MBDs), heterochromatin protein 1 (HP1) [Eissenberg and Elgin, 2000], silencing information regulator (Sir) [Shore, 2000], high mobility group (HMG) proteins [Ner and Travers, 1994] and PcG/TrxG proteins etc. The binding of these factors can alter chromatin structure. Some chromatin factors alter chromatin structure via histone tail covalent modification. For example, PcG protein Ezh2 catalyzes histone H3 Lysine27 tri-methylation [Czermin et al., 2002]. Others can form higher order chromatin structures like linker H1, HP1 [Eissenberg and Elgin, 2000] and PcG proteins [Sewalt et al., 2002]. A common characteristic of most chromatin accessory proteins is their ability to form polymers which can spread a specific chromatin conformation into and subsequently silence flanking DNA sequences.

Gene Silencing Network

Broad interplay among different chromatin alterations has been reported. Interplay can happen between the same levels of chromatin alterations. For example, MeCP2 was reported to replace linker histone H1 *in vitro* [Nan et al., 1997]. Lys9 deacetylation is proposed as a prerequisite for methylation at the same lysine residue [Cheung et al., 2000]. However, often the interplay is between different chromatin alteration levels. Histone tail modification such as H3 Lys9 methylation determines chromatin accessory protein binding. MeCP2 bound on methylated DNA can induce histone H3 deacetylation and methylation [Fuks et al., 2003b]. Also, H3 methylation is reported to direct DNA methylation in yeast [Schramke and Allshire, 2003] and plants [Jackson et al., 2002]. Chromatin modifications such as DNA methylation and histone modifications can be recognized by specific binding proteins. The binding proteins can subsequently recruit other chromatin modification enzymes to the same gene. As a result multiple chromatin modifications and multiple chromatin alterations co-exist on the same gene to enforce chromatin structures, be it condensed or open. It is reasonable to hypothesize that DNA methylation, histone modifications, and chromatin factor binding are all parts of a gene repression network. They function separately and cooperatively to achieve an inaccessible chromatin conformation.

Variegated Epigenetic Silencing

Variegated epigenetic silencing is observed in *Drosophila*, yeast and transgenic mice and is designated as position effect variegation (PEV). The prototype PEV phenomenon was observed in *Drosophila*. When a normally active euchromatic gene is translocated into the boundary between heterochromatin and euchromatin, its expression is shut down in a portion of the cells due to the spreading of condensed heterochromatin into the translocated gene. PEV of the eye color *white* gene results in the eye color mosaicism in *Drosophilia*. Massive genetic screens led to the identification of a large amount of PEV modifier genes in *Drosophila* [Schotta et al., 2002]. Among them, some can induce chromatin modification and structural alteration [Schotta et al., 2002].

PEV is also observed in transgenic mice. Transgene expression in mouse lines with multicopy transgenes does not correlate with the number of integrated transgene copies indicating an important role for integration sites. Transgene "copy number independent and position dependent" expression can be avoided by inclusion of locus control regions (LCR). LCRs are gene regulatory elements in mammals and birds that can overcome the repressive effects associated with heterochromatic transgene locations to promote position-independent and copy number-dependent transgene expression [Kioussis and Festenstein, 1997]. Transgenes with partially deleted LCRs located at the edge of centromeric heterochromatin regions are subject to PEV [Festenstein et al., 1996; Milot et al., 1996]. Using these PEV mouse lines, Milot and others have shown that PEV is controlled by the balance of positive transcription factors (Sp1 and EKLF) and repressive chromatin modifiers (Suv39H1, HP1, and PcG proteins) [Festenstein et al., 1999; Lundgren et al., 2000; McMorrow et al., 2000]. Interestingly, while most of the PEV in transgenic mice occurs when transgenes are located in heterochromatin, a recent paper demonstrates that PEV also can be established in non-centromeric regions [Ayyanathan et al., 2003].

Model Systems Used in Retrovirus Silencing Studies

From the first retrovirus silencing experiment done over 20 years ago, different model systems have been used. In general the model systems used for retrovirus silencing can be categorized into two groups: infection models and transfection models. In infection models, retrovirus or retrovirus vectors are infected into primitive cell types. In transfection models, retrovirus DNA sequences with or without modifications are introduced into different cell types. Both of these models have provided useful information on retrovirus silencing. Understanding the advantages and disadvantages of these systems is crucial to compare the data from different publications. The proper usage of retrovirus silencing systems is also important for the development of silencing-resistant retrovirus vectors.

Infection Models

In the infection models, retrovirus or retrovirus vectors are infected into preimplantation embryos, EC, ES or HSC. Comparison of retrovirus vector expression in primitive cells and mature cells is commonly used to determine retrovirus silencing effects. The existence of provirus DNA is used to determine successful viral infection. When wildtype retrovirus is used, infectious viral particles and viral protein expression are used to detect retrovirus expression. Expression of retrovirus vectors is detected by the expression of reporter genes. Because retrovirus vector infected cells cannot make infectious retrovirus vectors, it can avoid the virus spread problem associated with wildtype retrovirus infection. The existence of easily detectable reporter genes gives retrovirus vectors another advantage. Retrovirus vector infection models provide a unique opportunity to study different types of retrovirus silencing phenomena and their mechanisms. An infection model is also important to evaluate novel silencing-resistant retrovirus vector design.

Multiple copy provirus integration is a potential problem in infection models. The existence of expressing retrovirus vector can mask silent vectors in the same cell. The silent retrovirus vector transduced cells are also mixed with an excess of untransduced cells. The isolation and identification of single copy vector transduced cell lines are crucial to retrovirus silencing study. The use of *in vitro* cultured primitive cells facilitates the isolation and identification of silent retrovirus. The existence of homologous endogenous retrovirus in the mouse genome is another potential problem. It makes the determination of viral integration difficult. This problem can be avoided by careful design of retrovirus vectors.

Recently, HSC were used as model systems mainly because of their potential use in blood disease gene therapy. Research on HSC is always hampered by the lack of identity of HSC cells [Trevisan and Iscove, 1995]. Because the absolute purification of HSC is not possible, it is difficult to distinguish retrovirus expression in transduced HSC cells from other progenitor cells. One way to avoid this problem is serial transplantation. Since only HSC cells can survive after serial transplantation, the expression of the retrovirus vectors has to derive from HSC cells [Challita and Kohn, 1994].

Transfection Models

To study the involvement of DNA methylation in retrovirus silencing, two different transfection systems were introduced. M. Lorincz transfected *in vitro* methylated or unmethylated retrovirus sequences into a specific integration site in murine erythroleukemia (MEL) cells [Lorincz et al., 2001; Lorincz et al., 2002]. While this system can clearly determine whether DNA methylation functions in retrovirus silencing, the disadvantage is that the result might heavily depend on the integration site being used. Transgenic *Drosophila* also can be used to investigate the involvement of DNA methylation in retrovirus silencing. *Drosophila* was long believed to be a methylation-free system [Urieli-Shoval et al., 1982]. D. Pannell in our laboratory established that retrovirus sequences function as silencers in transgenic *Drosophila* suggesting that retrovirus silencing can be established independent of DNA methylation [Pannell et al., 2000]. However, a small amount of DNA methylation is detected in *Drosophila* especially in early embryos [Lyko et al., 2000]. The function of this DNA methylation is not yet clear. Nevertheless, the DNA methylation in *Drosophila* is not likely the same as in mammals, because the methylase being used is homologous to Dnmt2 [Hung et al., 1999] while Dnmt2 is not functional as a methylase in mammals [Okano et al., 1998]. The use of these two models has provided great information on whether and how DNA methylation is involved in retrovirus silencing.

Transfection models were also used to identify and modify silencer elements in retrovirus vectors. When a retrovirus sequence was linked to an LCR β-globin reporter gene and microinjected into fertilized mouse eggs, silencing of the reporter gene by the retrovirus sequence was detected in fetal liver [McCune and Townes, 1994; Osborne et al., 1999]. Transgene DNA and its expression were evaluated in each embryo. The silencing effect was evaluated by the percentage of expressing embryos and their expression levels. Since transgene expression suffers from PEV, a strong reporter gene that can promote position independent transgene expression is important to this model, and therefore a human LCR β-globin cassette was used as the reporter gene for retrovirus silencers. The human β-globin LCR is composed of four DNase I-hypersensitive sites all of which are required for full LCR activity [Pasceri et al., 1998]. The reporter cassette we used contains a 3.0-kb LCR (5'HS2 to 5'HS4) linked to the human β-globin transgene and directs 16 to 71% expression at all integration sites [Ellis et al., 1997]. The transgenic mouse model is very useful for evaluation of modifications to retrovirus vector silencer elements and development of silencing resistant retrovirus vectors.

Retrovirus Vector Silencing Mechanisms

Genome Defence Hypothesis

During the course of evolution, cells have been exposed to foreign DNAs. Intragenomic parasites are one of the important pathways for cellular genomes to obtain new genes. But integration of intragenomic parasites can cause endogenous gene damage and misregulation, so is potentially harmful. Genome defence systems may have developed during evolution to ensure genome integrity [Doerfler, 1991]. The genome defense hypothesis contends that cytosine methylation arose in mammals to restrict expression of intragenomic parasites [Walsh and Bestor, 1999]. Evidence supporting the genome defense hypothesis stems from the finding that the LTRs of retrotransposable elements such as L1s, IAPs, and endogenous retroviruses are all heavily methylated at CpG dinucleotides [Yoder et al., 1997a; Yoder et al., 1997b]. Further evidence supporting a role for methylation in genome defense comes from the finding that the IAP class of endogenous retrovirus becomes reactivated in ES cells that are null for the maintenance methyltransferase Dnmtl [Walsh et al., 1998]. Notably DNA methylation is also involved in the restriction-modification system of bacterial defense function indicating that DNA methylation is an old player in cell defense [Bestor, 1990].

DNA Methylation and Retrovirus Silencing

Functional Role of DNA Methylation in Retrovirus Silencing

Silencing of retrovirus is often accompanied by high density cytosine methylation [Jahner et al., 1982; Stewart et al., 1982]. The methylation inhibiting drug 5-azacytidine (5-AzaC) has been used to reactivate silent provirus in EC cells [Stewart et al., 1982]. The Jaenisch laboratory demonstrated that *in vitro* methylated retrovirus DNA is not infectious in cells, but treatment of the recipient cells with 5-AzaC reactivated infectious virus [Simon et al., 1983]. A functional role of DNA methylation in retrovirus silencing and extinction has been demonstrated [Cherry et al., 2000; Lorincz et al., 2000; Lorincz et al., 2001]. Recently Swindle et al [Swindle et al., 2004] also showed a causal role of DNA methylation in establishment and maintenance of retrovirus silencing by mutating the LTR promoter region CpGs.

It is proposed that the binding of MBD proteins on methylated DNA recruits HDACs and subsequently represses gene expression [Ballestar and Wolffe, 2001; Jones et al., 1998; Nan et al., 1998]. MeCP2 is recruited to silent retrovirus sequence [Lorincz et al., 2001; Yao, 2004]. However, no reactivation was observed by the HDAC inhibitor TSA or sodium butyrate alone in silent ES cells [Yao, 2004]. Clearly mechanisms other than histone deacetylation exist in the DNA methylation dependent retrovirus silencing pathway.

DNA Methylation Independent Pathway of Retrovirus Silencing

While there is data supporting a causal role of DNA methylation in retrovirus silencing, several lines of data suggest other mechanisms may also exist. First, 5-AzaC reactivation of silent retrovirus is only partial. This suggests that other methylation-independent mechanisms are actively silencing some of retroviruses in these cells. Second, retrovirus silencing is established much earlier than DNA methylation. Silencing occurs in EC cells 2 days post-infection whereas methylation is not detectable until 8-16 days [Gautsch and Wilson, 1983; Niwa et al., 1983]. Third, retrovirus silencing can be established in *Drosophila.* Because *Drosophila* has barely detectable levels of DNA methylation, this result suggests that DNA methylation is not essential for retrovirus silencing [Pannell et al., 2000]. Fourth, among the identified silencers in retrovirus LTR, the promoter is the only CpG-rich silencer. DNA methylation induced retrovirus silencing cannot account for the other silencers function.

Dnmt3a and 3b are the only de novo methylases known to methylate exogenous transduced retrovirus. Identification of silent retrovirus in dnmt3ab knockout ES cells is the best way to test the existence of DNA methylation independent retrovirus silencing pathway. Indeed, silent single copy retrovirus vector transduced dnmt3ab-/- ES cell clones were identified in our lab. DNA methylation sensitive digestion confirmed no DNA methylation on the integrated retrovirus [Yao, 2004], demonstrating the existence of a DNA methylation independent retrovirus silencing pathway.

Relationship between DNA Methylation Dependent and Independent Pathways

Recently, multiple pathways functioning stepwise in retrovirus silencing have been proposed. MoMLV vector infected MEL cells that initially expressed were sorted and silencing was examined over time. It was found that an initial extinction event is reversible by the HDAC inhibitor trichostatin A (TSA). At later time points, reactivation of expression required 5-AzaC in conjunction with trichostatin A, suggesting that methylation is a secondary step associated with maintenance of silencing [Lorincz et al., 2000; Lorincz et al., 2002]. By studying retrovirus infected ES cell long-term culture extinction, S. Cherry also proposed that there are DNA-methylation independent pathways at early stages and dependent pathways at later stages in retrovirus silencing [Cherry et al., 2000]. However, retrovirus silencing occurs in do novo methylase null ES cells [Yao, 2004]. Therefore, both DNA methylation dependent and independent pathways are capable of maintaining retrovirus silencing. We propose that DNA methylation dependent and independent pathways function in parallel in gene repression.

Chromatin Conformation in Retrovirus Silencing

Retrovirus provirus is integrated into the host genome and inherited with other endogenous genes [Breindl et al., 1979]. Epigenetic mechanisms such as histone modification and chromatin accessory protein binding may therefore also function on integrated provirus DNA. In 1979, Breindl and Jaenisch published that the chromatin conformation of transduced retrovirus is altered in the permissive and non-permissive tissue of the same mouse line by comparing their sensitivity to DNase digestion. In the non-permissive tissue in which the infected retrovirus is silent, the provirus is more resistant to DNase I digestion suggesting a condensed conformation [Breindl and Jaenisch, 1979]. DNase I hypersensitive site alteration was also observed in an endogenous allele mutated by Mo-MuLV insertion [Breindl et al., 1984].

Retrovirus sequence can silence the linked human LCRβ-globin transgene in transgenic mice. Using ChIP assay, we have shown that the retrovirus silenced transgene is marked by silent chromatin characterized by histone H3 deacetylation and linker histone H1 [Pannell et al., 2000]. These results demonstrate that histone modifications are involved in retrovirus silencing. A similar silent chromatin is formed on the silent retrovirus vectors in genuine infected ES cells as well as the silenced retrovirus in variegated ES cells [Yao, 2004]. As in wildtype ES cells, silent retrovirus vectors in *dnmt3ab-/-* cells are also marked by histone H3 deacetylation and linker histone H1 binding [Yao, 2004]. The existence of similar chromatin conformation on silent retrovirus regardless of DNA methylation shows that silent chromatin is a universal mark for retrovirus silencing.

No reactivation was observed by TSA treatment on retrovirus silenced transgene in transgenic mice [Pannell et al., 2000] and silenced retrovirus in transduced ES cells [Yao, 2004]. TSA treatment can increase the acetylation level of silenced transgene in transgenic mice, however TSA treatment cannot remove the linker histone H1 (unpublished data). These results suggest an important role for histone H1 in maintaining retrovirus silencing. The binding of histone H1 in silent *dnmt3ab-/-* cells is more abundant than in wildtype ES cells [Yao, 2004]. Since MeCP2 can replace histone H1 in binding linker DNA [Nan et al., 1997], we propose there is a compensatory competition between H1 and MeCP2 binding. When MeCP2 cannot bind, histone H1 will take over to maintain silent chromatin. This dynamic interplay of different pathways directs silent chromatin formation and maintenance.

Retrovirus Silencing Insulation

Retrovirus Vectors that Resist Silencing

The existence of retrovirus silencing in stem cells indicates the need to develop silencing-resistant retrovirus vectors. Development of such retrovirus vectors started with the isolation of EC cell permissive retrovirus strains and continued with the identification and removal of silencer elements from retrovirus vectors. Self-inactivating (SIN) retrovirus vectors are vectors with a deletion in the 3'LTR promoter region so that there are no functional promoters in any of the provirus LTRs after vector transduction. Although SIN retrovirus vectors were originally developed to improve vector safety, the extreme deletion of the retrovirus sequences in SIN vectors also reduces retrovirus silencing. Combining these efforts, many retrovirus vectors have been developed for stem cell gene transfer and gene therapy.

The identification and characterization of multiple silencer elements in retrovirus vectors paved the way for the construction of numerous modified vectors that lack these sequences. Silencer elements are DNA binding sites for trans-acting factors that directly or indirectly reduce transcription at promoters. At least four separate silencer elements located in the LTRs and the adjacent primer binding site (PBS) have been identified in retrovirus vectors. The LTR promoter is CpG rich serving as a potential silencer [Osborne et al., 1999; Swindle et al., 2004]. The Negative Control Region (NCR) was identified in the upstream region of the LTR which reduces virus expression [Flanagan et al., 1989]. Mutations in the direct repeat (DR) of the LTR enhancer were found to increase expression in stem cells [Hilberg et al., 1987]. The groups of Jaenisch and Goff found mutant PBS sites that increased expression from the LTR [Barklis et al., 1986; Colicelli and Goff, 1986]. Several negative trans-acting factors had been found to bind these silencer elements. Specifically, the multi-functional transcriptional regulator YY-1 binds to NCR [Flanagan et al., 1992], six distinct nuclear factors bind to each of the DRs [Speck and Baltimore, 1987], and the PBS is bound by factor A [Petersen et al., 1991]. In addition, the embryonic LTR-binding protein (ELP) [Tsukiyama and Niwa, 1992] was found to bind between the NCR and DR [Tsukiyama et al., 1989]. It has been shown that no single silencer element is sufficient to induce complete silencing [Osborne et al., 1999], suggesting that silencer elements additively increase the probability of silencing via multiple mechanisms.

Mo-MuLV is one of the earliest retrovirus isolated and studied. Given its well understood biology and its ability to efficiently infect a wide range of cells, Mo-MuLV has been widely used to derive retrovirus vectors. One problem for wildtype Mo-MuLV vectors is they are not permissive in ES or EC cells because of the poor functionality of Mo-MuLV

LTR enhancer in these cells [Tsukiyama et al., 1989]. PCMV is a mutant form of myeloproliferative sarcoma virus (MPSV) containing a deletion of one of the enhancer DRs, as well as a mutation in the ELP binding site and a point mutation that introduces an Sp1 binding site [Hilberg et al., 1987]. Although the PCMV enhancer is functional in both EC and ES cells, PCMV derived vectors are still restricted in ES cells [Grez et al., 1990]. Combination of the PCMV LTR with a PBS from an integration-defective Mo-MuLV mutant dl587rev [Colicelli and Goff, 1987] which does not bind factor A, yields the murine embryonic stem cell virus (MESV) [Grez et al., 1990]. Hawley et al further refined the MESV vector to increase titers by including an extended packaging signal from the LN vector and an upstream gag sequence from Moloney murine sarcoma virus, (MoMSV) resulting in MSCV [Hawley, 1994]. MSCV retains silencer elements such as the NCR, one of the DRs, and the CpG rich promoter.

Substitution of the MoMLV enhancer has been another approach to improve retrovirus vector expression in primitive cells. The MPSV (myeloproliferative sarcoma virus) enhancer had been shown to have increased expression in EC cells [Hilberg et al., 1987]. Substitution of the Mo-MuLV LTR with the MPSV LTR in conjunction with the dl587rev PBS yielded the MD (MPSV, dl587rev) vector [Challita et al., 1995]. Unfortunately, the expression from the MD vector did not significantly increase. However, the MND (MPSV, NCR, dl587rev) vector derived from a deletion of the NCR in the MD vector did show substantial improvement with expression in 43% of infected EC cells [Challita et al., 1995]. The MND vector also showed significantly increased expression in murine hematopoietic stem cells (HSCs) [Robbins et al., 1998].

Finally, SIN retrovirus vectors have steadily gained popularity due to the decreased probability that these vectors will undergo recombination leading to the generation of a rescued infectious virus. The first Mo-MuLV-based SIN retrovirus vector used a 299bp deletion in the 3'LTR that removes most of the DR and the promoter, but retains the NCR and wild-type PBS [Yu et al., 1986]. More recently the HSC1 vector was constructed in our laboratory by deletion of the entire NCR, DR, and part of the CpG rich promoter sequence in MSCV which retains the mutant PBS. HSC1 vector directs neo expression in 71% of transduced EC cells, offering a further improvement still [Osborne et al., 1999]. The HSC1 vector is particularly attractive both for its ability to express in non-permissive environments and its SIN properties.

SIN-Lentivirus vectors were created for safety reasons and may express better in primitive cell types due to the deletion of potential silencer sequences from the LTRs [Miyoshi et al., 1998]. Indeed, SIN-Lentivirus vector has been shown to transduce more efficiently into a broad range of cell types with higher expression levels compared with vectors with intact LTRs [Miyoshi et al., 1998]. Notably, the experience and knowledge gained in developing retrovirus vectors are well transferred into Lentivirus vector development.

In summary, progressive deletion or mutation of silencer sequences has lead to the development of retroviral vectors that have an increased probability of expression in primitive cell types. However, none of the current vectors are able to express at all integration sites despite mutations in all known retroviral silencer elements, suggesting that novel silencer elements remain to be discovered and may reside in the essential sequences of retrovirus vectors. In this consideration, insulation of the retrovirus silencer elements should be the next step in development of silencing resistant vectors.

Insulator Elements

The eukaryotic nucleus is organized into active and inactive domains [Weintraub and Groudine, 1976]. Within each chromosomal domain, each gene is embedded within a chromosomal environment of other DNA sequences that have the potential to affect its expression. Boundary elements such as insulators are needed to keep different domains and genes functioning separately. Insulators are defined as DNA elements located between different functional chromosomal domains that keep each of these domains functioning properly. In practice, insulators are defined by two assays: Enhancer Effect Blocking Assay [Kellum and Schedl, 1991] and Position Effect Protection Assay [Kellum and Schedl, 1992] (Figure 3A). Insulators can block the enhancing effect of an enhancer on a promoter. Transgenes in different integration sites suffer from PEV. Flanking transgenes with insulators can block position effects and promote position-independent transgene expression. Insulator elements themselves should have no enhancer, silencer or promoter activity. In other words, insulator elements should be neutral to make sure the blocking effect observed in these two assays is the result of true insulation. Insulator elements have silencer blocking activity. Otte and collegues showed that insulators block silencing effects of LexA silencing factor fusion proteins recruited to episomal reporter genes in human cells [van der Vlag et al., 2000]. Insulator silencer blocking activity was also tested in transgenic *Drosophila* [Barolo and Levine, 1997; Cai and Levine, 1995]. Insulator elements are found in *Drosophila* (*gypsy*, scs/scs', Fab-7, Fab-8, etc.) yeast (HMR, STAR etc.), sea urchin (sns), birds (cHS4) and mammals (H19/Ig2f, BEAD-1, HS5, etc.) [Bell and Felsenfeld, 1999]. Among them, chicken HS4 (cHS4), *Drosophila gypsy* and Scs are three well studied insulator elements.

Insulator elements were first discovered in *Drosophilia.* Scs/Scs' (special chromatin structures) insulators are located at the boundaries of the heat-shock *hsp70* gene and protect it from stimulation by neighboring genes [Udvardy et al., 1985]. These elements were shown to protect *white* transgenes against chromosomal position effects to produce uniform eye color [Kellum and Schedl, 1991]. Later the same group also showed that scs/scs' can block enhancer effects [Kellum and Schedl, 1992]. BEAF-32A and BEAF-32B (a boundary-element-associated factor of 32kDa) were found to bind to scs' [Hart et al., 1997; Zhao et al., 1995]. Zest-white-5 (Zw-5) binds to the scs insulator [Gaszner et al., 1999]. BEAF was found to locate at hundreds of interbands and many puff boundaries on polytene chromosomes [Zhao et al., 1995].

Gypsy insulator is another *Drosophila* insulator that has been extensively studied. It was originally found in the 5' non-translated region of *Drosophila gypsy* retrotransposon [Geyer and Corces, 1992]. Insertion of the *gypsy* retrotransposon in the *yellow* gene blocks enhancers located upstream of the insertion site causing abnormal eye pigmentation. A 430-bp fragment of the 5' non-translated region was found to be responsible for this blocking. *Gypsy* insulator was also reported to protect a DNA replication origin from position effects [Lu and Tower, 1997] and to block polycomb group (PcG) protein induced gene repression [Mallin et al.,

1998]. Interestingly, when multiple *gypsy* insulators were placed between an enhancer and promoter, or the enhancer was flanked with two *gypsy* insulators no insulation was observed [Cai and Shen, 2001; Melnikova et al., 2002; Muravyova et al., 2001]. Su(Hw) (Suppressor Hairy-wing) protein is required for *gypsy* insulator enhancer-blocking activity in *Drosophila*, and the *mod(mdg4)* gene is also involved [Gause et al., 2001; Gdula et al., 1996; Ghosh et al., 2001]. These two proteins direct *gypsy* insulator DNA to the nuclear periphery [Gerasimova et al., 2000]. No Su(Hw) homologue has been found in mammalian cells, but *gypsy* insulator was found to be bound by mammalian nuclear matrix proteins and function as a matrix attachment region (MAR) element [Nabirochkin et al., 1998]. Most recently the *gypsy* insulator was also reported to serve as a MAR in *Drosophila* [Byrd and Corces, 2003]. MAR elements are binding sites for nuclear scaffold proteins *in vitro* and are proposed to attach chromatin to the nuclear matrix *in vivo* [Dickinson et al., 1992]. It is proposed that some MAR elements can shield transgenes from the influence of flanking genes [Allen et al., 2000; Bonifer et al., 1994].

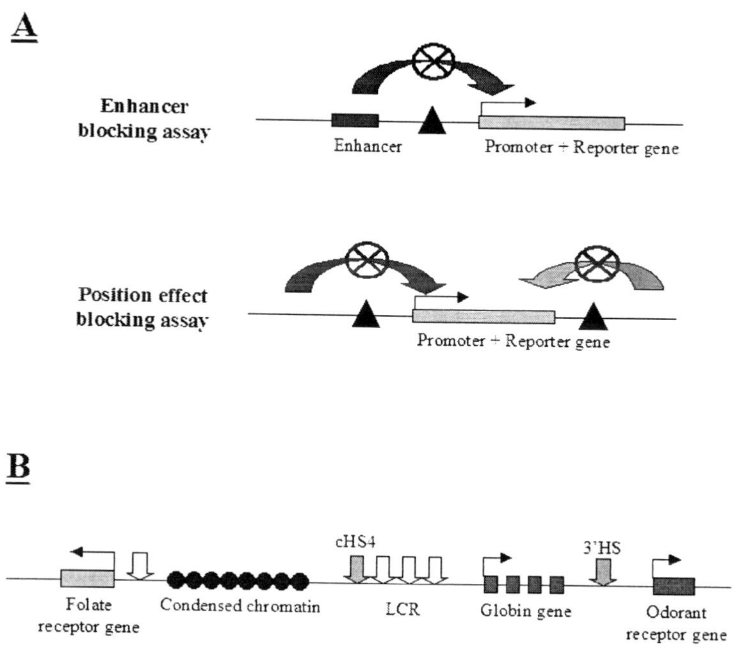

Figure 3. Insulator element. A. Two assays used to define insulator elements. In the enhancer blocking assay, the insulator element (triangle) located between the enhancer and promoter will block the enhancer effect. In the position effect blocking assay, flanking the transgene with the insulator elements will protect the transgene from position effects from flanking endogenous genes. B, Location of chicken HS4 insulator in the globin gene locus. cHS4 and 3'HS are two identified insulator elements functioning as boundary elements between different chromatin domains [Recillas-Targa et al., 2002]. See text for detail.

The first and the best-characterized vertebrate insulator element is the cHS4 insulator of the chicken β-globin gene (Figure 3B). The β-globin locus in chicken erythrocytes contains a 33-kb domain of open chromatin and a 16-kb stretch of heterochromatin [Litt et al., 2001a; Litt et al., 2001b]. Upstream of this 16-kb heterochromatin domain is the folate receptor gene which is regulated distinctly from that of globin genes [Prioleau et al., 1999]. The 1.2kb cHS4 insulator is located at the boundary between the active globin domain and the heterochromatin domain. cHS4 insulator showed position effect protection activity [Chung et al., 1993] and enhancer blocking activity [Chung et al., 1997]. It blocks the spreading of histone deacetylation from heterochromatin but not DNA methylation [Mutskov et al., 2002]. A 250-bp CpG island that contains the constitutive DNase I hypersensitive site was found to retain most of the insulating activity. This 250-bp core element is used as a dimer to achieve full insulation [Chung et al., 1997]. Further dissection of the 250-bp core element found that only the footprint II bound by CTCF is required for enhancer blocking activity [Bell et al., 1999]. CTCF is a ubiquitous zinc-finger DNA-binding protein which recognizes multiple DNA sequences [Burcin et al., 1997; Klenova et al., 1998]. CTCF is involved in multiple functions such as gene silencing, gene activation, X-inactivation and insulator activity [Lutz et al., 2000]. Position effect protection activity of the cHS4 core element does not require CTCF binding and resides in the other binding sites in the core element [Recillas-Targa et al., 2002]. The proteins responsible for position effect protection are not known.

Insulation of Retrovirus and Lentivirus Silencing

Expression of integrated retrovirus may be influenced by the flanking chromosomal environment of the integration site [Verma and Somia, 1997]. On the other hand, insertion of retrovirus also can induce insertional gene activation or repression [Breindl et al., 1984]. All retrovirus vectors contain LTRs and *gag* sequences that harbor known [Challita et al., 1995] and unidentified [Osborne et al., 1999] silencers. These elements can silence the transgene harbored in the retrovirus vector and potentially genomic genes flanking the integration site. Similarly, constant high expression of the retrovirus vector can activate flanking genes. Insulators are *cis*-acting elements and theoretically can be used to block retrovirus silencing spread and protect a transgene harbored inside a retrovirus vector. In addition, insulator elements also can be used to block silencing or activating effects of retrovirus sequences and protect the genes flanking the integration sites.

The monomer 1.2-kb cHS4 insulator has been used in retrovirus and Lentivirus vectors with limited success. Insertion of cHS4 into a Mo-MuLV construct 3'LTR can be faithfully duplicated into both LTRs after retroviral integration. The insulated retrovirus promotes an increased probability of expression at random chromosomal integration sites in MEL cells. Unfortunately, improved reporter gene eGFP expression was not observed when used to infect ES cells [Rivella et al., 2000]. Similar retrovirus vectors based on the MSCV vector were constructed by D. Emery [Emery et al., 2000]. cHS4 in one orientation in MSCV LTRs supports high percentage GFP positive cells and high GFP expression level in both 3T3 and bone marrow derived blood cells. However extinction of retrovirus expression was observed in long-term reconstituted mice. Further experiments showed that cHS4 insulators in the

LTRs dramatically decrease gene expression from an internal pgk promoter. In addition, severe differentiation extinction was observed regardless of the existence of cHS4 insulator [Yannaki et al., 2002]. At the same time, the same vector harboring a γ-globin cassette was able to successfully reduce gene silencing [Emery et al., 2002]. Notably, the γ-globin expression cassette used in this vector has its own enhancer. Similarly, cHS4 also failed to protect a Lentivirus vector when located in the LTRs in transduced CD34+ blood cells [Ramezani et al., 2003]. However, when the insertion of cHS4 in LTRs was combined with insertion of a matrix attachment region (MAR) element, higher and uniform gene expression was achieved [Ramezani and Hawley, 2003]. The same vector was also reported to support better gene expression in human ES cells although long term culture extinction was not avoided [Ma et al., 2003].

Use of insulators in retrovirus or Lentivirus vectors is mostly unsuccessful due to the lack of knowledge of retrovirus silencer blocking. Current insulator assays define the enhancer blocking activity and position effect protection activity of insulator elements. However, retrovirus silencer blocking activity is different from these two activities. Retrovirus silencing is a negative effect on the transgene from retrovirus sequences. Blocking a retrovirus silencer is not the same as blocking enhancer effects. Also, silencer blocking is not necessarily equal to position effect protection. In retrovirus vectors, the transgene is flanked by retrovirus silencers from both the 5' and 3'LTR. Retrovirus silencers may affect the transgene more than the flanking genomic DNA sequences simply because they are closer to the transgene. Our and other reports all show that blocking position effects by flanking retrovirus vectors with insulators is not sufficient to avoid silencing [Emery et al., 2000; Rivella et al., 2000; Yao et al., 2003]. An assay specific for retrovirus silencer blocking is needed to identify insulator elements suitable to be used in retrovirus vectors. In our work, an LCR β-globin expression cassette is used as a reporter transgene to evaluate the blocking of retrovirus silencer effects in transgenic mouse fetal liver cells [Yao et al., 2003]. Retrovirus sequences can silence this LCR β-globin transgene when linked directly [Osborne et al., 1999; Pannell et al., 2000]. Insulators with retrovirus silencer blocking activity will be able to protect transgene expression when inserted between the transgene and retrovirus sequence. Using this assay we screened different insulator elements and found that cHS4 has the ability to block retrovirus silencers [Yao et al., 2003]. The identification of retrovirus silencer blocking insulator elements is important to the design of insulated retrovirus vectors to avoid retrovirus silencing.

Using our retrovirus silencer blocking assay, we also showed retrovirus silencer blocking activity is independent of CTCF binding [Yao et al., 2003]. CTCF binding is the common feature for vertebrate insulator element enhancer blocking activities [Bell et al., 1999]. Therefore, this result further supports retrovirus silencer blocking activity is different from enhancer blocking activity. When the cHS4 dimer core insulator is used to flank the LCR β-globin reporter gene, full levels of expression were observed in the transgenic mice [Yao et al., 2003]. This result suggests the transgene in a retrovirus vector needs to be flanked by insulators to achieve full protection. However, flanking the reporter gene in retrovirus vectors with the cHS4 dimer core insulator failed to produce high titer virus suggesting another non-homologous insulator is needed to combine with cHS4 [Yao et al., 2003].

No silencer blocking activity was observed with *gypsy* and Scs insulators. However the *gypsy* insulator was observed to increase transgene expression when located in the retrovirus LTRs. cHS4, on the other hand, does not increase transgene expression in the same configuration. This result indicates that the *gypsy* insulator does not increase transgene expression by blocking retrovirus silencers [Yao et al., 2003]. *Gypsy* was reported to function as a MAR element in both *Drosophila* [Byrd and Corces, 2003] and mammalian cells [Nabirochkin et al., 1998]. The increase of transgene expression by *gypsy* in retrovirus vectors may be due to its ability to tether the transgene to the nuclear matrix. Regardless of the mechanism, the increase of transgene expression by *gypsy* insulator can be used to design high expressing retrovirus vectors.

An ideal insulated retrovirus vector would be able to protect both the internal transgene and the integration site flanking endogenous genes. Currently all the insulated retrovirus vectors use insulators located in LTRs and aim to block position effects from the flanking sequences and to prevent induction of insertional activation. However, some of the retrovirus silencers especially those located in the 5'LTR and *gag* sequences are left between the two insulators and adjacent to the transgene. To completely protect the transgene, two different insulators should be used, one located in the U3 region of the LTRs, the other between 5'LTR-PBS and the internal transgene. In this configuration, the transgene will be flanked by two insulators to achieve high expression and the flanking genomic genes will also be partially insulated from the integrated retrovirus vector.

Retrovirus Silencing Model

Although retrovirus vector silencing has long been documented, there is still argument about whether retrovirus silencing is solely due to position effects. Although the integration site flanking chromatin may be involved in the decision, the silencing of retrovirus vectors is mainly due to effects induced by the silencer elements in retrovirus vector sequences. Multiple silencer elements have been identified in retrovirus LTR and *gag* sequences [Challita et al., 1995; Osborne et al., 1999]. Osborne et al also demonstrate that deletions in the SIN retrovirus vectors are not enough to remove all the silencer elements [Osborne et al., 1999]. Transgene experiments show that both retrovirus and Lentivirus sequences are dominant silencers in both transgenic mice and *Drosophilia* [Pannell et al., 2000]. Direct evidence comes from blocking of retrovirus silencers by insulators. After the genomic position effect is blocked, retrovirus sequences are still silenced in transgenic mice [Yao et al., 2003]. Our and other results also show that flanking the retrovirus vector with insulators failed to avoid silencing on both retrovirus and Lentivirus vectors [Emery et al., 2000; Ramezani et al., 2003; Yannaki et al., 2002; Yao et al., 2003].

Present models account for transgene expression variegation by proposing that the cell equates the balance between positive and negative factors located in both the transgene and the flanking sequences [Dillon and Festenstein, 2002]. We propose that this can be readily applied to retrovirus silencing. The silencer elements in retrovirus vectors are dominant elements that recruit negative factors in stem cells. The transgenes harbored in the vectors bear positive elements such as enhancer, promoter or LCR sequences. The integration site flanking sequences can be either positive or negative. The final expressing state of the transgene is the sum of all the negative and positive effects of all of the recruited factors. We proposed that a negative integration environment permits the establishment of silent chromatin on the integrated retrovirus vectors. On the other hand, when a provirus locates in a constantly active domain the chromatin environment will not permit the silencing machinery to establish silent chromatin on the retrovirus vector. As a result, the integrated retrovirus will continue to express even after differentiation, but only to low levels because of continued binding by the stem cell silencing factors. The variegated retrovirus vector may reflect the recruitment of silencing proteins by retrovirus silencers into an otherwise active chromatin environment that is near heterochromatin. During differentiation of ES cells, some genomic domains are progressively silenced [Rasmussen, 2003]. Correspondingly, provirus located in these regions may also be shut down resulting in extinction (Figure 4.).

Since the Mo-MuLV enhancer is almost non-functional in ES cells [Tsukiyama et al., 1989], the balance of the positive and negative factors lean to the negative side. This explains why Mo-MuLV can induce stronger silencing compared to MSCV [Hawley, 1994]. This balance model also can explain why the inclusion of LCR or MAR elements in retrovirus vectors can limit silencing effects [Emery et al., 2002; Ma et al., 2003; Ramezani et al., 2003]. This model emphasizes the silencers as the main determinant of the retrovirus silencing, but does not underestimate the importance of the integration site flanking sequences.

Future Directions

Role of Integration Site in Determining Retrovirus Silencing

Our model proposes that the integration site is involved to some degree in the decision to silence retrovirus in stem cells. To test this model, the relationship between retrovirus vector expression status and its integration site needs to be investigated. The effect of the integrated retrovirus vector on the flanking genomic genes also needs to be studied. The inverse PCR technique has been combined with bioinformatics research to successfully identify large number of integration sites in retrovirus and Lentivirus infected bulk cell populations [Schroder et al., 2002; Wu et al., 2003]. After identification of the integration sites and their chromatin features, the relationship between retrovirus silencing and the integration environment can be studied.

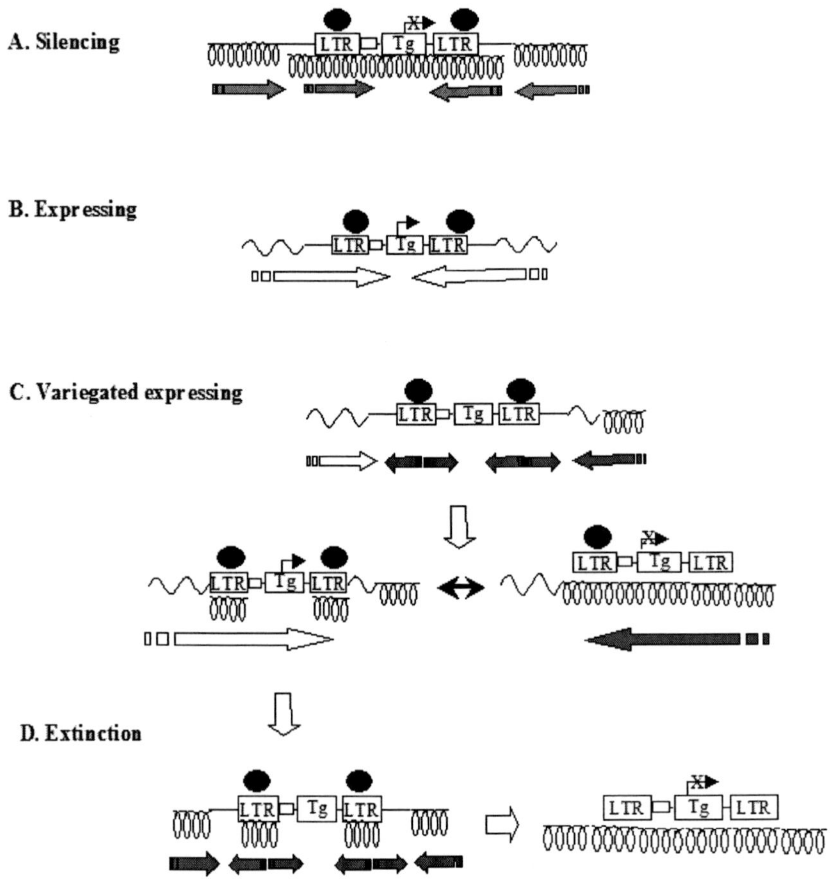

Figure 4. Retrovirus silencing hypothetical model. A). When the retrovirus vector is integrated into heterochromatin, the formation of silent chromatin on the LTRs and the spread of flanking heterochromatin establish silencing. B). When the retrovirus vector is integrated into extremely active euchromatin, the spread of active effects from the flanking gene inhibits the establishment of silent chromatin on LTRs. C). When the retrovirus vector integrates into euchromatin, variegated expression will be established by the balance of the spreading of silent effects from LTRs and the spread of active effects from euchromatin. D). During differentiation, the flanking euchromatin turns into heterochromatin and the transgene will be extinguished by the spreading of silencing effects from both LTRs and flanking heterochromatin. The black circle represents silencing factors.

Retrovirus Silencing in Somatic Stem Cells

Because of the ethical problems associated with using human ES cells, somatic stem cells have been the center of attention for gene therapy. The finding of somatic stem cell plasticity [Bjornson et al., 1999; Martin-Rendon and Watt, 2003] further emphasizes the importance of somatic stem cell studies. Although retrovirus and Lentivirus silencing have been documented in HSC, the understanding of this silencing and its mechanisms are limited. The lack of identity of HSC makes it difficult to monitor single HSC clones and their differentiation. As a result, the identification of different silencing phenomena and investigation of their mechanisms are restricted. In addition, whether other somatic stem cells retain retrovirus silencing pathways that may interfere with gene therapy needs to be addressed. The neural stem cell (NSC) has the potential to be used to study retrovirus silencing. Theoretically, a single neurosphere can be derived from a single NSC. Therefore, isolation of single NSC derived neurospheres can be used to identify retrovirus silencing or variegation. NSC have the potential to differentiate into three types of neural cells: astrocytes, oligodendrocytes and neurons [McKay, 1997]. Differentiation of single NSC derived neurospheres can be used to study retrovirus extinction.

Final Conclusions

In conclusion, retrovirus and Lentivirus silencing are heterogeneous epigenetic phenomena that involve induction of inheritable silent chromatin on integrated retroviruses. While every integrated retrovirus has the potential to be silenced, silent chromatin only can be established in a permissive genomic environment. Retrovirus silencing is achieved by cooperation of multiple cellular gene repression mechanisms to induce either static or dynamic silent chromatin conformations resulting in retrovirus silencing and variegated expression. Silent chromatin serves as the inheritable epigenetic mark that can be passed through DNA replication during cell division. The silent chromatin formed on the retrovirus vector sequences can spread into the flanking genes. Spread of retrovirus silencing can be blocked by insulator elements. Finally, insulator elements with retrovirus silencer blocking activity can be used to design silencing resistant retrovirus and Lentivirus vectors.

References

Allen, G. C., Spiker, S., and Thompson, W. F. (2000). Use of matrix attachment regions (MARs) to minimize transgene silencing. *Plant Mol Biol 43*, 361-376.

Ayyanathan, K., Lechner, M. S., Bell, P., Maul, G. G., Schultz, D. C., Yamada, Y., Tanaka, K., Torigoe, K., and Rauscher, F. J., 3rd (2003). Regulated recruitment of HP1 to a euchromatic gene induces mitotically heritable, epigenetic gene silencing: a mammalian cell culture model of gene variegation. *Genes Dev 17*, 1855-1869.

Bacheler, L., Jaenisch, R., and Fan, H. (1979). Highly inducible cell lines derived from mice genetically transmitting the Moloney murine leukemia virus genome. *J Virol 29*, 896-906.

Ballestar, E., and Wolffe, A. P. (2001). Methyl-CpG-binding proteins. Targeting specific gene repression. *Eur J Biochem 268*, 1-6.

Barklis, E., Mulligan, R. C., and Jaenisch, R. (1986). Chromosomal position or virus mutation permits retrovirus expression in embryonal carcinoma cells. *Cell 47*, 391-399.

Barolo, S., and Levine, M. (1997). hairy mediates dominant repression in the Drosophila embryo. *Embo J 16*, 2883-2891.

Bell, A. C., and Felsenfeld, G. (1999). Stopped at the border: boundaries and insulators. *Curr Opin Genet Dev 9*, 191-198.

Bell, A. C., West, A. G., and Felsenfeld, G. (1999). The protein CTCF is required for the enhancer blocking activity of vertebrate insulators. *Cell 98*, 387-396.

Berger, S. L. (2002). Histone modifications in transcriptional regulation. *Curr Opin Genet Dev 12*, 142-148.

Bestor, T. H. (1990). DNA methylation: evolution of a bacterial immune function into a regulator of gene expression and genome structure in higher eukaryotes. *Philos Trans R Soc Lond B Biol Sci 326*, 179-187.

Bhattacharjee, R. N., Banks, G. C., Trotter, K. W., Lee, H. L., and Archer, T. K. (2001). Histone H1 phosphorylation by Cdk2 selectively modulates mouse mammary tumor virus transcription through chromatin remodeling. *Mol Cell Biol 21*, 5417-5425.

Bird, A. P., and Wolffe, A. P. (1999). Methylation-induced repression--belts, braces, and chromatin. *Cell 99*, 451-454.

Bjornson, C. R., Rietze, R. L., Reynolds, B. A., Magli, M. C., and Vescovi, A. L. (1999). Turning brain into blood: a hematopoietic fate adopted by adult neural stem cells in vivo. *Science 283*, 534-537.

Bonifer, C., Yannoutsos, N., Kruger, G., Grosveld, F., and Sippel, A. E. (1994). Dissection of the locus control function located on the chicken lysozyme gene domain in transgenic mice. *Nucleic Acids Res 22*, 4202-4210.

Breindl, M., Doehmer, J., Willecke, K., Dausman, J., and Jaenisch, R. (1979). Germ line integration of Moloney leukemia virus: identification of the chromosomal integration site. *Proc Natl Acad Sci U S A 76*, 1938-1942.

Breindl, M., Harbers, K., and Jaenisch, R. (1984). Retrovirus-induced lethal mutation in collagen I gene of mice is associated with an altered chromatin structure. *Cell 38*, 9-16.

Breindl, M., and Jaenisch, R. (1979). Conformation of Moloney murine leukaemia proviral sequences in chromatin from leukaemic and nonleukaemic cells. *Nature 277*, 320-322.

Buchschacher, G. L., Jr., and Wong-Staal, F. (2000). Development of lentiviral vectors for gene therapy for human diseases. *Blood 95*, 2499-2504.

Burcin, M., Arnold, R., Lutz, M., Kaiser, B., Runge, D., Lottspeich, F., Filippova, G. N., Lobanenkov, V. V., and Renkawitz, R. (1997). Negative protein 1, which is required for function of the chicken lysozyme gene silencer in conjunction with hormone receptors, is identical to the multivalent zinc finger repressor CTCF. *Mol Cell Biol 17*, 1281-1288.

Byrd, K., and Corces, V. G. (2003). Visualization of chromatin domains created by the gypsy insulator of Drosophila. *J Cell Biol 162*, 565-574.

Cai, H., and Levine, M. (1995). Modulation of enhancer-promoter interactions by insulators in the Drosophila embryo. *Nature 376*, 533-536.

Cai, H. N., and Shen, P. (2001). Effects of cis arrangement of chromatin insulators on enhancer-blocking activity. *Science 291*, 493-495.

Cao, R., Wang, L., Wang, H., Xia, L., Erdjument-Bromage, H., Tempst, P., Jones, R. S., and Zhang, Y. (2002). Role of histone H3 lysine 27 methylation in Polycomb-group silencing. *Science 298*, 1039-1043.

Carlone, D. L., and Skalnik, D. G. (2001). CpG binding protein is crucial for early embryonic development. *Mol Cell Biol 21*, 7601-7606.

Challita, P. M., and Kohn, D. B. (1994). Lack of expression from a retroviral vector after transduction of murine hematopoietic stem cells is associated with methylation in vivo. *Proc Natl Acad Sci U S A 91*, 2567-2571.

Challita, P. M., Skelton, D., el-Khoueiry, A., Yu, X. J., Weinberg, K., and Kohn, D. B. (1995). Multiple modifications in cis elements of the long terminal repeat of retroviral vectors lead to increased expression and decreased DNA methylation in embryonic carcinoma cells. *J Virol 69*, 748-755.

Chen, W. G., Chang, Q., Lin, Y., Meissner, A., West, A. E., Griffith, E. C., Jaenisch, R., and Greenberg, M. E. (2003). Derepression of BDNF transcription involves calcium-dependent phosphorylation of MeCP2. *Science 302*, 885-889.

Cherry, S. R., Biniszkiewicz, D., van Parijs, L., Baltimore, D., and Jaenisch, R. (2000). Retroviral expression in embryonic stem cells and hematopoietic stem cells. *Mol Cell Biol 20*, 7419-7426.

Cheung, P., Allis, C. D., and Sassone-Corsi, P. (2000). Signaling to chromatin through histone modifications. *Cell 103*, 263-271.

Chicurel, M. (2000). Virology. Probing HIV's elusive activities within the host cell. *Science 290*, 1876-1879.

Chung, J. H., Bell, A. C., and Felsenfeld, G. (1997). Characterization of the chicken beta-globin insulator. *Proc Natl Acad Sci U S A 94*, 575-580.

Chung, J. H., Whiteley, M., and Felsenfeld, G. (1993). A 5' element of the chicken beta-globin domain serves as an insulator in human erythroid cells and protects against position effect in Drosophila. *Cell 74*, 505-514.

Colicelli, J., and Goff, S. P. (1986). Isolation of a recombinant murine leukemia virus utilizing a new primer tRNA. *J Virol 57*, 37-45.

_____ . (1987). Identification of endogenous retroviral sequences as potential donors for recombinational repair of mutant retroviruses: positions of crossover points. *Virology 160*, 518-522.

Crane-Robinson, C., Myers, F. A., Hebbes, T. R., Clayton, A. L., and Thorne, A. W. (1999). Chromatin immunoprecipitation assays in acetylation mapping of higher eukaryotes. *Methods Enzymol 304*, 533-547.

Czermin, B., Melfi, R., McCabe, D., Seitz, V., Imhof, A., and Pirrotta, V. (2002). Drosophila enhancer of Zeste/ESC complexes have a histone H3 methyltransferase activity that marks chromosomal Polycomb sites. *Cell 111*, 185-196.

Dickinson, L. A., Joh, T., Kohwi, Y., and Kohwi-Shigematsu, T. (1992). A tissue-specific MAR/SAR DNA-binding protein with unusual binding site recognition. *Cell 70*, 631-645.

Dillon, N., and Festenstein, R. (2002). Unravelling heterochromatin: competition between positive and negative factors regulates accessibility. *Trends Genet 18*, 252-258.

Doerfler, W. (1991). Patterns of DNA methylation--evolutionary vestiges of foreign DNA inactivation as a host defense mechanism. A proposal. *Biol Chem Hoppe Seyler 372*, 557-564.

Dull, T., Zufferey, R., Kelly, M., Mandel, R. J., Nguyen, M., Trono, D., and Naldini, L. (1998). A third-generation lentivirus vector with a conditional packaging system. *J Virol 72*, 8463-8471.

Eissenberg, J. C., and Elgin, S. C. (2000). The HP1 protein family: getting a grip on chromatin. *Curr Opin Genet Dev 10*, 204-210.

Ellis, J., Pasceri, P., Tan-Un, K. C., Wu, X., Harper, A., Fraser, P., and Grosveld, F. (1997). Evaluation of beta-globin gene therapy constructs in single copy transgenic mice. *Nucleic Acids Res 25*, 1296-1302.

Emery, D. W., Yannaki, E., Tubb, J., Nishino, T., Li, Q., and Stamatoyannopoulos, G. (2002). Development of virus vectors for gene therapy of beta chain hemoglobinopathies: flanking with a chromatin insulator reduces gamma-globin gene silencing in vivo. *Blood 100*, 2012-2019.

Emery, D. W., Yannaki, E., Tubb, J., and Stamatoyannopoulos, G. (2000). A chromatin insulator protects retrovirus vectors from chromosomal position effects. *Proc Natl Acad Sci U S A 97*, 9150-9155.

Festenstein, R., Sharghi-Namini, S., Fox, M., Roderick, K., Tolaini, M., Norton, T., Saveliev, A., Kioussis, D., and Singh, P. (1999). Heterochromatin protein 1 modifies mammalian PEV in a dose- and chromosomal-context-dependent manner. *Nat Genet 23*, 457-461.

Festenstein, R., Tolaini, M., Corbella, P., Mamalaki, C., Parrington, J., Fox, M., Miliou, A., Jones, M., and Kioussis, D. (1996). Locus control region function and heterochromatin-induced position effect variegation. *Science 271*, 1123-1125.

Flanagan, J. R., Becker, K. G., Ennist, D. L., Gleason, S. L., Driggers, P. H., Levi, B. Z., Appella, E., and Ozato, K. (1992). Cloning of a negative transcription factor that binds to the upstream conserved region of Moloney murine leukemia virus. *Mol Cell Biol 12*, 38-44.

Flanagan, J. R., Krieg, A. M., Max, E. E., and Khan, A. S. (1989). Negative control region at the 5' end of murine leukemia virus long terminal repeats. *Mol Cell Biol 9*, 739-746.

Fuks, F., Hurd, P. J., Deplus, R., and Kouzarides, T. (2003a). The DNA methyltransferases associate with HP1 and the SUV39H1 histone methyltransferase. *Nucleic Acids Res 31*, 2305-2312.

Fuks, F., Hurd, P. J., Wolf, D., Nan, X., Bird, A. P., and Kouzarides, T. (2003b). The methyl-CpG-binding protein MeCP2 links DNA methylation to histone methylation. *J Biol Chem 278*, 4035-4040.

Gao, G., Guo, X., and Goff, S. P. (2002). Inhibition of retroviral RNA production by ZAP, a CCCH-type zinc finger protein. *Science 297*, 1703-1706.

Gaszner, M., Vazquez, J., and Schedl, P. (1999). The Zw5 protein, a component of the scs chromatin domain boundary, is able to block enhancer-promoter interaction. *Genes Dev 13*, 2098-2107.

Gause, M., Morcillo, P., and Dorsett, D. (2001). Insulation of enhancer-promoter communication by a gypsy transposon insert in the Drosophila cut gene: cooperation between suppressor of hairy-wing and modifier of mdg4 proteins. *Mol Cell Biol 21*, 4807-4817.

Gautsch, J. W. (1980). Embryonal carcinoma stem cells lack a function required for virus replication. *Nature 285*, 110-112.

Gautsch, J. W., and Wilson, M. C. (1983). Delayed de novo methylation in teratocarcinoma suggests additional tissue-specific mechanisms for controlling gene expression. *Nature 301*, 32-37.

Gdula, D. A., Gerasimova, T. I., and Corces, V. G. (1996). Genetic and molecular analysis of the gypsy chromatin insulator of Drosophila. *Proc Natl Acad Sci U S A 93*, 9378-9383.

Georgel, P. T., Horowitz-Scherer, R. A., Adkins, N., Woodcock, C. L., Wade, P. A., and Hansen, J. C. (2003). Chromatin compaction by human MeCP2. Assembly of novel secondary chromatin structures in the absence of DNA methylation. *J Biol Chem 278*, 32181-32188.

Gerasimova, T. I., Byrd, K., and Corces, V. G. (2000). A chromatin insulator determines the nuclear localization of DNA. *Mol Cell 6*, 1025-1035.

Geyer, P. K., and Corces, V. G. (1992). DNA position-specific repression of transcription by a Drosophila zinc finger protein. *Genes Dev 6*, 1865-1873.

Ghosh, D., Gerasimova, T. I., and Corces, V. G. (2001). Interactions between the Su(Hw) and Mod(mdg4) proteins required for gypsy insulator function. *Embo J 20*, 2518-2527.

Graziano, V., Gerchman, S. E., Schneider, D. K., and Ramakrishnan, V. (1994). Histone H1 is located in the interior of the chromatin 30-nm filament. *Nature 368*, 351-354.

Grewal, S. I., and Moazed, D. (2003). Heterochromatin and epigenetic control of gene expression. *Science 301*, 798-802.

Grez, M., Akgun, E., Hilberg, F., and Ostertag, W. (1990). Embryonic stem cell virus, a recombinant murine retrovirus with expression in embryonic stem cells. *Proc Natl Acad Sci U S A 87*, 9202-9206.

Guenechea, G., Gan, O. I., Dorrell, C., and Dick, J. E. (2001). Distinct classes of human stem cells that differ in proliferative and self-renewal potential. *Nat Immunol 2*, 75-82.

Hamaguchi, I., Woods, N. B., Panagopoulos, I., Andersson, E., Mikkola, H., Fahlman, C., Zufferey, R., Carlsson, L., Trono, D., and Karlsson, S. (2000). Lentivirus vector gene expression during ES cell-derived hematopoietic development in vitro. *J Virol 74*, 10778-10784.

Hart, C. M., Zhao, K., and Laemmli, U. K. (1997). The scs' boundary element: characterization of boundary element-associated factors. *Mol Cell Biol 17*, 999-1009.

Hata, K., Okano, M., Lei, H., and Li, E. (2002). Dnmt3L cooperates with the Dnmt3 family of de novo DNA methyltransferases to establish maternal imprints in mice. *Development 129*, 1983-1993.

Hawley, R. G. (1994). High-titer retroviral vectors for efficient transduction of functional genes into murine hematopoietic stem cells. *Ann N Y Acad Sci 716*, 327-330.

Hilberg, F., Stocking, C., Ostertag, W., and Grez, M. (1987). Functional analysis of a retroviral host-range mutant: altered long terminal repeat sequences allow expression in embryonal carcinoma cells. *Proc Natl Acad Sci U S A 84*, 5232-5236.

Horn, P. J., Carruthers, L. M., Logie, C., Hill, D. A., Solomon, M. J., Wade, P. A., Imbalzano, A. N., Hansen, J. C., and Peterson, C. L. (2002). Phosphorylation of linker histones regulates ATP-dependent chromatin remodeling enzymes. *Nat Struct Biol 9*, 263-267.

Hung, M. S., Karthikeyan, N., Huang, B., Koo, H. C., Kiger, J., and Shen, C. J. (1999). Drosophila proteins related to vertebrate DNA (5-cytosine) methyltransferases. *Proc Natl Acad Sci U S A 96*, 11940-11945.

Iida, T., Suetake, I., Tajima, S., Morioka, H., Ohta, S., Obuse, C., and Tsurimoto, T. (2002). PCNA clamp facilitates action of DNA cytosine methyltransferase 1 on hemimethylated DNA. *Genes Cells 7*, 997-1007.

Illmensee, K., and Mintz, B. (1976). Totipotency and normal differentiation of single teratocarcinoma cells cloned by injection into blastocysts. *Proc Natl Acad Sci U S A 73*, 549-553.

Iwakuma, T., Cui, Y., and Chang, L. J. (1999). Self-inactivating lentiviral vectors with U3 and U5 modifications. *Virology 261*, 120-132.

Jackson, J. P., Lindroth, A. M., Cao, X., and Jacobsen, S. E. (2002). Control of CpNpG DNA methylation by the KRYPTONITE histone H3 methyltransferase. *Nature 416*, 556-560.

Jaenisch, R. (1980). Retroviruses and embryogenesis: microinjection of Moloney leukemia virus into midgestation mouse embryos. *Cell 19*, 181-188.

Jaenisch, R., Fan, H., and Croker, B. (1975). Infection of preimplantation mouse embryos and of newborn mice with leukemia virus: tissue distribution of viral DNA and RNA and leukemogenesis in the adult animal. *Proc Natl Acad Sci U S A 72*, 4008-4012.

Jahner, D., Stuhlmann, H., Stewart, C. L., Harbers, K., Lohler, J., Simon, I., and Jaenisch, R. (1982). De novo methylation and expression of retroviral genomes during mouse embryogenesis. *Nature 298*, 623-628.

Jones, P. L., Veenstra, G. J., Wade, P. A., Vermaak, D., Kass, S. U., Landsberger, N., Strouboulis, J., and Wolffe, A. P. (1998). Methylated DNA and MeCP2 recruit histone deacetylase to repress transcription. *Nat Genet 19*, 187-191.

Karymov, M. A., Tomschik, M., Leuba, S. H., Caiafa, P., and Zlatanova, J. (2001). DNA methylation-dependent chromatin fiber compaction in vivo and in vitro: requirement for linker histone. *Faseb J 15*, 2631-2641.

Kellum, R., and Schedl, P. (1991). A position-effect assay for boundaries of higher order chromosomal domains. *Cell 64*, 941-950.

Kellum, R., and Schedl, P. (1992). A group of scs elements function as domain boundaries in an enhancer-blocking assay. *Mol Cell Biol 12*, 2424-2431.

Kim, V. N., Mitrophanous, K., Kingsman, S. M., and Kingsman, A. J. (1998). Minimal requirement for a lentivirus vector based on human immunodeficiency virus type 1. *J Virol 72*, 811-816.

Kioussis, D., and Festenstein, R. (1997). Locus control regions: overcoming heterochromatin-induced gene inactivation in mammals. *Curr Opin Genet Dev 7*, 614-619.

Klenova, E. M., Fagerlie, S., Filippova, G. N., Kretzner, L., Goodwin, G. H., Loring, G., Neiman, P. E., and Lobanenkov, V. V. (1998). Characterization of the chicken CTCF genomic locus, and initial study of the cell cycle-regulated promoter of the gene. *J Biol Chem 273*, 26571-26579.

Klug, C. A., Cheshier, S., and Weissman, I. L. (2000). Inactivation of a GFP retrovirus occurs at multiple levels in long-term repopulating stem cells and their differentiated progeny. *Blood 96*, 894-901.

Laker, C., Meyer, J., Schopen, A., Friel, J., Heberlein, C., Ostertag, W., and Stocking, C. (1998). Host cis-mediated extinction of a retrovirus permissive for expression in embryonal stem cells during differentiation. *J Virol 72*, 339-348.

Lange, C., and Blankenstein, T. (1997). Loss of retroviral gene expression in bone marrow reconstituted mice correlates with down-regulation of gene expression in long-term culture initiating cells. *Gene Ther 4*, 303-308.

Leonhardt, H., Page, A. W., Weier, H. U., and Bestor, T. H. (1992). A targeting sequence directs DNA methyltransferase to sites of DNA replication in mammalian nuclei. *Cell 71*, 865-873.

Lever, M. A., Th'ng, J. P., Sun, X., and Hendzel, M. J. (2000). Rapid exchange of histone H1.1 on chromatin in living human cells. *Nature 408*, 873-876.

Lewis, J. D., Meehan, R. R., Henzel, W. J., Maurer-Fogy, I., Jeppesen, P., Klein, F., and Bird, A. (1992). Purification, sequence, and cellular localization of a novel chromosomal protein that binds to methylated DNA. *Cell 69*, 905-914.

Li, E., Bestor, T. H., and Jaenisch, R. (1992). Targeted mutation of the DNA methyltransferase gene results in embryonic lethality. *Cell 69*, 915-926.

Liang, G., Chan, M. F., Tomigahara, Y., Tsai, Y. C., Gonzales, F. A., Li, E., Laird, P. W., and Jones, P. A. (2002). Cooperativity between DNA methyltransferases in the maintenance methylation of repetitive elements. *Mol Cell Biol 22*, 480-491.

Litt, M. D., Simpson, M., Gaszner, M., Allis, C. D., and Felsenfeld, G. (2001a). Correlation between histone lysine methylation and developmental changes at the chicken beta-globin locus. *Science 293*, 2453-2455.

Litt, M. D., Simpson, M., Recillas-Targa, F., Prioleau, M. N., and Felsenfeld, G. (2001b). Transitions in histone acetylation reveal boundaries of three separately regulated neighboring loci. *Embo J 20*, 2224-2235.

Liu, K., Wang, Y. F., Cantemir, C., and Muller, M. T. (2003). Endogenous assays of DNA methyltransferases: Evidence for differential activities of DNMT1, DNMT2, and DNMT3 in mammalian cells in vivo. *Mol Cell Biol 23*, 2709-2719.

Lois, C., Hong, E. J., Pease, S., Brown, E. J., and Baltimore, D. (2002). Germline transmission and tissue-specific expression of transgenes delivered by lentiviral vectors. *Science 295*, 868-872.

Lorincz, M. C., Schubeler, D., Goeke, S. C., Walters, M., Groudine, M., and Martin, D. I. (2000). Dynamic analysis of proviral induction and De Novo methylation: implications for a histone deacetylase-independent, methylation density-dependent mechanism of transcriptional repression. *Mol Cell Biol 20*, 842-850.

Lorincz, M. C., Schubeler, D., and Groudine, M. (2001). Methylation-mediated proviral silencing is associated with MeCP2 recruitment and localized histone H3 deacetylation. *Mol Cell Biol 21*, 7913-7922.

Lorincz, M. C., Schubeler, D., Hutchinson, S. R., Dickerson, D. R., and Groudine, M. (2002). DNA methylation density influences the stability of an epigenetic imprint and Dnmt3a/b-independent de novo methylation. *Mol Cell Biol 22*, 7572-7580.

Lu, L., and Tower, J. (1997). A transcriptional insulator element, the su(Hw) binding site, protects a chromosomal DNA replication origin from position effects. *Mol Cell Biol 17*, 2202-2206.

Lundgren, M., Chow, C. M., Sabbattini, P., Georgiou, A., Minaee, S., and Dillon, N. (2000). Transcription factor dosage affects changes in higher order chromatin structure associated with activation of a heterochromatic gene. *Cell 103*, 733-743.

Lutz, M., Burke, L. J., Barreto, G., Goeman, F., Greb, H., Arnold, R., Schultheiss, H., Brehm, A., Kouzarides, T., Lobanenkov, V., and Renkawitz, R. (2000). Transcriptional repression by the insulator protein CTCF involves histone deacetylases. *Nucleic Acids Res 28*, 1707-1713.

Lyko, F., Ramsahoye, B. H., and Jaenisch, R. (2000). DNA methylation in Drosophila melanogaster. *Nature 408*, 538-540.

Ma, Y., Ramezani, A., Lewis, R., Hawley, R. G., and Thomson, J. A. (2003). High-level sustained transgene expression in human embryonic stem cells using lentiviral vectors. *Stem Cells 21*, 111-117.

Mallin, D. R., Myung, J. S., Patton, J. S., and Geyer, P. K. (1998). Polycomb group repression is blocked by the Drosophila suppressor of Hairy-wing [su(Hw)] insulator. *Genetics 148*, 331-339.

Martinowich, K., Hattori, D., Wu, H., Fouse, S., He, F., Hu, Y., Fan, G., and Sun, Y. E. (2003). DNA methylation-related chromatin remodeling in activity-dependent BDNF gene regulation. *Science 302*, 890-893.

Martin-Rendon, E., and Watt, S. M. (2003). Exploitation of stem cell plasticity. *Transfus Med 13*, 325-349.

May, C., Rivella, S., Callegari, J., Heller, G., Gaensler, K. M., Luzzatto, L., and Sadelain, M. (2000). Therapeutic haemoglobin synthesis in beta-thalassaemic mice expressing lentivirus-encoded human beta-globin. *Nature 406*, 82-86.

McCune, S. L., and Townes, T. M. (1994). Retroviral vector sequences inhibit human beta-globin gene expression in transgenic mice. *Nucleic Acids Res 22*, 4477-4481.

McInerney, J. M., Nawrocki, J. R., and Lowrey, C. H. (2000). Long-term silencing of retroviral vectors is resistant to reversal by trichostatin A and 5-azacytidine. *Gene Ther 7*, 653-663.

McKay, R. (1997). Stem cells in the central nervous system. *Science 276*, 66-71.

McMorrow, T., van den Wijngaard, A., Wollenschlaeger, A., van de Corput, M., Monkhorst, K., Trimborn, T., Fraser, P., van Lohuizen, M., Jenuwein, T., Djabali, M., *et al.* (2000). Activation of the beta globin locus by transcription factors and chromatin modifiers. *Embo J 19*, 4986-4996.

McNairn, A. J., and Gilbert, D. M. (2003). Epigenomic replication: linking epigenetics to DNA replication. *Bioessays 25*, 647-656.

Melnikova, L., Gause, M., and Georgiev, P. (2002). The gypsy insulators flanking yellow enhancers do not form a separate transcriptional domain in Drosophila melanogaster: the enhancers can activate an isolated yellow promoter. *Genetics 160*, 1549-1560.

Mikkola, H., Woods, N. B., Sjogren, M., Helgadottir, H., Hamaguchi, I., Jacobsen, S. E., Trono, D., and Karlsson, S. (2000). Lentivirus gene transfer in murine hematopoietic progenitor cells is compromised by a delay in proviral integration and results in transduction mosaicism and heterogeneous gene expression in progeny cells. *J Virol 74*, 11911-11918.

Milne, T. A., Briggs, S. D., Brock, H. W., Martin, M. E., Gibbs, D., Allis, C. D., and Hess, J. L. (2002). MLL targets SET domain methyltransferase activity to Hox gene promoters. *Mol Cell 10*, 1107-1117.

Milot, E., Strouboulis, J., Trimborn, T., Wijgerde, M., de Boer, E., Langeveld, A., Tan-Un, K., Vergeer, W., Yannoutsos, N., Grosveld, F., and Fraser, P. (1996). Heterochromatin effects on the frequency and duration of LCR-mediated gene transcription. *Cell 87*, 105-114.

Milutinovic, S., Zhuang, Q., Niveleau, A., and Szyf, M. (2003). Epigenomic stress response. Knockdown of DNA methyltransferase 1 triggers an intra-S-phase arrest of DNA replication and induction of stress response genes. *J Biol Chem 278*, 14985-14995.

Misteli, T., Gunjan, A., Hock, R., Bustin, M., and Brown, D. T. (2000). Dynamic binding of histone H1 to chromatin in living cells. *Nature 408*, 877-881.

Miyoshi, H., Blomer, U., Takahashi, M., Gage, F. H., and Verma, I. M. (1998). Development of a self-inactivating lentivirus vector. *J Virol 72*, 8150-8157.

Moazed, D. (2001). Common themes in mechanisms of gene silencing. *Mol Cell 8*, 489-498.

Muravyova, E., Golovnin, A., Gracheva, E., Parshikov, A., Belenkaya, T., Pirrotta, V., and Georgiev, P. (2001). Loss of insulator activity by paired Su(Hw) chromatin insulators. *Science 291*, 495-498.

Mutskov, V. J., Farrell, C. M., Wade, P. A., Wolffe, A. P., and Felsenfeld, G. (2002). The barrier function of an insulator couples high histone acetylation levels with specific protection of promoter DNA from methylation. *Genes Dev 16*, 1540-1554.

Nabirochkin, S., Ossokina, M., and Heidmann, T. (1998). A nuclear matrix/scaffold attachment region co-localizes with the gypsy retrotransposon insulator sequence. *J Biol Chem 273*, 2473-2479.

Naldini, L., Blomer, U., Gallay, P., Ory, D., Mulligan, R., Gage, F. H., Verma, I. M., and Trono, D. (1996). In vivo gene delivery and stable transduction of nondividing cells by a lentiviral vector. *Science 272*, 263-267.

Naldini, L., and Verma, I. M. (2000). Lentiviral vectors. *Adv Virus Res 55*, 599-609.

Nan, X., Campoy, F. J., and Bird, A. (1997). MeCP2 is a transcriptional repressor with abundant binding sites in genomic chromatin. *Cell 88*, 471-481.

Nan, X., Meehan, R. R., and Bird, A. (1993). Dissection of the methyl-CpG binding domain from the chromosomal protein MeCP2. *Nucleic Acids Res 21*, 4886-4892.

Nan, X., Ng, H. H., Johnson, C. A., Laherty, C. D., Turner, B. M., Eisenman, R. N., and Bird, A. (1998). Transcriptional repression by the methyl-CpG-binding protein MeCP2 involves a histone deacetylase complex. *Nature 393*, 386-389.

Narlikar, G. J., Fan, H. Y., and Kingston, R. E. (2002). Cooperation between complexes that regulate chromatin structure and transcription. *Cell 108*, 475-487.

Ner, S. S., and Travers, A. A. (1994). HMG-D, the Drosophila melanogaster homologue of HMG 1 protein, is associated with early embryonic chromatin in the absence of histone H1. *Embo J 13*, 1817-1822.

Nielsen, S. J., Schneider, R., Bauer, U. M., Bannister, A. J., Morrison, A., O'Carroll, D., Firestein, R., Cleary, M., Jenuwein, T., Herrera, R. E., and Kouzarides, T. (2001). Rb targets histone H3 methylation and HP1 to promoters. *Nature 412*, 561-565.

Niwa, O., Yokota, Y., Ishida, H., and Sugahara, T. (1983). Independent mechanisms involved in suppression of the Moloney leukemia virus genome during differentiation of murine teratocarcinoma cells. *Cell 32*, 1105-1113.

Okano, M., Bell, D. W., Haber, D. A., and Li, E. (1999). DNA methyltransferases Dnmt3a and Dnmt3b are essential for de novo methylation and mammalian development. *Cell 99*, 247-257.

Okano, M., Xie, S., and Li, E. (1998). Dnmt2 is not required for de novo and maintenance methylation of viral DNA in embryonic stem cells. *Nucleic Acids Res 26*, 2536-2540.

Osborne, C. S., Pasceri, P., Singal, R., Sukonnik, T., Ginder, G. D., and Ellis, J. (1999). Amelioration of retroviral vector silencing in locus control region beta-globin-transgenic mice and transduced F9 embryonic cells. *J Virol 73*, 5490-5496.

Pannell, D., Osborne, C. S., Yao, S., Sukonnik, T., Pasceri, P., Karaiskakis, A., Okano, M., Li, E., Lipshitz, H. D., and Ellis, J. (2000). Retrovirus vector silencing is de novo methylase independent and marked by a repressive histone code. *Embo J 19*, 5884-5894.

Pasceri, P., Pannell, D., Wu, X., and Ellis, J. (1998). Full activity from human beta-globin locus control region transgenes requires 5'HS1, distal beta-globin promoter, and 3' beta-globin sequences. *Blood 92*, 653-663.

Peries, J., Debons-Guillemin, M. C., Canivet, M., Emanoil-Ravicovitch, R., Tavitian, A., and Boiron, M. (1977). [Multiplication of murine C-type viruses in mouse teratocarcinoma cell lines]. *Nouv Rev Fr Hematol Blood Cells 18*, 383-390.

Peters, A. H., O'Carroll, D., Scherthan, H., Mechtler, K., Sauer, S., Schofer, C., Weipoltshammer, K., Pagani, M., Lachner, M., Kohlmaier, A., *et al.* (2001). Loss of the Suv39h histone methyltransferases impairs mammalian heterochromatin and genome stability. *Cell 107*, 323-337.

Petersen, R., Kempler, G., and Barklis, E. (1991). A stem cell-specific silencer in the primer-binding site of a retrovirus. *Mol Cell Biol 11*, 1214-1221.

Pfeifer, A., Ikawa, M., Dayn, Y., and Verma, I. M. (2002). Transgenesis by lentiviral vectors: lack of gene silencing in mammalian embryonic stem cells and preimplantation embryos. *Proc Natl Acad Sci U S A 99*, 2140-2145.

Pirrotta, V. (1998). Polycombing the genome: PcG, trxG, and chromatin silencing. *Cell 93*, 333-336.

Prioleau, M. N., Nony, P., Simpson, M., and Felsenfeld, G. (1999). An insulator element and condensed chromatin region separate the chicken beta-globin locus from an independently regulated erythroid-specific folate receptor gene. *Embo J 18*, 4035-4048.

Quinonez, R., and Sutton, R. E. (2002). Lentiviral vectors for gene delivery into cells. *DNA Cell Biol 21*, 937-951.

Ramezani, A., and Hawley, R. G. (2003). Human immunodeficiency virus type 1-based vectors for gene delivery to human hematopoietic stem cells. *Methods Mol Med 76*, 467-492.

Ramezani, A., Hawley, T. S., and Hawley, R. G. (2003). Performance- and safety-enhanced lentiviral vectors containing the human interferon-beta scaffold attachment region and the chicken beta-globin insulator. *Blood 101*, 4717-4724.

Rasmussen, T. P. (2003). Embryonic stem cell differentiation: A chromatin perspective. *Reprod Biol Endocrinol 1*, 100.

Rea, S., Eisenhaber, F., O'Carroll, D., Strahl, B. D., Sun, Z. W., Schmid, M., Opravil, S., Mechtler, K., Ponting, C. P., Allis, C. D., and Jenuwein, T. (2000). Regulation of chromatin structure by site-specific histone H3 methyltransferases. *Nature 406*, 593-599.

Recillas-Targa, F., Pikaart, M. J., Burgess-Beusse, B., Bell, A. C., Litt, M. D., West, A. G., Gaszner, M., and Felsenfeld, G. (2002). Position-effect protection and enhancer blocking by the chicken beta-globin insulator are separable activities. *Proc Natl Acad Sci U S A 99*, 6883-6888.

Reid, J. L., Iyer, V. R., Brown, P. O., and Struhl, K. (2000). Coordinate regulation of yeast ribosomal protein genes is associated with targeted recruitment of Esa1 histone acetylase. *Mol Cell 6*, 1297-1307.

Rivella, S., Callegari, J. A., May, C., Tan, C. W., and Sadelain, M. (2000). The cHS4 insulator increases the probability of retroviral expression at random chromosomal integration sites. *J Virol 74*, 4679-4687.

Robbins, P. B., Skelton, D. C., Yu, X. J., Halene, S., Leonard, E. H., and Kohn, D. B. (1998). Consistent, persistent expression from modified retroviral vectors in murine hematopoietic stem cells. *Proc Natl Acad Sci U S A 95*, 10182-10187.

Roth, S. Y., Denu, J. M., and Allis, C. D. (2001). Histone acetyltransferases. *Annu Rev Biochem 70*, 81-120.

Russo, V. E. A., Martienssen, R.A. and Riggs, A.D. (1996). Epigenetic Mechanisms of Gene Regulation. In *Cold Spring Harbor Laboratory Press*, pp. 1-5.

Santos, F., Hendrich, B., Reik, W., and Dean, W. (2002). Dynamic reprogramming of DNA methylation in the early mouse embryo. *Dev Biol 241*, 172-182.

Schilz, A. J., Brouns, G., Knoss, H., Ottmann, O. G., Hoelzer, D., Fauser, A. A., Thrasher, A. J., and Grez, M. (1998). High efficiency gene transfer to human hematopoietic SCID-repopulating cells under serum-free conditions. *Blood 92*, 3163-3171.

Schotta, G., Ebert, A., Krauss, V., Fischer, A., Hoffmann, J., Rea, S., Jenuwein, T., Dorn, R., and Reuter, G. (2002). Central role of Drosophila SU(VAR)3-9 in histone H3-K9 methylation and heterochromatic gene silencing. *Embo J 21*, 1121-1131.

Schramke, V., and Allshire, R. (2003). Hairpin RNAs and retrotransposon LTRs effect RNAi and chromatin-based gene silencing. *Science 301*, 1069-1074.

Schroder, A. R., Shinn, P., Chen, H., Berry, C., Ecker, J. R., and Bushman, F. (2002). HIV-1 integration in the human genome favors active genes and local hotspots. *Cell 110*, 521-529.

Sewalt, R. G., Lachner, M., Vargas, M., Hamer, K. M., den Blaauwen, J. L., Hendrix, T., Melcher, M., Schweizer, D., Jenuwein, T., and Otte, A. P. (2002). Selective interactions between vertebrate polycomb homologs and the SUV39H1 histone lysine methyltransferase suggest that histone H3-K9 methylation contributes to chromosomal targeting of Polycomb group proteins. *Mol Cell Biol 22*, 5539-5553.

Shen, X., and Gorovsky, M. A. (1996). Linker histone H1 regulates specific gene expression but not global transcription in vivo. *Cell 86*, 475-483.

Shore, D. (2000). The Sir2 protein family: A novel deacetylase for gene silencing and more. *Proc Natl Acad Sci U S A 97*, 14030-14032.

Simon, D., Stuhlmann, H., Jahner, D., Wagner, H., Werner, E., and Jaenisch, R. (1983). Retrovirus genomes methylated by mammalian but not bacterial methylase are non-infectious. *Nature 304*, 275-277.

Sirven, A., Pflumio, F., Zennou, V., Titeux, M., Vainchenker, W., Coulombel, L., Dubart-Kupperschmitt, A., and Charneau, P. (2000). The human immunodeficiency virus type-1 central DNA flap is a crucial determinant for lentiviral vector nuclear import and gene transduction of human hematopoietic stem cells. *Blood 96*, 4103-4110.

Sirven, A., Ravet, E., Charneau, P., Zennou, V., Coulombel, L., Guetard, D., Pflumio, F., and Dubart-Kupperschmitt, A. (2001). Enhanced transgene expression in cord blood CD34(+)-derived hematopoietic cells, including developing T cells and NOD/SCID mouse repopulating cells, following transduction with modified trip lentiviral vectors. *Mol Ther 3*, 438-448.

Speck, N. A., and Baltimore, D. (1987). Six distinct nuclear factors interact with the 75-base-pair repeat of the Moloney murine leukemia virus enhancer. *Mol Cell Biol 7*, 1101-1110.

Speers, W. C., Gautsch, J. W., and Dixon, F. J. (1980). Silent infection of murine embryonal carcinoma cells by Moloney murine leukemia virus. *Virology 105*, 241-244.

Stancheva, I., Collins, A. L., Van den Veyver, I. B., Zoghbi, H., and Meehan, R. R. (2003). A mutant form of MeCP2 protein associated with human Rett syndrome cannot be displaced from methylated DNA by notch in Xenopus embryos. *Mol Cell 12*, 425-435.

Stewart, C. L., Stuhlmann, H., Jahner, D., and Jaenisch, R. (1982). De novo methylation, expression, and infectivity of retroviral genomes introduced into embryonal carcinoma cells. *Proc Natl Acad Sci U S A 79*, 4098-4102.

Strahl, B. D., and Allis, C. D. (2000). The language of covalent histone modifications. *Nature 403*, 41-45.

Swindle, C. S., Kim, H. G., and Klug, C. A. (2004). Mutation of CpGs in the Murine Stem Cell Virus Retroviral Vector Long Terminal Repeat Represses Silencing in Embryonic Stem Cells. *J Biol Chem 279*, 34-41.

Tachibana, M., Sugimoto, K., Fukushima, T., and Shinkai, Y. (2001). Set domain-containing protein, G9a, is a novel lysine-preferring mammalian histone methyltransferase with hyperactivity and specific selectivity to lysines 9 and 27 of histone H3. *J Biol Chem 276*, 25309-25317.

Teich, N. M., Weiss, R. A., Martin, G. R., and Lowy, D. R. (1977). Virus infection of murine teratocarcinoma stem cell lines. *Cell 12*, 973-982.

Trevisan, M., and Iscove, N. N. (1995). Phenotypic analysis of murine long-term hemopoietic reconstituting cells quantitated competitively in vivo and comparison with more advanced colony-forming progeny. *J Exp Med 181*, 93-103.

Tsukiyama, T., and Niwa, O. (1992). Isolation of high affinity cellular targets of the embryonal LTR binding protein, an undifferentiated embryonal carcinoma cell-specific repressor of Moloney leukemia virus. *Nucleic Acids Res 20*, 1477-1482.

Tsukiyama, T., Niwa, O., and Yokoro, K. (1989). Mechanism of suppression of the long terminal repeat of Moloney leukemia virus in mouse embryonal carcinoma cells. *Mol Cell Biol 9*, 4670-4676.

Uchida, N., Sutton, R. E., Friera, A. M., He, D., Reitsma, M. J., Chang, W. C., Veres, G., Scollay, R., and Weissman, I. L. (1998). HIV, but not murine leukemia virus, vectors mediate high efficiency gene transfer into freshly isolated G0/G1 human hematopoietic stem cells. *Proc Natl Acad Sci U S A 95*, 11939-11944.

Udvardy, A., Maine, E., and Schedl, P. (1985). The 87A7 chromomere. Identification of novel chromatin structures flanking the heat shock locus that may define the boundaries of higher order domains. *J Mol Biol 185*, 341-358.

Urieli-Shoval, S., Gruenbaum, Y., Sedat, J., and Razin, A. (1982). The absence of detectable methylated bases in Drosophila melanogaster DNA. *FEBS Lett 146*, 148-152.

van der Vlag, J., den Blaauwen, J. L., Sewalt, R. G., van Driel, R., and Otte, A. P. (2000). Transcriptional repression mediated by polycomb group proteins and other chromatin-associated repressors is selectively blocked by insulators. *J Biol Chem 275*, 697-704.

Verma, I. M., and Somia, N. (1997). Gene therapy -- promises, problems and prospects. *Nature 389*, 239-242.

Vogelauer, M., Wu, J., Suka, N., and Grunstein, M. (2000). Global histone acetylation and deacetylation in yeast. *Nature 408*, 495-498.

Voo, K. S., Carlone, D. L., Jacobsen, B. M., Flodin, A., and Skalnik, D. G. (2000). Cloning of a mammalian transcriptional activator that binds unmethylated CpG motifs and shares a CXXC domain with DNA methyltransferase, human trithorax, and methyl-CpG binding domain protein 1. *Mol Cell Biol 20*, 2108-2121.

Walsh, C. P., and Bestor, T. H. (1999). Cytosine methylation and mammalian development. *Genes Dev 13*, 26-34.

Walsh, C. P., Chaillet, J. R., and Bestor, T. H. (1998). Transcription of IAP endogenous retroviruses is constrained by cytosine methylation. *Nat Genet 20*, 116-117.

Weintraub, H., and Groudine, M. (1976). Chromosomal subunits in active genes have an altered conformation. *Science 193*, 848-856.

Wolffe, A. P., and Guschin, D. (2000). Review: chromatin structural features and targets that regulate transcription. *J Struct Biol 129*, 102-122.

Wolffe, A. P., and Matzke, M. A. (1999). Epigenetics: regulation through repression. *Science 286*, 481-486.

Woods, N. B., Fahlman, C., Mikkola, H., Hamaguchi, I., Olsson, K., Zufferey, R., Jacobsen, S. E., Trono, D., and Karlsson, S. (2000). Lentiviral gene transfer into primary and secondary NOD/SCID repopulating cells. *Blood 96*, 3725-3733.

Wu, X., Li, Y., Crise, B., and Burgess, S. M. (2003). Transcription start regions in the human genome are favored targets for MLV integration. *Science 300*, 1749-1751.

Xie, S., Wang, Z., Okano, M., Nogami, M., Li, Y., He, W. W., Okumura, K., and Li, E. (1999). Cloning, expression and chromosome locations of the human DNMT3 gene family. *Gene 236*, 87-95.

Yannaki, E., Tubb, J., Aker, M., Stamatoyannopoulos, G., and Emery, D. W. (2002). Topological constraints governing the use of the chicken HS4 chromatin insulator in oncoretrovirus vectors. *Mol Ther 5*, 589-598.

Yao, S., Osborne, C. S., Bharadwaj, R. R., Pasceri, P., Sukonnik, T., Pannell, D., Recillas-Targa, F., West, A. G., and Ellis, J. (2003). Retrovirus silencer blocking by the cHS4 insulator is CTCF independent. *Nucleic Acids Res 31*, 5317-5323.

Yao, S., Sukonnik, T., Kean, T., Bharadwaj R.R., Pasceri P. and Ellis J. (2004). Retrovirus silencing, variegation, extinction and memory are controlled by a dynamic interplay of multiple epigenetic modifications. *Mol Ther 10, 27-36.*

Yoder, J. A., Soman, N. S., Verdine, G. L., and Bestor, T. H. (1997a). DNA (cytosine-5)-methyltransferases in mouse cells and tissues. Studies with a mechanism-based probe. *J Mol Biol 270*, 385-395.

Yoder, J. A., Walsh, C. P., and Bestor, T. H. (1997b). Cytosine methylation and the ecology of intragenomic parasites. *Trends Genet 13*, 335-340.

Yu, S. F., von Ruden, T., Kantoff, P. W., Garber, C., Seiberg, M., Ruther, U., Anderson, W. F., Wagner, E. F., and Gilboa, E. (1986). Self-inactivating retroviral vectors designed for transfer of whole genes into mammalian cells. *Proc Natl Acad Sci U S A 83*, 3194-3198.

Zentilin, L., Qin, G., Tafuro, S., Dinauer, M. C., Baum, C., and Giacca, M. (2000). Variegation of retroviral vector gene expression in myeloid cells. *Gene Ther 7*, 153-166.

Zhao, K., Hart, C. M., and Laemmli, U. K. (1995). Visualization of chromosomal domains with boundary element-associated factor BEAF-32. *Cell 81*, 879-889.

Zufferey, R., Donello, J. E., Trono, D., and Hope, T. J. (1999). Woodchuck hepatitis virus posttranscriptional regulatory element enhances expression of transgenes delivered by retroviral vectors. *J Virol 73*, 2886-2892.

Zufferey, R., Dull, T., Mandel, R. J., Bukovsky, A., Quiroz, D., Naldini, L., and Trono, D. (1998). Self-inactivating lentivirus vector for safe and efficient in vivo gene delivery. *J Virol 72*, 9873-9880.

Zufferey, R., Nagy, D., Mandel, R. J., Naldini, L., and Trono, D. (1997). Multiply attenuated lentiviral vector achieves efficient gene delivery in vivo. *Nat Biotechnol 15*, 871-875.

In: Trends in Stem Cell Research
Editor: Erik V. Greer, pp. 67-100

ISBN 1-59454-315-1

Chapter III

Strategies for the Therapeutic Use of Adult Stem Cells in Regenerative Medicine: Cell Biological and Technological Approaches

Katrin Zeilinger[1*], Mariah S. Hout[2], and Jörg C. Gerlach[1,2]
[1]Clinic for Surgery, Experimental Surgery, Charité - Campus Virchow, University Medicine, Berlin, Germany, [2]McGowan Institute for Regenerative Medicine, University of Pittsburgh, Pittsburgh, USA

Abstract

Stem cell research has raised hopes of addressing the difficulties inherent in whole organ transplantation: the limited availability of donor tissue and the side effects of long-term immunosuppression. Exciting work has been performed within the last several years, providing evidence that tissue-specific stem cells capable of self-renewal and differentiation are present in some adult organs, enabling life-long organ regeneration. Moreover, plasticity of adult stem cells has been described. This, in conjunction with described cell fusion phenomena challenges traditional concepts of cell biology. The underlying mechanisms of adult stem cell growth and differentiation are still unclear. However, there is increasing evidence that these processes are greatly influenced by the surrounding milieu. The development of culture systems offering a complex, tissue-specific microenvironment for the cells could greatly enhance further *in vitro* studies.

In this article, we review recent findings on adult stem cell research with a focus on cell biological and technological strategies for the possible clinical use of such cells. Three-dimensional four-compartment capillary membrane bioreactors with integral oxygenation and decentralized mass exchange support high-density cell co-culture and tissue reorganization under *in vivo*-like conditions. Initial results on liver progenitor cell culture in these bioreactors illustrate the potential of innovative culture technology in the field of stem cell-based regenerative medicine.

Introduction

The exciting reports published recently on the extraordinary potential of embryonic and adult stem cells have triggered worldwide intensive research in this field. Research efforts are driven in particular by the hope that stem cells could serve as a tool for developing innovative cell-based medical therapies. Although solid organ transplantation is widely used clinically to treat many diseases, including kidney, heart, and liver disease, the limited availability of donor organs, the frequency of graft failure and immune rejection, and the long-term side effects of immunosuppression, limit the benefits of transplantation medicine.

The use of stem cells capable of proliferating and differentiating into specific cell types could serve as an alternative to using whole organs or tissue to replace damaged tissue. Further advancement of stem cell-related research may produce new therapies in transplantation medicine such as cell transplantation and extracorporeal bioreactor application for the temporary support of a failing organ. In selected indications initial efforts are already being made to replace organ transplantation with adult cell transplantation. However, stem cell transplantation may be more suitable then transplantation of the fully differentiated adult cells, because stem cells may better retain the ability to form cell-cell junctions and communicate with host cells. In addition, the problems of immune reactions could be avoided if autologous stem cells are used, taking tissue from one part of the body to repair another. At present, this form of therapy is only performed with some solid tissue, such as autologous skin grafts for burn victims. The concept behind autologous stem cell therapy is to isolate adult stem cells, use *in vitro* technologies to multiply and differentiate them into cells of the injured tissue and, subsequently, use them for transplantation.

Cell-based therapies involve the transplantation of healthy cells to replace damaged cells, to treat diseases such as liver disease [1], heart disease [2], diabetes [3], Parkinson's disease [4], and muscular dystrophy [5]. Beneficial effects of those therapies have been shown. However, the studies also revealed some of the potential problems of stem cell therapy, including lack of cell purity, reduced cell functionality, and poor cell compatibility.

To realize the full therapeutic potential of stem cells, *in vitro* methods to maintain the unspecialized state of the stem cells, promote the proliferation of the stem cells, and direct the differentiation of the stem cells to the desired specialized cells are necessary. To prevent harmful side effects, a special focus must be placed on the safety issues of stem-cell-based therapies.

* Dr. Katrin Zeilinger, AG Experimentelle Chirurgie, Charité Campus Virchow-Klinikum, Augustenburger Platz 1, 13353 Berlin, Germany. Phone: 0049-30-450559009; Fax: 0049-30-450559909; email: katrin.zeilinger@charite.de

Stem Cells

Stem cells are defined as "undifferentiated cells", capable of a) proliferation, b) self-maintenance, c) the production of a large number of differentiated, functional progeny, d) regenerating tissue after injury, and e) flexibility in the use of these options [6].

Stem cells develop from totipotent zygotic cells of the embryo (capable of developing new organisms) via pluripotent blastocystic/morula cells of the embryo (with the potency to develop organs and tissues), to committed progenitor cells of the body (which differentiate into defined cell types).

Cells from the blastocyst form the pluripotent embryonic stem cells, which are the subject of embryonic stem cell research. Tissue-specific stem cells, which can be found in the fetus and in the adult, form the basis of adult stem cell research or (more precisely) somatic stem cell research. The potency of adult stem cells ranges from multipotency (ability to differentiate into several different cell types) to unipotency (ability to differentiate into a single cell type). Multipotent cells capable of differentiating into various cell types of the organism are located, for example, in bone marrow and cord blood. In the skin, multipotent stem cells are capable of generating all of the lineages of the skin, including multiple hair follicles, sebaceous glands, and epidermis [7]. Further sources for multipotent stem cells, such as fat tissue, are also being investigated.

Committed progenitor cells, such as hepatocytic or neuronal progenitor cells, belong to the class of unipotent cells with a defined progeny. Between multipotent and unipotent cells, various intermediate stages are possible, for example bipotent liver progenitor cells capable of differentiating into hepatocytes or biliary cells.

Both, embryonic and adult stem cells are considered potential instruments to address the major problems of organ transplantation: the limited availability of donor tissue and the immunological reactions following solid non-autologous organ transplantation.

In cell transplantation, two approaches are being investigated to avoid immunosuppression: therapeutic cloning of human embryos using autologous donor cells and the use of autologous adult stem cells.

Embryonic Stem Cells

In 1998 the culture of human pluripotent embryonic stem cells was enabled *in vitro*, stimulating speculation on the growth of organs from embryonic cells [8,9,10].

Although these cells have been shown to differentiate into single specific cell types, such as neural cells [11,12,13,14], pancreatic islets [15], and heart muscle cells [16], no method has yet been developed to enable directed organ development from embryonic stem cells *in vitro*. The molecular processes that direct the differentiation of embryonic stem cells toward specific cell lineages remain largely unknown. Initial gene technological processes remain a challenge [17].

The safety and stability of therapies based on human embryonic stem cells must be demonstrated if the clinical potential of these cells is to be realized.

A major problem is the regulation of proliferation, considering the tumorigenic potential of these cells [18]. Insufficient purification of transplanted cell preparations could lead to the formation of tumors in the patient. After experimental injection of undifferentiated embryonic stem cells into the brain of rats, five of 19 animals developed teratoma-like structures [19].

Karyotype abnormalities detected in cultured human embryonic stem cell lines involve a further safety risk [20,21]. The existence of chromosomal abnormalities in human embryonic stem cells destined for therapies could be detrimental, since karyotype changes are often associated with carcinogenesis. Additionally, human embryonic stem cells are currently co-cultivated with mouse feeder cells to maintain their undifferentiated state [8,22]. Reliable methods for removing mouse feeder cells from stem cell preparations intended for clinical application are required.

Adult stem cell therapy can be performed with autologous cells taken from the patient to prevent clinical rejection and avoid immunosuppression. Alternatively, human embryos must be cloned ("therapeutic cloning") to generate autologous embryonic stem cells, which requires nuclear reprogramming to switch off adult genes and switch on embryonic ones. The mechanisms involved in these processes are to date largely unknown. Although cloning of animals has been achieved, most of the cloned embryos die before birth, and the surviving ones generally develop health problems. Cloning studies in mice indicate that nuclear reprogramming may induce anomalous gene activity in the growing organism [23]. Cloning of human embryos up to the six-cell stage has been reported [24], and, more recently, the derivation of a pluripotent embryonic stem cell line from a cloned human blastocyst has been achieved [25]. The cells generated embryoid bodies *in vitro* and formed teratomas *in vivo* in immunodeficient mice. However, from the experience with animal cloning, the possibility of abnormal developments of these cells cannot be exluded.

Apart from scientific aspects, the use of human embryonic stem cells for research and clinical therapies raises ethical questions, regarding the attainment of the cells, their in-vitro processing, and their clinical usage. Considerable effort and cost will be required to demonstrate that embryonic stem cell therapies are safe enough to be approved for clinical use.

Adult Stem Cells

Adult stem cells, also known as tissue-derived or somatic stem cells, reside in specific structures within many organs, where they enable life-long organ regeneration. Adult stem cells are capable of self-renewal and give rise to committed progenitor cells that differentiate along specific pathways. Thus, cells that die due to aging or because of injury are replaced in most organs by adult stem cells.

Adult hematopoietic stem cells from bone marrow have been used in clinical therapy since 1959 to regenerate the immune system and blood system of cancer patients after chemotherapy, and to treat autoimmune diseases [26].

Organ-specific adult stem cells have now been described in many organs [27, 28], including bone marrow [29], skin [7, 30, 31, 32, 33, 34, 35, 36, 37, 38, 39], liver [40, 41, 42, 43, 44, 45, 46, 47], pancreas [48], muscle [49], and neural tissue [50, 51, 52, 53, 54, 55, 56,

57, 58, 59, 60]. Adult stem cells can be found in specific microanatomic tissue structures. Interestingly, some of these structures have been known for centuries. As far back as 1865, Hering described channels in the liver tissue [61] that can now be related to the origin of liver regeneration by liver progenitors [62], from which mature bile epithelial cells and hepatocytes are derived.

The potential of adult stem cells is enormous. Cells of the intestinal tract, for example, regenerate the mucous lining of the intestine at three-day intervals. In the case of acute liver disease, liver tissue is generated at a rate of one gram per day. The skin regenerates from the basal cell to the uppermost layer within approximately four weeks.

In addition to autologous cells, derived for example from the patient´s own bone marrow, cells derived from homologous tissue from an adult human donor or from fetal tissue may also be considered for cell-based therapies. One possibility for this approach is the use of cells from adult organ donors. For example, neural progenitors survive for a number of hours in the dead body and could be cultivated after death [63]. However, immunosuppression would be necessary in the same way as with solid organ transplantation. Thus, the development of autologous stem cell sources for therapy is a challenge.

Adult Stem Cell Plasticity

Traditional stem cell biology holds that the functional potential of cells – self-renewal and tissue repair - is restricted to the tissue from which they are derived. Within the last few years, several reports have been published challenging this concept. *In vivo* studies carried out in 2001 by Lagasse et al. [64] describe an extraordinary plasticity of adult stem cells [65].

In independent studies Shen et al. [66] and Dabeva et al. [67] demonstrated that pancreas-derived cells differentiate into mature liver cells expressing hepatocyte-specific markers. Furthermore, Malouf et al. [68] demonstrated that adult liver-derived stem cells respond to the tissue microenvironment of the adult heart *in vivo* and differentiate into mature cardiac myocytes. The differentiation of stem cells from fat tissue into mesenchymal cells, such as muscle, cartilage and bone was shown by Zuk et al. [69]. However, the possibility that the fat cells may have contained blood stem cells from the circulation could not be excluded. Bjornson et al. [70] showed that stem cell plasticity is not restricted to one direction. They enabled neuronal stem cells to convert *in vivo* to a variety of blood cells (including myeloid and lymphoid cells) as well as early hematopoietic cells. Studies by Galli et al. [71] indicate the skeletal myogenic differentiation potential of neural stem cells if exposed to myoblasts. In a further study, cells from skeletal muscle were shown to generate hematopoietic progeny in irradiated mice [72]. The skin may also harbour multipotent adult stem cells capable of differentiating into both neural and mesenchymal cells, as shown by Toma et al. [73].

Perhaps most impressive is the broad developmental capacity of bone marrow-derived adult stem cells [74,75]. Until recently, hematopoietic stem cells contained in the adult bone marrow were considered to be the source of blood cells only. Recently, transplanted cells isolated from the bone marrow of rodents and humans have been shown to differentiate into hepatic stem cells and mature hepatocytes in the liver [64,76,77,78,79,80]. A possible explanation for these findings seems to be the close association between hematopoiesis and the hepatic environment during fetal hepatic development. However, independent groups also demonstrated the capacity of bone marrow cells to differentiate into various other organ-specific cells, including neuronal cells [81,82], lung, skin and intestinal cells [74], blood vessel cells [83,84], and liver vasculature cells [85].

In 2002, studies by Verfaillie et al. [86,87] on the pluripotent potential of adult bone marrow stem cells evoked significant excitement. Verfaillie and her group were able to demonstrate that adult progenitor cells from bone marrow were able to differentiate *in vitro* into multiple different cell types with mesoderm, neuroectoderm and endoderm characteristics. The cells showed no signs of senescence or loss of their differentiation potential. Differentiation into a neuronal phenotype was achieved *in vitro* by co-culturing multipotent adult progenitor cells with astrocytes, suggesting that neuronal differentiation may require astrocyte-derived factors [88]. After implantation of single multipotent adult progenitor cells from bone marrow into mouse blastocysts, the cells differentiated into both neuronal and glial cell types, indicating that the cells exhibit properties similar to embryonic stem cells [89]. Single human bone marrow-derived mesodermal progenitor cells were also shown to differentiate into osteoblasts, chondrocytes, adipocytes, myocytes, endothelial cells, and hepatocyte-like cells [90,91]. Recent results from Verfaillie and her group indicate that cells with multipotent progenitor characteristics can be isolated not only from bone marrow, but also from brain and muscle tissue [92]. The authors showed that cells derived from all three tissues were able to differentiate into endothelial, neuronal, glial, and hepatic cell types.

The mechanisms involved in the presumed capacity of adult stem cells to differentiate into multiple cell types of specific organs or tissues are not yet clear. In initial reports on stem cell plasticity, transdifferentiation of the cells, involving reprogramming of genes to adopt the new cell type was assumed. However, more recent studies indicate that the generation of specific cell types derived from adult stem cells may be due to cell fusion with native cells located within the target organ rather than to transdifferentiation. The development of specific genetic techniques made it possible to distinguish between fusion and transdifferentiation processes in studies of transplanted stem cells. For example, fusion between cardiomyocytes and donor cells was observed after systemic delivery of genetically labelled adult heart-derived stem cells [93]. Studies in murine models of liver regeneration provide evidence that the repopulation of liver by bone-marrow-derived hepatocytes is also due to cell fusion between donor and recipient cells [94,95]. Similar processes were also described for other organs, like neural cells [96].

Clinical Usage of Bone Marrow Stem Cell Plasticity

The chameleonic potential of multipotent adult stem cells and the possibility of obtaining autologous multipotent stem cells for clinical therapy from the patient's own bone marrow could make these cells ideal candidates for the therapy of inherited or degenerative diseases. This prompted the initiation of a number of clinical trials based on bone marrow stem cells, primarily in the field of cardiac therapy.

First clinical trials on the usage of bone marrow stem cells for the treatment of cardiac infarction were stimulated by experimental data on myocardial regeneration through hematopoietic stem cells reported by Orlic et al. [97,98]. They tested the ability of mouse hematopoietic stem cells to differentiate into cardiac myocytes in heart muscle that had been damaged by experimental heart infarct. The cells migrated into the damaged heart muscle region and began to reconstitute the myocardium. In accordance with these findings, several independent studies on biopsies from human heart recipients revealed the presence of recipient-derived myocytes within transplanted human hearts [99,100,101]. The detection of donor-derived cardiomyocytes in bone marrow transplantation patients suggests that these cells originate from bone marrow [102]. Besides stem cell antigens, markers for myocytes and endothelial cells were detected, indicating tissue-specific differentiation of the cells in the transplanted hearts. These findings raised the hope that bone marrow may serve as a source for cardiac precursor cells and may enhance the regeneration of injured heart *in vivo*.

Indeed, two recent clinical trials showed a beneficial effect of bone marrow stem cell therapy for myocardial infarction. Strauer et al. [103] investigated ten patients treated by intracoronary injection of autologous, mononuclear bone marrow cells in addition to standard therapy. After three months, a significantly larger decrease of the infarct region, improved contractility and increased myocardial perfusion of the infarct region were observed compared with a control group given the standard treatment. In another study performed by Assmus et al. [104] a significant improvement in myocardial function, viability and coronary perfusion was detected after infusion of bone marrow-derived or circulating blood-derived progenitor cells into the infarct artery in patients with myocardial infarction. Both studies indicated that bone marrow cell therapy of myocardial infarcts is feasible and safe, and that it may beneficially affect myocardial regeneration.

However, these hopeful clinical results were called into question by a further clinical study that was interrupted prematurely due to serious cardiac side effects of bone marrow stem cell transplantation into injured hearts [105]. In addition, two recent experimental studies indicate that bone marrow stem cells fail to develop into heart muscle cells. Balsam et al. [106] isolated bone marrow cells from transgenic mice expressing green fluorescent protein and injected the labelled cells into the ischemic myocardium of wild-type mice. Marker analysis after 10 and 30 days showed that the fluorescence-labelled cells expressed only haematopoietic markers, and no evidence of myocyte generation from the bone marrow-derived cells was found. In a simultaneously published report, Murry et al. [107] studied the fate of haematopoietic stem cells after injection into normal and injured mouse hearts using a cardiomyocyte-specific reporter transgene. They also found no indications of the generation of bone marrow-derived cardiomyocytes or an increase in cardiomyocytes compared to sham-engrafted hearts.

Part of the discrepancies between these results and the results of earlier studies could be explained by differences in the experimental approach, e.g. type of injury, mode and site of cell delivery, or intrinsic properties of the stem cells used. An explanation for the improvement in cardiac function observed by Orlic *et al.* is that myocardial regeneration may have been stimulated by the growth of new blood vessels in the diseased area rather than by stem cell differentiation into myocardial cells. Such growth of new blood vessels has been found in transplantation studies of the heart [108]. Kocher et al. [109] demonstrated that human bone marrow-derived angioblasts may be used to reconstitute the vasculature after heart injury. In the same line of evidence, inhibition of bone marrow stem cell recruitment was shown to prevent tumor angiogenesis in mice [110].

The controversial results of experimental and clinical bone marrow stem cell therapy clearly show the need for further research to get a closer insight into the processes directing the migration and differentiation behaviour of these cells in the organism.

Although the studies described above indicate a large clinical potential of stem cells, the transition from basic laboratory research to clinical applications of stem cell-based therapies implicates several safety challenges [111]. Experimental testing in large-animal model systems has to precede clinical applications in patients in order to avoid premature clinical studies, but precautions must be taken before transferring the data from animal models into humans, with regard to possible interspecies differences. Thus, exhaustive pre-clinical and clinical testing of the behaviour of stem cells *in vivo* is necessary to prevent harmful side effects of stem cell-based therapy. Major requirements are the development of standardized assays and the selection of appropriate patient populations for early phase trials.

Stem Cells as Carriers for Drug Targeting

The plasticity of bone marrow stem cells could also be employed to target potential therapeutic factors to injured organs or tissues, as demonstrated by Priller et al. [112]. They found that gene-modified hematopoietic cells labelled with green fluorescent protein can mobilize from the blood after transplantation and enter the brain where they differentiate into microglia at sites of tissue damage, providing a vehicle for local gene delivery. To monitor the route and distribution of stem cells after transplantation, magnetic resonance tracking of magnetically labelled cells could also be used as an alternative to fluorescence marking [113].

Another approach could be the use of factors that stimulate bone marrow cells *in vivo* and, via the blood system, set them in motion toward damaged organs where they may enable regeneration [114]. Such drug-like substances may also be derived from stem cell cultures as products of cells cultivated in bioreactor systems.

Characterization of Stem Cells

The definite identification and characterization of stem cells is an essential requirement for *in vitro* studies and for future clinical applications of these cells for transplantation and gene therapy. Methods for isolation and characterization of epithelial stem cells are mostly based on morphological, physical and antigenic characteristics, such as the cells' adhesiveness, the expression of specific cell surface markers, or their proliferative capacity. A further important criterion is the functional competence of the cells, namely their capacity to differentiate into specific cell types. For example the multipotency of hematopoietic stem cells can be verified *in vivo* using cell-transfer experiments in animals.

However, the specific markers required for isolation of a certain type of stem cell are not available in all cases. If stem cells are co-cultured with non-parenchymal feeder cells, the distinction of proliferating stem cells from other growing cell types may be problematic. Assessing the functional competence of stem cells is often difficult due to a lack of suitable functional assays.

For example, the characterization of skin stem cells is complicated by the fact that a wide spectrum of epidermal keratinocytes can regenerate multilayered stratified epidermis. Although the presence of hair follicle stem cells residing in the bulge area of the hair follicle was shown in 1990 by Cotsarelis et al. [30], their genotypic and phenotypic characteristics were unclear.

Recently, two distinct approaches to surmount these difficulties were reported for skin stem cells. Both methods are based on the generation of transgenic mice designed to mark hair follicle bulge cells in cutaneous epithelium, thus permitting their subsequent isolation using fluorescence-activated cell sorting. Tumbar et al. [115] developed a strategy for fluorescently labelling slow-cycling cells marking the skin stem cell niche of the bulge in a cell type-specific fashion. The analysis of the transcriptional profile of labelled cells after isolation showed that many of the mRNAs niche encoding surface receptors and secreted proteins that allow for communication with their environment were preferentially expressed in the stem cell niche. Morris et al. [116] used a keratin-15 promoter to target bulge cells with either the green fluorescent protein marker or an inducible Cre recombinase construct, to facilitate cell characterization and isolation. The authors were able to demonstrate that isolated adult bulge stem cells are able to regenerate hair follicles, interfollicular epidermis and sebaceous glands, indicating a multilineage differentiation capability of the cells.

Similar approaches could also be useful to more closely characterize the localization and behaviour of stem cells from other organs and tissues.

In Vitro Culture of Stem Cells

The reported studies provide evidence that the differentiation of stem cells into a certain cell type is highly dependent on the cell environment. Under *in vivo* conditions, the differentiation of adult stem cells is apparently directed by the organ-specific milieu. To allow for controlled proliferation and differentiation of stem cells *in vitro*, culture systems offering a complex, tissue-specific microenvironment are required [117,118].

Co-culture models of adult stem cells with autologous mesenchymal cells, from the same or from other organs, may provide a valuable tool to enable physiological cell-cell contacts and the supply of specific cell-derived factors without adding synthetic additives. For example, bone marrow stromal cells were shown to support *in vitro* hepatocyte growth and differentiation, presumably by the activation of specific transcription factors [119]. Extracellular matrix components produced by stromal cells were identified as important growth regulators for human hematopoietic progenitors [120]. Jiang et al. showed that co-culture with astrocytes favours differentiation of mouse multipotent adult progenitor cells into neuroectodermal cell types [88]. Successful induction of embryonic stem cell differentiation into mature functional T-lymphocytes was achieved by co-culturing the cells with a stromal cell line [121].

The maintenance of primitive human hematopoietic stem cells is decreased by direct contact with bone marrow stromal cells, while separation from the stroma layer through membranes permeable for soluble mediators positively influenced cell maintenance [122]. Similarly umbilical cord hematopoietic cells maintained in stroma non-contact culture models or in stroma-free cultures supplied with preconditioned medium from stromal cultures preserved long-term engrafting capacities, i.e. the ability of engrafting secondary or tertiary hosts [123]. This phenomenon was attributed to adhesion-mediated proliferation inhibition by the stromal cell layer. Co-culture experiments using the AFT024 cell line that supports primitive human hematopoiesis performed better in contact cultures compared to primary feeder cells, suggesting that these cells produce factors that prevent contact growth inhibition of human hematopoietic progenitors [124].

Further identification of the factors or combinations of factors involved in promoting stem cell growth and differentiation is needed to allow for directed regulation and control of the stem cell behaviour *in vitro* and *in vivo*.

Bioreactors as Tools for Stem Cell-Based Research and Therapy

Several *in vitro* investigations on stem cells can be performed using conventional two-dimensional culture models. However, studies focusing for example on the impact of exogenous factors could benefit from the use of custom-built bioreactors that enable three-dimensional (3D) high-density cultures of stem cells possibly in co-culture with stromal cells. The development of specific 3D bioreactor technology appears important in the context of developing individual culture models for specific cells, as well as for subsequent translation of the findings to develop animal models, establish cell banks, and prepare clinical trials. Cell culture in a 3D environment is important not only in accommodating an organ-like mass but also in enhancing physiological cell-cell contacts, accumulation of extracellular matrices, and local growth factor delivery. For example, much greater biochemical effects on the maturation of fetal mouse hepatocytes were achieved in a 3D bioreactor when compared to monolayer cultures [125,126]. Thus, bioreactor technology development may facilitate studies on the behaviour of cells and the mechanisms of tissue formation, under reproducible and controllable *in vitro* conditions. In addition, large-scale stem cell expansion for clinical use can be realized using high-density culture bioreactors. To enable safe clinical applications, the stem cells and their progeny must be stable during *in vitro* maintenance and, after transplantation, into existing physiological tissues, at least until they are replaced by the body's own tissue repair process, or permanently if self-repair is not possible [127].

The aim of advanced bioreactor technology development for stem cell culture is to allow cells to re-build functional tissue with organ functions *ex vivo*. Possible applications include

- Cell maintenance for *in vitro* research or for clinical therapy (such as extracorporeal liver support based on primary liver cells)
- Cell expansion for the production of substances *for in vitro* or *in vivo* usage (such as mediator production for organ regeneration)
- Cell expansion and/or differentiation for clinical cell transplantation (such as hematopoietic stem cell growth for therapy of injured tissue)

Many different types of bioreactors spanning a wide range of geometric constructions, flow arrangements and mass exchange conditions have been developed and tested *in vitro* (table 1) and *in vivo* (tables 2 and 3).

Table 1. Bioreactor constructions tested in vitro

Technology	Cell Type	Company/ published studies
Early perfusion chambers	Chick heart fibroblasts, human malignant epithelial cells, Chinese hamster cells, hybridomas	Christiansen et al, 1953[128]; Rose, 1954[129]; Freed, 1963[130]; Katinger, 1985[131]
Commercially available perfusion chambers	Bone marrow-derived osteoblasts	Minucells [132]
Commercially available systems for non-adherent cells	Hybridomas	VectraCell gas-permeable bags [133]; Rotary Cell Culture System [134]; Wave Bioreactor [135]; CELLine [136]; miniPERM Bioreactor [137]; CellMax [138]; Tecnomouse [139]
Commercially available system for bone marrow expansion	Hematopoetic stem cells	AastromReplicell [140]
Hollow fiber-based bioreactors	Mouse fibroblasts, human choriocarcinoma cells, Reuber hepatoma cells, human hepatocytes	Knazek et al, 1972 [141]; Wolf et al, 1975[142]; Hager et al, 1978[143,144]; 1983
Hollow fiber-based bioreactor with integral oxygenation	Human leukemic cell lines	Gloeckner et al, 2001 [145]
Coaxial hollow fiber-based bioreactor with integral oxygenation	Rat hepatocytes	Macdonald et al, 2001 [146]
Hollow fiber-based bioartificial liver with integral oxygenation (Gerlach CellModule)	Porcine and human liver cells	Gerlach et al, 1994 [147]; 1996 [148]; 2003 [149,150]; Sauer et al, 2002[151]
Flat sheet and hollow fiber-based bioartificial liver with integral oxygenation	Porcine hepatocytes	Flendrig et al, 1997 [152]
Hollow fiber-based renal tubule assist device (Humes RAD)	Porcine renal tubule cells	Humes et al, 1999 [153]
Flat membrane bioreactor with integral oxygenation	Porcine hepatocytes	De Bartolo et al, 2000 [154]
Flat plate bioartificial liver with integral oxygenation	Porcine hepatocytes	Shito et al, 2001 [155]
Micropatterned borosilicate wafers	Rat hepatocytes and 3T3 fibroblasts	Bhatia et al, 1997 [156]
Biodegradable polymer bioreactor constructed via 3D printing	Rat liver cells	Kim et al, 1998[157]
Microfabricated bioreactor constructed via ion etching of silicon wafers	Rat hepatocytes	Powers et al, 2002 [158,159]
Titanium mesh bioreactor	Rat bone marrow stromal osteoblasts	Bancroft et al, 2003 [160]
Bioreactor containing hydrated polyester fibers and porcine autologous biomatrix	Porcine hepatocytes (1010)	Ambrosino et al, 2002 [161]
Bioreactor containing nonwoven polyurethane matrix with integral oxygenation	Rat or pig hepatocytes	Linti et al, 2002[162]

Table 2. Bioreactor constructions tested in animal models

Technology	Cell Type	Published studies
Hollow fiber-based bioartificial liver with integral oxygenation (Gerlach CellModule)	Porcine liver cells	Gerlach et al, 1993[163]; 1994[164]; 2001[165]; Janke et al, 1997[166]
Spirally wound flat sheet and hollow fiber-based bioartificial liver with integral oxygenation	Porcine hepatocytes	Flendrig et al, 1999[167]
Flat plate bioartificial liver with integral oxygenation	Porcine hepatocytes	Shito et al, 2003[168]
Hollow fiber-based renal tubule assist device	Human renal tubule cells	Humes et al, 2002[169]

Table 3. Bioreactor constructions tested in clinical trials

Technology	Cell Type	Published studies
Hollow fiber-based bioartificial liver (Vitagen ELAD)	Human hepatoblastoma cell line (C3A)	Sussman et al, 1992 [170]
Hollow fiber-based bioartificial liver with hepatocytes attached to dextran microcarriers (Circe Biomedical HepatAssist)	Porcine hepatocytes	Watanabe et al, 1997 [171]
Hollow fiber-based renal tubule assist device	Human renal tubule cells	Humes et al, 2002 [172]
Hollow fiber-based bioartificial liver perfused with whole blood (Excorp Medical BLSS)	Porcine hepatocytes	Mazariegos et al, 2002 [173]
Hollow fiber-based bioartificial liver with integral oxygenation (Gerlach CellModule)	Porcine and human liver cells	Mundt et al, 2002 [174]; Irgang et al, 2003[175]; Sauer et al, 2003 [176, 177]

Selected bioreactor motifs, such as multi-axial hollow fibres [146] and flat plate bioreactors [155] have been modelled using simple engineering equations that relate operating parameters to conditions within the local cellular environment, e.g. plasma flow rates in cell chambers, oxygen tension profiles, and specific biological performances such as metabolic activity or substance production. These idealized geometries are excellent tools for investigating the impact of the microenvironment upon cell behaviour and function; however, extension of these designs toward clinical dimensions has been a challenge.

Four-Compartment Bioreactor Technology for Stem Cell Growth and Expansion

We focus on bioreactor developments that address the cellular needs of 3D tissue density conditions. The objective of our research was to develop a four-compartment 3D stem cell bioreactor technology platform that will allow investigators to effectively study embryonic, fetal, and adult stem cells in a tightly controlled environment. A major feature of the technology developed is the option of scaling up the technology from small analytical devices for research purposes towards large systems for expansion of cells and for their use in clinical bioartificial liver support.

Our working hypothesis states that to allow for tissue assembly by the cells themselves, maintenance, and proliferation/differentiation of stem cells, a step from the conventional two- or three-compartment systems towards a four-compartment technology is necessary. Furthermore we hypothesize that stem cells themselves can create their own typical microenvironment in such *in vitro* culture models, and they may benefit from interactions with non-parenchymal cells co-cultured in the systems.

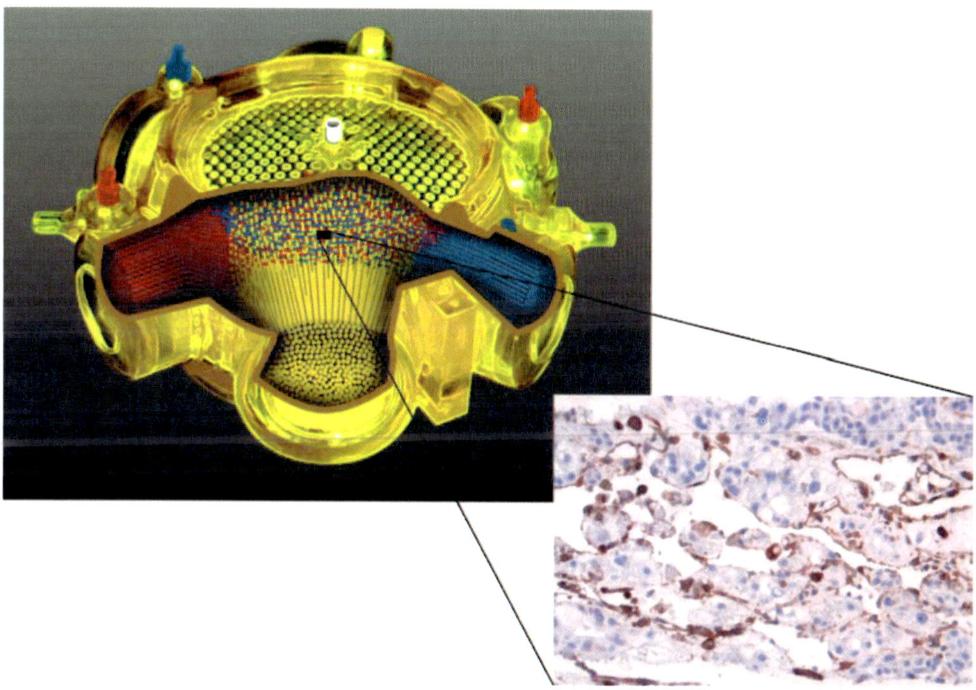

Figure 1. 3D multi-compartment bioreactor for high-density culture of primary liver cells. Medium perfusion is performed via two independent capillary systems each connected to the perfusion circuit via separate ports for inflow and outflow. A further capillary system is provided for oxygenation. Cells are inoculated into the extra-capillary space through multiple tube endings, which are open at different locations within the cell compartment. The histological analysis of primary human liver cell cultures (insert) shows that parenchymal and non-parenchymal liver cells build up tissue-like structures in bioreactors. Endothelial cells characterized by vimentin immunoreactivity (red staining) form vascular channels, which allow for efficient transport of nutrients/metabolites within the cell compartment.The histological picture was kindly provided by Prof. Dr. Sebastian Bachmann and Gudrun Holland, Department of Anatomy, Charité University Medicine Berlin.

The four-compartment bioreactor (Figure 1) consists of two independent bundles of hollow fiber microfiltration membranes for the transport of culture medium (medium compartments 1,2), interwoven with one bundle of hollow fiber oxygenation membranes for the transport of oxygen, carbon dioxide, and/or other gases (gas compartment 3). The hollow fiber membranes are potted within a polymer housing, and the cells are cultured in the interstitial space between the capillaries (cell compartment 4). The four-compartment design allows integral oxygenation of the cells and efficient transfer of nutrients to the cells and removal of waste products from the cells. Additionally, the interwoven fibers provide a 3D scaffold allowing the cells to form 3D tissue-like structures within the cell compartment.

Primary Liver Cell and Liver Stem Cell Culture in Four-Compartment Bioreactors

Initial applications of our bioreactor technology focused on large-scale primary liver cell culture for extracorporeal liver support. The bioreactors currently used for primary liver cell culture and clinical liver support are hollow fiber-based and exhibit two functional compartments [171,178]. Xu et al. pointed out that four bioreactor compartments are necessary to enable integral oxygenation and distributed mass exchange with low solute gradients [179].

We developed a scale for the described bioreactor technology specific for clinical liver support that accommodates 400 – 800g of primary cells [165]. The design focused on transforming the previously published bioreactor constructions into a technology with up to five compartments. Our approach was to mimic the native structure of an organ using an artificial capillary bed. A construction principle was developed to form independently perfused subunits from woven synthetic capillaries, each with a different function, and arranged in parallel [180]. The lack of erythrocytes was addressed by flowing oxygen through one compartment. Mass exchange was addressed by perfusing two independent capillary systems with plasma and media. Small artificial hollow fiber capillary subunits, in which interwoven membranes represent the organs secondary level of vascular supply, are simultaneously perfused [181]. To scale up the bioreactor, additional subunits are added resulting in a capillary system [147,182] similar to the natural organ's microvasculature. Primary liver cells isolated from porcine or human organs [183,184,185,186] are inoculated into the extracapillary space (cell compartment) [187].

The bioreactor is integrated into a perfusion device with separate pump systems for medium circulation and medium substitution, a heating unit for maintaining the temperature of the bioreactor chamber, and a gas mixing unit for the regulation of O2/CO2 perfusion (Figure 2).

Under conditions of capillary perfusion, parenchymal and non-parenchymal tissue formation is possible by spontaneous organization to cellular aggregates and cellular attachment to the surface of the capillaries [188]. The cells spontaneously re-formed the primary capillary structure of liver sinusoids thus supporting a larger cell mass at a higher density [189]. Maintenance of the differentiated functions of primary liver cells was demonstrated over a period of several weeks [148,190]. Primary human liver cells, obtained from preservation-injured organs, showed morphological and biochemical signs of recovery and regeneration in bioreactors [149,191].

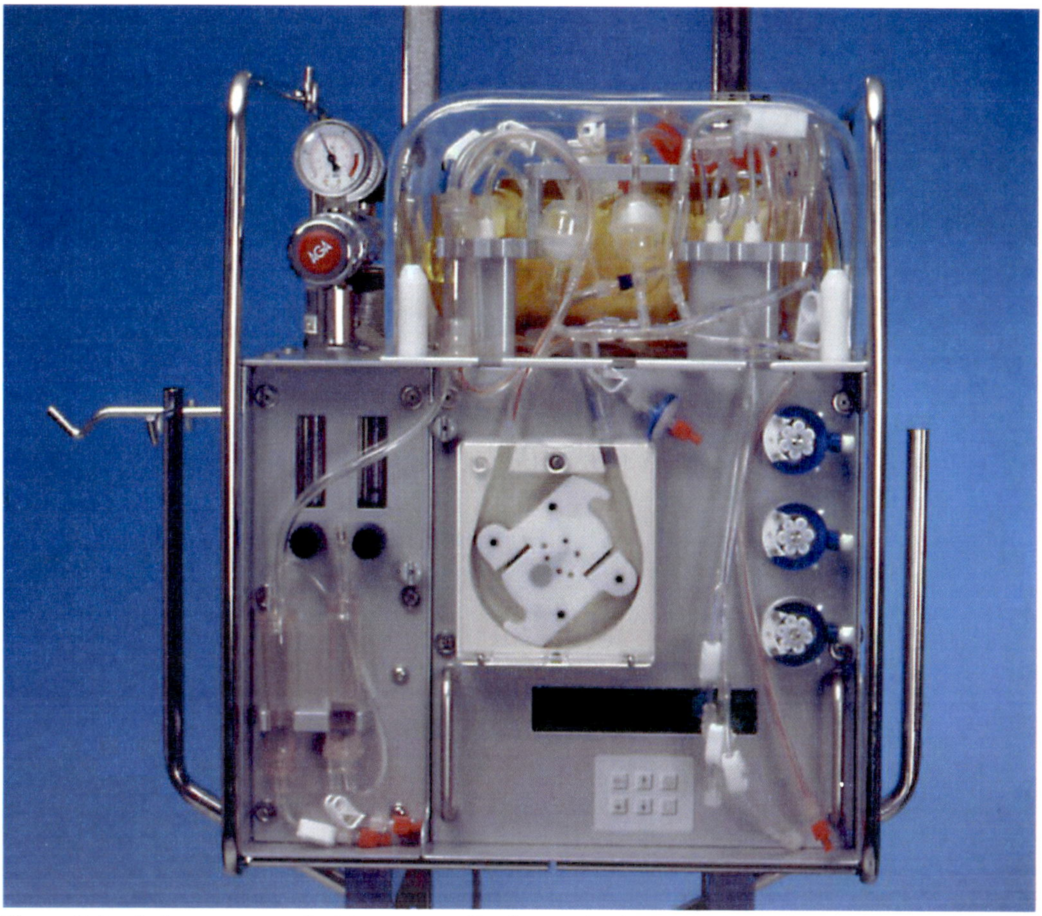

Figure 2: Perfusion apparatus for bioreactor cell culture. The device disposes of individual pump systems for medium circulation and medium supply, equipment for gas regulation, and a heating unit for maintaining a constant temperature in the bioreactor chamber.

The bioreactor has been successfully used in a device to provide extracorporeal liver support [192,193,194]. Animal studies showed encouraging results for the clinical use of the bioreactor [118,163,164,166,195,196]. Using porcine cells, the device was used as a bridge to liver transplantation in a phase-I study with eight patients in acute liver failure, coma stage III-IV [176]. To date the patient and organ survival rate is 100% with an observation period of 3 years [175]. A pilot study in Berlin, Germany, is using the device with primary human liver cells harvested from organs explanted for transplantation but subsequently discarded due to steatosis, cirrhosis or mechanical injury [151,177]. These results demonstrate the feasibility of employing primary human liver cells in clinical applications with specific bioreactor systems [197]. However, the limiting issue of liver support remains the availability of primary human cells. Thus although an attractive technology with therapeutic potential is available, the availability of the cell source will be crucial to enable further clinical studies.

This problem prompted us to work on methods of using adult liver stem cells as a source for generating functional, differentiated liver cells. Reports on the existence of liver-specific adult stem cells [40,41,42,43,44,45,46,47] raised the hope that these cells could, in the future, enable tissue repair processes in regenerative medicine. Several studies described the differentiation process of "small hepatocytes/oval cells" that are presumably identical with the stem cells of the adult liver [198,199,200] (Figure 3). To characterize the differentiation state of liver progenitor cells *in vitro* and on biopsies taken from intact or injured liver, a number of different markers have been used, like α-Fetoprotein [201,202,203,204], stem cell factor (SCF) and its receptor c-kit (CD 117) [205,206,207,208,209,210,211], CD 34 [205,212] and cytokeratin (CK) 19 [41,204,213,214], as indicated in figure 3.

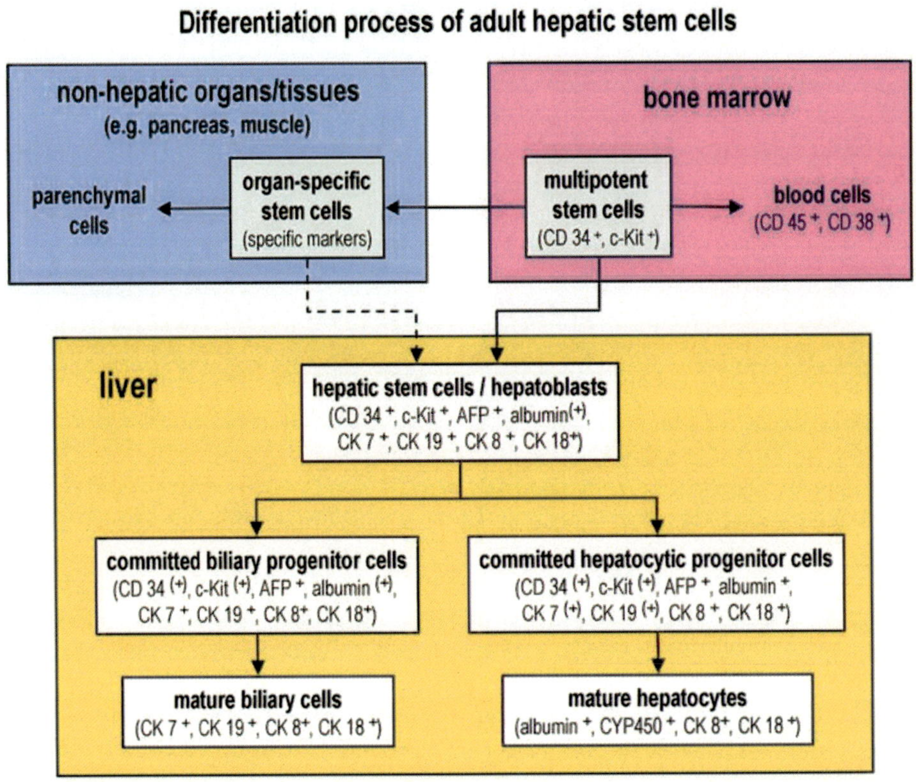

Figure 3: Schematic figure of the process of liver stem cell differentiation, showing the possible bone marrow origin of adult liver stem cells. Markers described for the characterization of liver stem cells in different differentiation stages are indicated.

Studies on the growth and differentiation behaviour of fetal liver stem cells have demonstrated the importance of growth factors and cytokines [215]. Hepatocyte growth factor (HGF), for example, is generally known to stimulate hepatocytes to replicate *in vitro*, and oncostatin M, a multifunctional cytokine of the interleukin-6 family, has been shown to induce maturation of fetal mouse hepatocytes *in vitro* [216,217,218], while it attenuates fetal liver hematopoiesis [219].

The generally used approach is to add these and other factors at a defined concentration and combination to the culture. The developed 3D high-density multicompartment bioreactor could favour the creation of such a microenvironment, by allowing for physiological cell-

cell-contacts, extracellular matrix provision and signal exchange, as well as controllable nutrient supply and oxygenation.

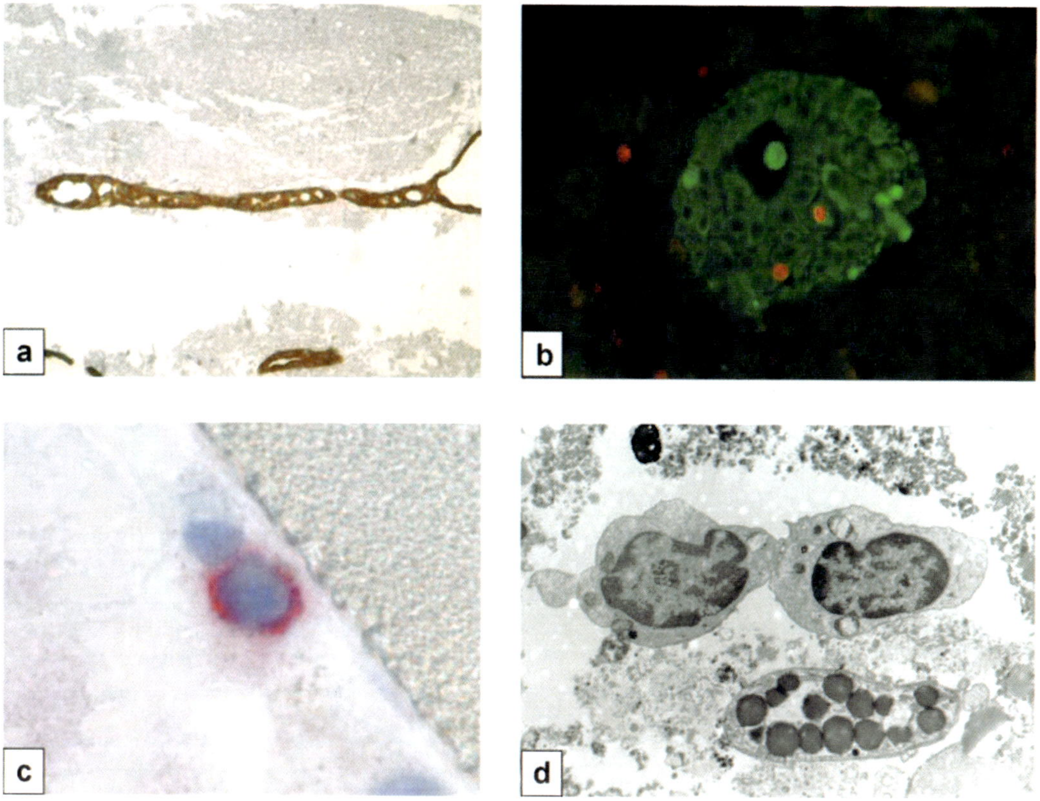

Figure 4: Immunohistochemical and ultrastructural observations on liver stem cells in bioreactor cultures of primary human liver cells. a) Bile ductular structure resembling the Canals of Hering, showing CK 19 immunoreactivity (magnification x 200); b) cell proliferation originating from bile ductular structures, as shown by double-immunofluorescence labeling with CK 19 (green) and the proliferation marker Ki-67 (red), (magnification x 400); c) small cell expressing CD 34, adjacent to an artificial capillary (magnification x 800); d) cluster of three small cells (transmission electron microscopy, magnification x 2,784). The pictures were kindly provided by Prof. Dr. Sebastian Bachmann, Petra Schrade and Kerim Mutig, Department of Anatomy, Charité University Medicine Berlin.

Initial histological and immune histochemical studies performed on bioreactor co-cultures of primary human hepatocytes and non-parenchymal cells indicate the presence of proliferating cells and of cells expressing typical morphological and antigenic properties of adult liver stem cells [150]. Light and electron microscopic analyses of biopsies taken from the cell compartment after one to five weeks of bioreactor culture showed the formation of parenchyma-like tissue associations. Hepatocyte aggregates were penetrated by vascular channels characterized by CD 31 immunoreactivity and bile duct-like channels expressing CK 19. Biliary structures formed complex tubular systems, which increased in number with culture time. The detection of the proliferation marker ki-67 indicated cell growth originating from these structures. The observed proliferation of bile ducts shows similarities to the so-called "ductular reaction" associated with hepatic necrosis or chronic cholestasis [220,221,222]. The origin of bile duct growth after liver injury may be liver stem cell proliferation. Studies by Theise et al. [62] indicate that adult stem cells in the human liver are harbored within the Canals of Hering, which are located in the periportal area between bile ducts and parenchymal bile canaliculi. The ultrastructural analysis of bioreactor cultures (Figure 4) revealed the presence of channel-like structures showing characteristics of the Canals of Hering in the cultures. Furthermore, small cells expressing the stem cell antigens CD34 and c-kit, often associated in cell couples, were detected in the biopsies suggesting the presence of early, undifferentiated stem cells in the cultures.

Further studies are required to more closely investigate the proliferation and differentiation processes of liver stem cells in bioreactors on molecular biological, morphological and functional levels.

To address these needs, the following criteria were applied in the adaptation of bioreactors for the specific demands of stem cell culture and research:

- Small cell compartment volume thus minimizing the required starting mass of stem cells and the required amount of reagents
- Efficient mass transfer and integral oxygenation allowing high density cell culture
- Ability to introduce gaseous mediators such as nitrous oxide through the oxygenation membranes
- Non-destructive access to the culture medium compartments for biochemical analyses
- Non-destructive access to the cell compartment for gene-arrays and PCR sampling
- Non-destructive access to the cell compartment for harvesting of proliferating cell populations
- Ability to conduct on-line (video) microscopy for continuous observation of the cell behaviour during the culture period

These technical features were realized by constructing small-scale bioreactor prototypes, which are based on the multi-compartment technology developed for primary cell culture, but allow for the culture of small cell amounts (up to 1 resp. 10 ml of cells).

With the integration of different compartments for medium inflow and outflow, oxygenation and cell maintenance, these bioreactors allow for 3D culture of various stem cell populations in a perfused environment, providing decentralized mass exchange at physiological gradients and integral oxygenation, under conditions closely reflecting the *in vivo* situation, where stem cells reside and interact in microvascularised structures.

Removal of living cells from the system for cell passaging or further studies can be attained through enzymatic or non-enzymatic rinsing of the cell compartment, which is directly accessible via a separate port, or by opening a lid and mechanical removal. This provides the option for cell transfer from the system into larger bioreactors for cell expansion at a clinical scale for potential therapeutic use.

Figure 5: Small-scale multi-compartment bioreactors for cell volumes of up to 10 ml (left side) or 1 ml (right side). The structure corresponds to the model shown in figure 1, exhibiting two independent capillary systems for medium perfusion, one further capillary bundle for oxygenation, and a separate access to the cell compartment for cell inoculation, as well as for cell removal from the bioreactor.

Thus, the technologies developed can be used as tools for various studies on *in vitro* maintenance, proliferation and differentiation of stem cells, as well as for the generation of a cell pool that may serve as a continuous source for stem cells to be used *in vitro* (for differentiation into different specific cell types) or *in vivo* (for cell transplantation).

Outlook

The possible advantages of processing autologous adult stem cells for transplantation medicine are obvious: no limit to availability, reduced waiting times, and, since the body recognizes its own cells, no life-long immunosuppression with its adverse side-effects.

To provide a better understanding of the mechanisms contributing to stem cell growth and differentiation, basic research must explore these processes on cell biologic and molecular levels. Biotechnologists and engineers face the task of developing new technology platforms in medical biotechnology.

Bioreactor technology facilitates: investigations of stem cell maintenance, proliferation, and directed differentiation; investigations of stem cell behaviour after 3D reorganization (possibly in co-culture with differentiated cells) at high cell densities comparable to the cell densities of biological tissues; *in vitro* safety and quality investigations of stem cell cultures; and standardization of experiments.

Bone marrow-derived stem cells and the possibility of producing "circulating stem cells" that exhibit considerable plasticity *in vivo* may be the most interesting source of stem cells for cell-based therapies. Therefore, the development of bioreactors should also consider bone marrow cells. Such systems could enable the use of autologous bone marrow cells for the differentiation/production of liver cells, heart muscle cells and other cell types, which would represent a milestone in the development of the field.

References

[1] Strom, SC; Chowdhury, JR; Fox, IJ. Hepatocyte transplantation for the treatment of human disease. *Semin Liver Dis,* 1999, 19, 39-48.

[2] Couzin, J; Vogel, G. Cell therapy. Renovating the heart. *Science,* 2004, 304, 192-194.

[3] Kaczorowski, DJ; Patterson, ES; Jastromb, WE; Shamblott, MJ. Glucose-responsive insulin-producing cells from stem cells. *Diabetes Metab Res Rev,* 2002, 18, 442-450.

[4] Isacson, O. The production and use of cells as therapeutic agents in neurodegenerative diseases. *Lancet Neurol,* 2003, 2, 417-424.

[5] Deasy, BM; Huard, J. Gene therapy and tissue engineering based on muscle-derived stem cells. *Curr Opin Mol Ther,* 2002, 4, 382-389.

[6] Potten, CS; Loeffler, M. Stem cells: attributes, cycles, spirals, pitfalls and uncertainties. Lessons for and from the crypt. *Development*, 1990, 110, 1001-1020.

[7] Oshima, H; Rochat, A; Kedzia, C; Kobayashi, K; Barrandon, Y. Morphogenesis and renewal of hair follicles from adult multipotent stem cells. *Cell*, 2001, 104(2), 233-245.

[8] Thomson, JA; Itskovitz-Eldor, J; Shapiro, SS; Waknitz, MA; Swiergiel, JJ; Marshall, VS; Jones, JM. Embryonic Stem Cell Lines Derived from Human Blastocysts. *Science,* 1998, 282, 1145-1147.

[9] Gearhart, J. New Potential for Human Embryonic Stem Cells. *Science,* 1998, 282, 1061.

[10] Shamblott, MJ; Axelman, J; Wang, S; Bugg, EM; Littlefield, JW; Donovan, PJ; Blumenthal, PD; Huggins, GR; Gearhart, JD. Derivation of pluripotent stem cells from cultured human primordial germ cells. *Proc Natl Acad Sci,* 1998, 95,13726.

[11] Brustle, O; Jones, KN; Learish, RD; Karram, K; Choudhary, K; Wiestler, OD; Duncan, ID; McKay, RD. Embryonic stem cell-derived glial precursors: a source of myelinating transplants. *Science*, 1999, 285, 754-756.

[12] Liu, S; Qu, Y; Stewart, TJ; Howard, MJ; Chakrabortty, S; Holekamp, TF; McDonald, JW. Embryonic stem cells differentiate into oligodendrocytes and myelinate in culture and after spinal cord transplantation. *Proc Natl Acad Sci*, 2000, 97(11), 6126-6131.

[13] Zhang, S-C; Wernig, M; Duncan, ID; Brüstle, O; Thomson, JA. In vitro differentiation of transplantable neural precursors from human embryonic stem cells. *Nat Biotechnol,* 2001, 19 (12), 1129 -1133.

[14] Reubinoff, BE; Itsykson, P; Turetsky, T; Pera, MF; Reinhartz, E; Itzik, A; Ben-Hur, T. Neural progenitors from human embryonic stem cells. *Nat Biotechnol,* 2001, 19 (12), 1134-1140.

[15] Lumelsky, N; Blondel, O; Laeng, P; Velasco, I; Ravin, R; McKay, R. Differentiation of embryonic stem cells to insulin-secreting structures similar to pancreatic islets. *Science,* 2001,292, 1389-1394.

[16] Kehat, I; Kenyagin-Karsenti, D; Snir, M; Segev, H; Amit, M; Gepstein, A; Livne, E; Binah, O; Itskovitz-Eldor, J; Gepstein, L. Human embryonic stem cells can differentiate into myocytes with structural and functional properties of cardiomyocytes. *J Clin Invest*, 2001, 108(3), 407-414.

[17] Wakayama, T; Tabar, V; Rodriguez, I; Perry, ACF; Studer, L; Mombaerts, P. Differentiation of Embryonic Stem Cell Lines Generated from Adult Somatic Cells by Nuclear Transfer. *Science*, 2001, 292, 740-743.

[18] Langa, F; Kress, C; Colucci-Guyon, E; Khun, H; Vandormael-Pournin, S; Huerre, M; Babinet, C. Teratocarcinomas induced by embryonic stem (ES) cells lacking vimentin: an approach to study the role of vimentin in tumorigenesis. *J Cell Sci*, 2000, 113, 3463-3472.

[19] Bjorklund, LM; Sanchez-Pernaute, R; Chung, S; Andersson, T; Chen, IY; McNaught, KS; Brownell, AL; Jenkins, BG; Wahlestedt, C; Kim, KS; Isacson, O. Embryonic stem cells develop into functional dopaminergic neurons after transplantation in a Parkinson rat model. *Proc Natl Acad Sci USA,* 2002, 99(4), 2344-2349.

[20] Draper, JS; Smith, K; Gokhale, P; Moore, HD; Maltby, E;Johnson, J; Meisner, L; Zwaka, TP; Thomson, JA; Andrews, PW. Recurrent gain of chromosomes 17q and 12 in cultured human embryonic stem cells. *Nat Biotechnol*, 2004, 22(1), 53-54.

[21] Rosler, ES; Fisk, GJ; Ares, X; Irving, J; Miura, T; Rao, MS; Carpenter, MK. Long-term culture of human embryonic stem cells in feeder-free conditions. *Dev Dyn*, 2004, 229(2), 259-274.

[22] Reubinoff, BE; Pera, MF; Fong, CY; Trounson, A; Bongso, A. Embryonic stem cell lines from human blastocysts: somatic differentiation in vitro. *Nat Biotechnol*, 2000, 18, 399-404.

[23] Humpherys, D; Eggan, K; Akutsu, H; Hochedlinger, K; Rideout, WM 3rd; Biniszkiewicz, D; Yanagimachi, R; Jaenisch, R. Epigenetic Instability in ES Cells and Cloned Mice. *Science*, 2001, 293, 95-97.

[24] Cibelli, JB; Kiessling, AA; Cunniff, K; Richards, C; Lanza, RP; West, MD. Somatic cell nuclear transfer in humans: pronuclear and early embryonic development. *J Regenerative Medicine,* 2001, 2, 25-31.

[25] Hwang, WS; Ryu, YJ; Park, JH; Park, ES; Lee, EG; Koo, JM; Jeon, HY; Lee, BC; Kang, SK; Kim, SJ; Ahn, C; Hwang, JH; Park, KY; Cibelli, JB; Moon, SY; Hwang, WS. Evidence of a pluripotent human embryonic stem cell line derived from a cloned blastocyst. *Science*, 2004, 303(5664), 1669-1674.

[26] Saba, N; Abraham, R; Keating, A. Overview of autologous stem cell transplantation. *Crit Rev Oncol Hematol,* 2000, 36(1), 27-48.

[27] Fuchs, E; Segre, JA. Stem Cells: A New Lease on Life. *Cell*, 2000, 100, 143-155.

[28] Slack, JMW. Stem Cells in Epithelial Tissues. *Science*, 2000, 287, 1431-1433.

[29] Kuci, S; Wessels, JT; Buhring, HJ; Schilbach, K; Schumm, M; Seitz, G; Loffler, J; Bader, P; Schlegel, PG; Niethammer, D; Handgretinger, R. Identification of a novel class of human adherent CD34- stem cells that give rise to SCID-repopulating cells. *Blood*, 2003, 101, 869-876.

[30] Cotsarelis, G; Sun, TT; Lavker, RM. Label-retaining cells reside in the bulge area of pilosebaceous unit: implications for follicular stem cells, hair cycle and skin carcinogenesis. *Cell,* 1990, 6, 1329-1337.

[31] Toma, JG; Akhavan, M; Fernandes, KJL; Barnabé-Heider, F; Sadikot, A; Kaplan, DR; Miller, FD. Isolation of multipotent adult stem cells from the dermis of mammalian skin. *Nat Cell Biol*, 2001, 3, 778-784.

[32] Epstein, E Jr. Split stories converge at desmoglein 1. *Nat Med*, 2000, 6(11), 1213-1214.

[33] Huelsken, J; Vogel, R; Erdmann, B; Cotsarelis, G; Birchmeier, W. ß-Catenin controls hair follicle morphogenesis and stem cell differentiation in the skin. *Cell,* 2001, 105, 533-545.

[34] Jones, PH; Watt, FM. Separation of human epidermal stem cells from transit amplifying cells on the basis of differences in integrin function and expression. *Cell*, 1993, 73, 713-724.

[35] Jones, PH; Harper, S; Watt, FM. Stem cell patterning and fate in human epidermis. *Cell*, 1995, 80, 83-93.

[36] Lavker, RM; Sun, T-T. Epidermal stem cells: Properties, markers and location. *Proc Natl Acad Sci*, 2000, 97, 13473-13475.

[37] Li, A; Simmons, PJ; Kaur, P. Identification and isolation of candidate human keratinocyte stem cells based on cell surface phenotype. *Proc Natl Acad Sci*, 1998, 95, 3902-3907.

[38] Tani, H; Morris, RJ; Kaur, P. Enrichment for murine keratinocyte stem cells based on cell surface phenotype. *Proc Natl Acad Sci,* 2000, 97, 10960-10965.

[39] Taylor, G; Lehrer, MS; Jensen, PJ; Sun, T-T; Lavker, RM. Involvement of follicular stem cells in forming not only the follicle but also the epidermis. *Cell*, 2000, 102, 451-461.

[40] Suzuki, A; Zheng, Y-W; Kondo, R; Kusakabe, M; Takada, Y; Fukao, K; Nakauchi, H; Taniguchi, H. Flow-cytometric separation and enrichment of hepatic progenitor cells in the developing mouse liver. *Hepatology*, 2000, 32, 1230-1239.

[41] Crosby, HA; Hubscher, S; Fabris, L; Joplin, R; Sell, S; Kelly, D; Strain, AJ. Immunolocalization of putative human liver progenitor cells in livers from patients with end-stage primary biliary cirrhosis and sclerosing cholangitis using the monoclonal antibody ov-6. *Am J Pathol,* 1998, 152, 777-779.

[42] Reid, LM. Stem cell/lineage biology and lineage-dependent extracellular matrix chemistry: keys to tissue engineering of quiescent tissues such as liver. In: Ed Lanza RP, Langer R, Chick W, editors. *Principles* of *Tissue Engineering*. Georgetown, Texas, USA: R. G. Landes Company and Academic press, Inc.; 1997; 481-514.

[43] Sell, S. Heterogeneity and plasticity of hepatocyte lineage cells. *Hepatology*, 2001, 33(3), 738-750.

[44] Alison, M; Sarraf, C. Hepatic stem cells. *J Hepatol,* 1998, 29, 676-682.

[45] Sigal, SH; Brill, S; Fiorino, AS; Reid, LM. The liver as a stem cell and lineage system. *Am J Physiol*, 1992, 263, G139-G148.

[46] Brill, S; Holst, P; Sigal, S; Zvibel, I; Fiorino, A; Ochs, A; Somasundaran, U; Reid, LM. Hepatic progenitor populations in embryonic, neonatal and adult liver. *Proc Soc Exp Biol Med,* 1993, 261-269.

[47] Strain, AG; Crosby, HA. Hepatic Stem Cells. *Gut*, 2000, 46, 743-745.

[48] Ramiya, VK; Maraist, M; Arfors, KE; Schatz, DA; Peck, AB; Cornelius, JG. Reversal of insulin-dependent diabetes using islets generated in vitro from pancreatic stem cells. *Nat Med*, 2000, 6(3), 278-282.

[49] Lee, J-Y; Qu-Petersen, Z; Cao, B; Kimura, S; Jankowski, R; Cummins, J; Usas, A; Gates, C; Robbins, P; Wernig, A; Huard, J, Clonal isolation of muscle-derived cells capable of enhancing muscle regeneration and bone healing. *J Cell Biol.*, 2000 150(5), 1085-1100.

[50] Gage, FH. Mammalian neural stem cells. *Science,* 2000, 287,1433-1438.

[51] Bachoud-lévi, AC; Rémy, P; Ngyuyen, J-P; Brugières, P; Lefaucheur, J-P; Bourdet, C; Baudic, S; Gaura, V; Maison, P; Haddad, B; Boissé, M-F; Grandmougin,T; Jény, R; Bartolomeo, P; Dalla Barba, G; Degos, J-D; Lisovoski, F; Ergis, A-M; Pailhous, E; Cesaro, P; Hantraye, P; Peschanski, M. Motor and cognitive improvements in patients with Huntington´s disease after neural transplantation. *Lancet*, 2000, 356, 1975-1979.

[52] Piccini, P; Brooks, DJ; Bjorklund, A; Gunn, RN; Grasby, PM; Rimoldi, O; Brundin, P; Hagell, P; Rehncrona, S; Widner, H; Lindvall, O. Dopamine release from nigral transplants visualized in vivo in a Parkinson's patient. *Nat Neurosci*, 1999, 12, 1137-1140.

[53] Kondo, T; Raff, M. Oligodendrocyte precursor cells reprogrammed to become multipotential CNS stem cells. *Science*, 2000, 289(5485), 1754-1757.

[54] Clarke, DL; Johansson, C B; Wilbertz, J; Veress, B; Nilsson, E; Karlstrom, H; Lendahl, U; Frisen, J. Generalized potential of adult neural stem cells. *Science*, 2000, 288(5471), 1660-1663.

[55] Benedetti, S; Pirola, B; Pollo, B; Magrassi, L; Bruzzone, G; Rigamonti, D; Galli, R; Selleri, S; Di Meco, F; De Fraja, C; Vescovi, A; Cattaneo, E; Finocchiaro, G. Gene therapy of experimental brain tumors using neural progenitor cells. *Nat Med*, 2000, 6(4), 447-450.

[56] Han, SS; Fischer, I. Neural stem cells and gene therapy: prospects for repairing the injured spinal cord. *JAMA*, 2000, 283(17), 2300-2301.

[57] Doetsch, F; Caille, I; Lim, DA; Garcia Verdugo, JM; Alvarez Buylla, A. Subventricular zone astrocytes are neural stem cells in the adult mammalian brain. *Cell*, 1999, 97(6), 703-716.

[58] Johansson, CB; Momma, S; Clarke, DL; Risling, M; Lendahl, U; Frisen, J. Identification of a neural stem cell in the adult mammalian central nervous system. *Cell,* 1999, 96(1), 25-34.

[59] Rietze, RL; Valcanis, H; Brooker, GF; Thomas, T; Voss, AK; Bartlett, PF. Purification of a pluripotent neural stem cell from the adult mouse brain. *Nature,* 2001, 412, 736-739.

[60] Flax, JD; Aurora, S; Yang, C; Simonin, C; Wills, AM; Billinghurst, LL; Jendoubi, M; Sidman, RL; Wolfe, JH; Kim, SU; Snyder, EY. Engraftable human neural stem cells respond to developmental cues, replace neurons, and express foreign genes. *Nat Biotechnol,* 1998, 16(11), 1033-1090.

[61] Hering, E. Über den Bau der Wirbeltierleber, *Akad Wiss Wien,* Math Naturic KL 1865, 54, 496-515.

[62] Theise, ND; Saxena, R; Portmann, BC; Thung, SN; Yee, H, Chiriboga, L; Kumar, A; Crawford, JM. The canals of Hering and hepatic stem cells in humans. *Hepatology,* 1999, 30(6), 1425-1433.

[63] Palmer, TD; Schwartz, PH; Taupin, P; Kaspar, B; Stein, SA; Gage, FH. Cell culture: Progenitor cells from human brain after death. *Nature*, 2001, 411, 42-43.

[64] Lagasse, E; Connors, H; Al-Dhalimy, M; Reitsma, M, Dohse, M; Osborne, L; Wang, X; Finegold, M; Weissman, IL; Grompe, M. Purified hematopoietic stem cells can differentiate into hepatocytes in vivo. *Nat Med,*2000, 6, 1229-1234.

[65] Orkin, SH. Stem cell alchemy. *Nat Med,* 2000, 6, 1229-1234.

[66] Chia-Ning, Shen; Jonathan, M. W. Slack; David, Tosh. Molecular basis of transdifferentiation of pancreas to liver. *Nat Cell Biol*, 2000, 2(12), 879-887.

[67] Dabeva, MD; Hwang, S-G; Vasa, SRG; Hurston, E; Novikoff, PM; Hixson, DC; Gupta, S; Shafritz, DA. Differentiation of pancreatic epithelial progenitor cells into hepatocytes following transplantation into rat liver. *Proc Natl Acad Sci USA*, 1997, 94, 7356-7361.

[68] Malouf, NN; Coleman, WB; Grisham, JW; Lininger,RA; Madden, VJ; Sproul, M; Anderson, PAW. Adult-derived stem cells from the liver become myocytes in the heart *in vivo*. *Am J Pathol*, 2001, 158, 1929-1935.

[69] Zuk, PA; Zhu, M; Mizuno, H; Huang, J; Futrell, JW; Katz, AJ; Benhaim, P; Lorenz, HP; Hedrick, MH. Multilineage cells from human adipose tissue: implications for cell-based therapys. *Tissue Engineering 2001*, 7, 211-228).

[70] Bjornson, CRR; Rietze, RL; Reynolds, BA; Magli, MC; Vescovi, AL. Turning brain into blood: a hematopoietic fate adopted by adult neural stem cells in vivo. *Science,* 1999, 283, 534-537.

[71] Galli, R; Borello, U; Gritti, A; Minasi, MG; Bjornson, C; Coletta, M; Mora, M; De Angelis, MGC; Fiocco, R; Cossu, G; Vescovi, AL. Skeletal myogenic potential of human and mouse neural stem cells. *Nat Neurosci*, 2000, 3(19), 986-991.

[72] Jackson, KA; Mi, T; Goodell, MA. Hematopoietic potential of stem cells isolated from murine skeletal muscle. *Proc Nat Acad Sci*, 1999, 96 (25), 14482-14486.

[73] Toma, JG; Akhavan, M; Fernandes, KJL; Barnabé-Heider, F; Sadikot, A; Kaplan, DR; Miller, FD. Isolation of multipotent adult stem cells from the dermis of mammalian skin. *Nat Cell Biol,* 2001,3, 778-784.

[74] Krause, DS; Theise, ND; Collector, MI; Henegariu, O; Hwang, S; Gardener, R; Neutzel, S; Sharkis, SJ. Multi-organ, multi-lineage engraftment by a single bone marrow-derived stem cell. *Cell*, 2001, 105, 369-377.

[75] Davern, TJ. Bloody liver – prometheus revised (comment). *Hepatology*, 2001, 34(1), 212-214.

[76] Mitaka, T. Hepatic stem cells: From bone marrow cells to hepatocytes. *Biochem Biophys Res Commun*, 2001, 281, 1-5.

[77] Petersen, BE; Bowen, WC; Patrene, KD; Mars, WM; Sullivan, AK; Murase, N; Boggs, SS; Greenberger, JS; Goff, JP. Bone marrow as a potential source of hepatic oval cells. *Science*, 1999, 284, 1168-1170.

[78] Theise, ND; Badve, S; Saxena, R; Henegariu, O; Sell, S; Crawford, JM; Krause, DS. Derivation of hepatocytes from bone marrow cells in mice after radiation-induced myeloablation. *Hepatology*, 2000, 31(1), 235-240.

[79] Theise, ND; Nimmakayalu, M; Gardener, R; Illei, PB; Morgan, G; Teperman, L; Henegariu, O; Krause, DS. Liver from bone marrow in humans. *Hepatology*, 2000, 32(1), 11-16.

[80] Alison, MR; Poulsom, R; Jeffery, R; Dhillon, AP; Quaglia, A; Jacob, J; Novelli, M; Prentice, G; Williamson, J; Wright, NA. Hepatocytes from non-hepatic adult stem cells. *Nature,* 2000, 406.

[81] Brazelton, TR; Rossi, FM; Keshet, GI; Blau, HM. From marrow to brain: expression of neuronal phenotypes in adult mice. *Science*, 2000, 290, 1775-1779.

[82] Mezey, É; Chandross, KJ; Harta, G; Maki, RA; McKercher, SR. Turning blood into brain: cells bearing neuronal antigens generated in vivo from bone marrow. *Science*, 2000, 290, 1779-1782.

[83] Shimizu, K; Sugiyama, S; Aikawa, M; Fukumoto, Y; Rabkin, E; Libby, P; Mitchell, RN. Host bone-marrow cells are a source of donor intimal smooth- muscle-like cells in murine aortic transplant arteriopathy. *Nat Med*, 2001, 7, 738-741.

[84] Kaushal, S; Amiel, GE; Guleserian, KJ; Shapira, OM; Perry, T; Sutherland, FW; Rabkin, E; Moran, AM; Schoen, FJ; Atala, A; Soker, S; Bischoff, J; Mayer, JE Jr. Functional small-diameter neovessels created using endothelial progenitor cells expanded *ex vivo*. *Nat Med*, 2001, 7(9), 1035-1040.

[85] Gao, Z-H; McAlister, VC; Williams, M. Repopulation of liver endothelium by bone-marrow-derived cells. *Lancet*, 2001, 357, 932-933.

[86] Jiang, Y; Jahagirdar, BN; Reinhardt, RL; Schwartz, RE; Keene, CD; Ortiz-Gonzalez, XR; Reyes, M; Lenvik, T; Lund, T; Blackstad, M; Du, J; Aldrich, S; Lisberg, A; Low, WC; Largaespada, DA; Verfaillie, CM. Pluripotency of mesenchymal stem cells derived from adult marrow. *Nature*, 2002, 418(6893), 41-49.

[87] Verfaillie, CM; Schwartz, R; Reyes, M; Jiang, Y. Unexpected potential of adult stem cells. *Ann N Y Acad Sci.*, 2003, 996, 231-234.

[88] Jiang, Y; Henderson, D; Blackstad, M; Chen, A; Miller, RF; Verfaillie, CM. Neuroectodermal differentiation from mouse multipotent adult progenitor cells. *Proc Natl Acad Sci USA,* 2003, 100 Suppl 1, 11854-11860.

[89] Keene, CD; Ortiz-Gonzalez, XR; Jiang, Y; Largaespada, DA; Verfaillie, CM; Low, WC. Neural differentiation and incorporation of bone marrow-derived multipotent adult progenitor cells after single cell transplantation into blastocyst stage mouse embryos. *Cell Transplant,* 2003, 12(3), 201-213.

[90] Schwartz, RE; Reyes, M; Koodie, L; Jiang, Y; Blackstad, M; Lund, T; Lenvik, T; Johnson, S; Hu, WS; Verfaillie, CM. Multipotent adult progenitor cells from bone marrow differentiate into functional hepatocyte-like cells. *J Clin Invest*, 2002, 109(10), 1291-1302.

[91] Qi, H; Aguiar, DJ; Williams, SM; La Pean, A; Pan, W; Verfaillie,CM. Identification of genes responsible for osteoblast differentiation from human mesodermal progenitor cells. *Proc Natl Acad Sci USA,* 2003, 100(6), 3305-3310.

[92] Jiang, Y; Vaessen, B; Lenvik, T; Blackstad, M; Reyes, M; Verfaillie, CM. Multipotent progenitor cells can be isolated from postnatal murine bone marrow, muscle, and brain. *Exp Hematol,* 2002, 30(8), 896-904.

[93] Oh, H; Bradfute, SB; Gallardo, TD; Nakamura, T; Gaussin, V; Mishina, Y; Pocius, J; Michael, LH; Behringer, RR; Garry, DJ; Entman, ML; Schneider, MD. Cardiac progenitor cells from adult myocardium: homing, differentiation, and fusion after infarction. *Proc Natl Acad Sci USA,* 2003, 100(21), 12313-12318.

[94] Vassilopoulos, G; Wang, PR; Russell, DW. Transplanted bone marrow regenerates liver by cell fusion. *Nature,* 2003, 422(6934), 901-904.

[95] Wang, X; Willenbring, H; Akkari, Y; Torimaru, Y; Foster, M; Al-Dhalimy, M; Lagasse, E; Finegold, M; Olson, S; Grompe, M. Cell fusion is the principal source of bone-marrow-derived hepatocytes. *Nature,* 2003, 422(6934), 897-901.

[96] Alvarez-Dolado, M; Pardal, R; Garcia-Verdugo, JM; Fike, JR; Lee, HO; Pfeffer, K; Lois, C; Morrison, SJ; Alvarez-Buylla, A. Fusion of bone-marrow-derived cells with Purkinje neurons, cardiomyocytes and hepatocytes. *Nature,* 2003, 425(6961), 968-973.

[97] Orlic, D; Kajstura, J; Chimenti, S; Li, B; Anderson, SM; Jakoniuk, I; Pickel, J; McKay, R; Bodine, DM; Leri, A; Anversa, P. Transplanted hematopoietic stem cells repair myocardial infarcts. *Blood,* 2000, 96, 221a.

[98] Orlic, D; Kajstura, J; Chimenti, S; Bodine, DM; Leri, A; Anversa, P. Bone marrow cells regenerate infarcted myocardium. *Nature,* 2001, 410, 701-705.

[99] Quaini, F; Urbanek, K; Beltrami, AP; Finato, N; Beltrami, CA; Nadal-Ginard, B; Kajstura, J; Leri, A. M.D.; Anversa, P. Chimerism of the Transplanted Heart. *New Engl J Med,* 2002, 346, 5-15.

[100] Laflamme, MA; Myerson, D; Saffitz, JE; Murry, CE. Evidence for cardiomyocyte repopulation by extracardiac progenitors in transplanted human hearts. *Circ Res.*, 2002, 90(6), 634-640.

[101] Muller, P; Pfeiffer, P; Koglin, J; Schafers, HJ; Seeland, U; Janzen, I; Urbschat, S; Bohm, M. Cardiomyocytes of noncardiac origin in myocardial biopsies of human transplanted hearts. *Circulation,* 2002, 106(1), 31-35.

[102] Deb, A; Wang, S; Skelding, KA; Miller, D; Simper, D; Caplice, NM. Bone marrow-derived cardiomyocytes are present in adult human heart: A study of gender-mismatched bone marrow transplantation patients. *Circulation,* 2003, 107(9), 1247-1249.

[103] Strauer, BE; Brehm, M; Zeus, T; Kostering, M; Hernandez, A; Sorg, RV; Kogler, G; Wernet, P. Repair of infarcted myocardium by autologous intracoronary mononuclear bone marrow cell transplantation in humans. *Circulation,* 2002, 106(15), 1913-1918.

[104] Assmus, B; Schachinger, V; Teupe, C; Britten, M; Lehmann, R; Dobert, N; Grunwald, F; Aicher, A; Urbich, C; Martin, H; Hoelzer, D; Dimmeler, S; Zeiher, AM. Transplantation of progenitor cells and regeneration enhancement in acute myocardial infarction. (TOPCARE-AMI). *Circulation,* 2002, 106(24), 3009-3017.

[105] Kang, HJ; Kim, HS; Zhang, SY; Park, KW; Cho, HJ; Koo, BK; Kim, YJ; Soo Lee, D; Sohn, DW; Han, KS; Oh, BH; Lee, MM; Park, YB. Effects of intracoronary infusion of peripheral blood stem-cells mobilised with granulocyte-colony stimulating factor on left ventricular systolic function and restenosis after coronary stenting in myocardial

infarction: the MAGIC cell randomised clinical trial. *Lancet,* 2004, 363(9411), 751-756.

[106] Balsam, LB; Wagers, AJ; Christensen, JL; Kofidis, T; Weissman, IL; Robbins, RC. Haematopoietic stem cells adopt mature haematopoietic fates in ischaemic myocardium. *Nature*, 2004, 428, 668-673.

[107] Murry, CE; Soonpaa, MH; Reinecke, H; Nakajima, H; Nakajima, HO; Rubart, M; Pasumarthi, KB; Ismail Virag, J; Bartelmez, SH; Poppa, V; Bradford, G; Dowell, JD; Williams, DA; Field, LJ. Haematopoietic stem cells do not transdifferentiate into cardiac myocytes in myocardial infarcts. *Nature,* 2004, 428, 664-668.

[108] Dowell, JD; Rubart, M; Pasumarthi, KB; Soonpaa, MH; Field, LJ. Myocyte and myogenic stem cell transplantation in the heart. *J Cardiovasc Res,* 2003, 58, 336–350.

[109] Kocher, AA; Schuster, MD; Szabolcs, MJ; Takuma, S; Burkhoff, D; Wang, J; Homma, S; Edwards, NM; Itescu, S. Neovascularization of ischemic myocardium by human bone-marrow-derived angioblasts prevents cardiomyocyte apoptosis, reduces remodeling and improves cardiac function. *Nat Med*, 2001, 7, 430-436.

[110] Lyden, D; Hattori, K; Dias, S; Costa, C; Blaikie, P; Butros, L; Chadburn, A; Heissig, B; Marks, W; Witte, L; Wu, Y; Hicklin, D; Zhu, Z; Hackett, NR; Crystal, RG; Moore, MA; Hajjar, KA; Manova, K; Benezra, R; Rafii, S. Impaired recruitment of bone-marrow-derived endothelial and hematopoietic precursor cells blocks tumor angiogenesis and growth. *Nat Med*, 2001, 7, 1194-1201.

[111] Dawson, L; Bateman-House, AS; Mueller Agnew, D; Bok, H; Brock, DW; Chakravarti, A; Greene, M; King, PA; O'Brien, SJ; Sachs, DH; Schill, KE; Siegel, A; Solter, D; Suter, SM; Verfaillie, CM; Walters, LB; Gearhart, JD; Faden, RR. Safety issues in cell-based intervention trials. *Fertil Steril,* 2003, 80(5), 1077-1085.

[112] Priller, J; Flugel, A; Wehner, T; Boentert, M; Haas, CA; Prinz, M; Fernandez-Klett, F; Prass, K; Bechmann, I; De Boer, BA; Frotscher, M; Kreutzberg, GW; Persons, DA; Dirnagl, U. Targeting gene-modified hematopoietic cells to the central nervous system: Use of green fluorescent protein uncovers microglial engraftment. *Nat Med*, 2001, 7, 1356-1361.

[113] Bulte, JWM; Douglas, T; Witwer, B; Zhang, S-C; Strable, E; Lewis, BK; Zywicke, H; Miller, B; van Gelderen, P; Moskowitz, BM; Duncan, ID; Frank, JA. Magnetodendrimers allow endosomal magnetic labeling and in vivo tracking of stem cells. *Nat Biotechnol*, 2001, 19, 1141-1147.

[114] Orlic, D; Kajstura, J; Chimenti, S; Limana, F; Jakoniuk, I; Quaini, F; Nadal-Ginard, B; Bodine, DM; Leri, A; Anversa, P. Mobilized bone marrow cells repair the infarcted heart, improving function and survival. *Proc Natl Acad Sci*, 2001, 98(18), 10344-10349.

[115] Tumbar, T; Guasch, G, Greco, V; Blanpain, C; Lowry, WE; Rendl, M; Fuchs, E. Defining the epithelial stem cell niche in skin. *Science,* 2004, 303(5656), 359-363.

[116] Morris, RJ; Liu, Y; Marles, L; Yang, Z; Trempus, C; Li, S; Lin, JS; Sawicki, JA; Cotsarelis, C. Capturing and profiling adult hair follicle stem cells. *Nature Biotechnology,* 2004, 22, 411-417.

[117] Gerlach, J; Schnoy, N; Encke, J; Müller, C; Smith, M; Neuhaus, P. Improved hepatocyte *in vitro* maintenance in a culture model with woven multicompartment capillary systems: electron microscopy studies. *Hepatology*, 1995, 22, 546-552.

[118] Gerlach, J. Development of a Hybrid Liver Support System - a review. *Int J Art Org,* 1996, 19, 645-655.

[119] Mizuguchi, T; Hui, T; Palm, K; Sugiyama, N; Mitaka, T; Demetriou, AA; Rozga, J; Enhanced proliferation and differentiation of rat hepatocytes cultured with bone marrow stromal cells. *J Cell Physiol*, 2001, 189, 106-119.

[120] Verfaillie, CM; Gupta, P; Prosper, F; Hurley, R; Lundell, B; Bhatia, R. The hematopoietic microenvironment: stromal extracellular matrix components as growth regulators for human hematopoietic progenitors. *Hematology*, 1999, 4(4), 321-333.

[121] Schmitt, TM; de Pooter, RF; Gronski, MA; Cho, SK; Ohashi, PS; Zuniga-Pflucker, JC. Induction of T cell development and establishment of T cell competence from embryonic stem cells differentiated in vitro. *Nature Immunology,* 2004, 5, 410-417.

[122] Verfaillie, CM; Catanzaro, P. Direct contact with stroma inhibits proliferation of human long-term culture initiating cells. *Leukemia,* 1996, 10(3), 498-504.

[123] Lewis, ID; Almeida-Porada, G; Du, J; Lemischka, IR; Moore, KA; Zanjani, ED; Verfaillie, CM. Umbilical cord blood cells capable of engrafting in primary, secondary, and tertiary xenogeneic hosts are preserved after ex vivo culture in a noncontact system. *Blood,* 2001, 97(11), 3441-3449.

[124] Punzel, M; Gupta, P; Verfaillie, CM. The microenvironment of AFT024 cells maintains primitive human hematopoiesis by counteracting contact mediated inhibition of proliferation. *Cell Commun Adhes,* 2002, 9(3), 149-159.

[125] Miyoshi, H; Ehashi, T; Ema, H; Hsiang, CH; Nakauchi, H; Ohshima, N. Long-term culture of fetal liver cells using a three-dimentional porous polymer substrate. *ASAIO J*, 2000, 46, 397-402.

[126] Jiang, J; Kojima, N; Kinoshita, T; Miyajima, A; Yan, W; Akai, Y. Cultivation of fetal liver cells in a three-dimentional poly-L-lactic acid scaffold in the presence of oncostatin M. *Cell Transplant,* 2002, 11, 403-406.

[127] Auchincloss, H; Bonventre, JV. Transplanting cloned cells into therapeutic promise. *Nat Biotechnol,* 2002, 20 (7), 665.

[128] Christiansen, GS; Danes, B; Allen, L; Leinfelder, PJ. A culture chamber for the continuous biochemical and morphological study of living cells in tissue culture. *Exp Cell Res,* 1953, 5, 10-15.

[129] Rose, G. A separable and multipurpose tissue culture chamber. *Tex Rep Biol Med,* 1954, 12, 1074-1083.

[130] Freed, JJ. Cell culture perfusion chamber: Adaptation for microscopy of clonal growth. *Science,* 1963, 140, 1334-1335.

[131] Katinger, H. Principles of animal cell fermentation. *Dev Biol Stand,* 1987, 66, 195-209.

[132] Minucells and Minutissue Vertriebs GmbH, Bad Abbach, Germany.

[133] Diagnostic Chemicals Limited, Prince Edward Island, Canada.

[134] Synthecon, Inc., Houston, Texas, U.S.A.

[135] Wave Biotech LLC, Bridgewater, New Jersey, U.S.A.

[136] Integra Biosciences AG, Chur, Switzerland.

[137] Sartorius AG, Goettingen, Germany.

[138] Spectrum Laboratories, Inc., Rancho Dominguez, California, U.S.A.

[139] Integra Biosciences AG, Chur, Switzerland.

[140] Aastrom Biosciences, Inc., Ann Arbor, Michigan, U.S.A.

[141] Knazek, RA; Gullino, PM; Kohler, PO; Dedrick, RL. Cell culture on artificial capillaries: An approach to tissue growth in vitro. *Science,* 1972, 178, 65-66.

[142] Wolf, CF; Munkelt, BE. Bilirubin conjugation by an artificial liver composed of cultured cells and synthetic capillaries. *Trans Am Soc Artif Intern Organs,* 1975, 21, 16-27.

[143] Hager, JC; Carman, R; Stoller, R; Panol, G; Leduc, EH; Thayer, WR; Porter, LE; Galletti, PM; Calabresi, P. A prototype for a hybrid artificial liver. *Trans Am Soc Artif Intern Organs,* 1978, 24, 250-253.

[144] Hager, JC; Carman, R; Porter, LE; et al. Neonatal hepatocyte culture on artificial capillaries: a model for drug metabolism and the artificial liver. *ASAIO,* 1983, J 6, 26-35.

[145] Gloeckner, H; Lemke, HD. New miniaturized hollow-fiber bioreactor for in vivo like cell culture, cell expansion, and production of cell-derived products. *Biotechnol Prog,* 2001, 17, 828-831.

[146] MacDonald, JM; Wolfe, SP; Roy-Chowdhury, I; Kubota, H; Reid, LM. Effect of flow configuration and membrane characteristics on membrane fouling in a novel multicoaxial hollow-fiber bioartificial liver. *Ann NY Acad Sci*, 2001, 944, 334-343.

[147] Gerlach, JC; Encke, J; Hole, O; Muller, C; Ryan, CJ; Neuhaus, P. Bioreactor for a larger scale hepatocyte in vitro perfusion. *Transplantation,* 1994, 58, 984-988.

[148] Gerlach, JC; Fuchs, M; Smith, MD; Bornemann, R; Encke, J; Neuhaus, P; Riedel, E. Is a clinical application of hybrid liver support systems limited by an initial disorder in cellular amino acid and alpha-keto acid metabolism, rather than by later gradual loss of primary hepatocyte function? *Transplantation*, 1996, 62, 224-228.

[149] Gerlach, JC; Zeilinger, K; Grebe, A; Puhl, G; Pless, G; Sauer, I; Grunwald, A; Schnoy, N; Muller, C; Neuhaus, P. Recovery of preservation-injured primary human hepatocytes and non-parenchymal cells to tissuelike structures in large-scale bioreactors for liver support: An initial transmission electron microscopy study. *J Invest Surg*, 2003, 16, 83-92.

[150] Gerlach, JC; Mutig, K; Sauer, IM; Schrade, P; Efimova, E; Mieder, T; Naumann, G; Grunwald, A; Pless, G; Mas, A; Bachmann, S; Neuhaus, P; Zeilinger, K. Use of primary human liver cells originating from discarded grafts in a bioreactor for liver support therapy and the prospects of culturing adult liver stem cells in bioreactors: A morphological study. *Transplantation*, 2003, 76, 781-786.

[151] Sauer, IM; Zeilinger, K; Obermayer, N; Pless, G; Grunwald, A; Pascher, A; Mieder, T; Roth, S; Goetz, M; Kardassis, D; Mas, A; Neuhaus, P; Gerlach, JC. Primary human liver cells as source for modular extracorporeal liver support--a preliminary report. *Int J Artif Organs*, 2002, 25, 1001-1005.

[152] Flendrig, LM; la Soe, JW; Jorning, GG; Steenbeek, A; Karlsen, OT; Bovee, WM; Ladiges, NC; te Velde, AA; Chamuleau, RA. In vitro evaluation of a novel bioreactor

based on an integral oxygenator and a spirally wound nonwoven polyester matrix for hepatocyte culture as small aggregates. *J Hepatol,* 1997, 26, 1379-1392.

[153] Humes, HD; MacKay, SM; Funke, AJ; Buffington, DA. Tissue engineering of a bioartificial renal tubule assist device: In vitro transport and metabolic characteristics. *Kidney Int*, 1999, 55, 2502-2514.

[154] De Bartolo, L; Jarosch-Von Schweder, G; Haverich, A; Bader, A. A novel full-scale flat membrane bioreactor utilizing porcine hepatocytes: Cell viability and tissue-specific functions. *Biotechnol Prog*, 2000, 16, 102-108.

[155] Shito, M; Kim, NH; Baskaran, H; Tilles, AW; Tompkins, RG; Yarmush, ML; Toner, M. In vitro and in vivo evaluation of albumin synthesis rate of porcine hepatocytes in a flat-plate bioreactor. *Artif Organs*, 2001, 25, 571-578.

[156] Bhatia, SN, Yarmush, ML; Toner, M. Controlling cell interactions by micropatterning in co-cultures: Hepatocytes and 3T3 fibroblasts. *J Biomed Mater Res*, 1997, 34, 189-199.

[157] Kim, SS; Utsunomiya, H; Koski, JA; Wu, BM; Cima, MJ; Sohn, J; Mukai, K; Griffith, LG; Vacanti, JP. Survival and function of hepatocytes on a novel three-dimensional synthetic biodegradable polymer scaffold with an intrinsic network of channels. *Ann Surg*, 1998, 228, 8-13.

[158] Powers, MJ; Janigian, DM; Wack, KE; Baker, CS; Beer Stolz, D; Griffith, LG. Functional behavior of primary rat liver cells in a three-dimensional perfused microarray bioreactor. *Tissue Eng*, 2002, 8, 499-513.

[159] Powers, MJ; Domansky, K; Kaazempur-Mofrad, MR; Kalezi, A; Capitano, A; Upadhyaya, A; Kurzawski, P; Wack, KE; Stolz, DB; Kamm, R; Griffith, LG. A microfabricated array bioreactor for perfused 3D liver culture. *Biotechnol Bioeng*, 2002, 78, 257-269.

[160] Bancroft, GN; Sikavitsas, VI; Mikos, AG. Design of a flow perfusion bioreactor system for bone tissue-engineering applications. *Tissue Eng*, 2003, 9, 549-554.

[161] Ambrosino, G; Varotto, S; Basso, S; Galavotti, D; Cecchetto, A; Carraro, P; Naso, A; De Silvestro, G; Plebani, M; Giron, G; Abatangelo, G; Donato, D; Braga, GP; Cestrone, A; Marrelli, L; Trombetta, M; Lorenzelli, V; Picardi, A; Valente, ML; Palu, G; Colantoni, A; Van Thiel, D; Ricordi, C; D'Amico, DF. ALEX (artificial liver for extracorporeal xenoassistance): A new bioreactor containing a porcine autologous biomatrix as hepatocyte support. Preliminary results in an ex vivo experimental model. *Int J Artif Organs*, 2002, 25, 960-965.

[162] Linti, C; Zipfel, A; Schenk, M; Dauner, M; Doser, M; Viebahn, R; Becker, HD; Planck, H. Cultivation of porcine hepatocytes in polyurethane nonwovens as part of a biohybrid liver support system. *Int J Artif Organs*, 2002, 25, 994-1000.

[163] Gerlach, J; Jorres, A; Trost, O; Hole, O; Vienken, J; Courtney, JM; Gahl, GM; Neuhaus, P. Side effects of hybrid liver support therapy: TNF-alpha liberation in pigs, associated with extracorporeal bioreactors. *Int J Artif Organs*, 1993, 16, 604-608.

[164] Gerlach, J; Trost, T; Ryan, CJ; Meissler, M; Hole, O; Muller, C; Neuhaus, P. Hybrid liver support system in a short term application on hepatectomized pigs. *Int J Artif Organs*, 1994, 17, 549-553.

[165] Gerlach, JC; Botsch, M; Kardassis, D; Lemmens, P; Schon, M; Janke, J; Puhl, G; Unger, J; Kraemer, M; Busse, B; Bohmer, C; Belal, R; Ingenlath, M; Kosan, M; Kosan, B; Sultmann, J; Patzold, A; Tietze, S; Rossaint, R; Muller, C; Monch, E; Sauer, IM; Neuhaus, P. Experimental evaluation of a cell module for hybrid liver support. *Int J Artif Organs,* 2001, 24, 793-798.

[166] Janke, J; Gerlach, J; Kardassis, D; Bohmer, C; Rossaint, R. Effect of a hybrid liver support system on cardiopulmonary function in healthy pigs. *Int J Artif Organs*, 1997, 20, 570-575.

[167] Flendrig, LM; Calise, F; Di Florio, E; Mancini, A; Ceriello, A; Santaniello, W; Mezza, E; Sicoli, F; Belleza, G; Bracco, A; Cozzolino, S; Scala, D; Mazzone, M; Fattore, M; Gonzales, E; Chamuleau, RA. Significantly improved survival time in pigs with complete liver ischemia treated with a novel bioartificial liver. *Int J Artif Organs*, 1999, 22, 701-709.

[168] Shito, M; Tilles, AW; Tompkins, RG; Yarmush, ML; Toner, M. Efficacy of an extracorporeal flat-plate bioartificial liver in treating fulminant hepatic failure. *J Surg Res,* 2003, 111, 53-62.

[169] Humes, HD; Fissell, WH; Weitzel, WF; Buffington, DA; Westover, AJ; MacKay, SM; Gutierrez, JM. Metabolic replacement of kidney function in uremic animals with a bioartificial kidney containing human cells. *Am J Kidney Dis*, 2002, 39, 1078-1087.

[170] Sussman, NL; Chong, MG; Koussayer, T; He, DE; Shang, TA; Whisennand, HH; Kelly, JH. Reversal of fulminant hepatic failure using an extracorporeal liver assist device. *Hepatology*, 1992, 16, 60-65.

[171] Watanabe, FD; Mullon, CJ; Hewitt, WR; Arkadopoulos, N; Kahaku, E; Eguchi, S; Khalili, T; Arnaout, W; Shackleton, CR; Rozga, J; Solomon, B; Demetriou, AA. Clinical experience with a bioartificial liver in the treatment of severe liver failure. A phase I clinical trial. *Ann Surg,* 1997, 225, 484-491.

[172] Humes, HD; Fissell, WH; Weitzel, WF. The bioartificial kidney in the treatment of acute renal failure. *Kidney Int*, 2002, 61 Suppl, 121-125.

[173] Mazariegos, GV; Patzer, JF 2nd; Lopez, RC; Giraldo, M; Devera, ME; Grogan, TA; Zhu, Y; Fulmer, ML; Amiot, BP; Kramer, DJ. First clinical use of a novel bioartificial liver support system (BLSS). *Am J Transplant*, 2002, 2, 260-266.

[174] Mundt, A; Puhl, G; Muller, A; Sauer, I; Muller, C; Richard, R; Fotopoulou, C; Doll, R; Gabelein, G; Hohn, W; Hofbauer, R; Neuhaus, P; Gerlach, J. A method to assess biochemical activity of liver cells during clinical application of extracorporeal hybrid liver support. *Int J Artif Organs*, 2002, 25, 542-548.

[175] Irgang, M; Sauer, IM; Karlas, A; Zeilinger, K; Gerlach, JC; Kurth, R; Neuhaus, P; Denner, J. Porcine endogenous retroviruses: No infection in patients treated with a bioreactor based on porcine liver cells. *J Clin Virol*, 2003, 28, 141-154.

[176] Sauer, IM; Kardassis, D; Zeillinger, K; Pascher, A; Gruenwald, A; Pless, G; Irgang, M; Kraemer, M; Puhl, G; Frank, J; Muller, AR; Steinmuller, T; Denner, J; Neuhaus, P; Gerlach, JC. Clinical extracorporeal hybrid liver support - phase I study with primary porcine liver cells. *Xenotransplantation*, 2003, 10, 460-469.

[177] Sauer, IM; Zeilinger, K; Pless, G; Kardassis, D; Theruvath, T; Pascher, A; Goetz, M; Neuhaus, P; Gerlach, JC. Extracorporeal liver support based on primary human liver

cells and albumin dialysis - treatment of a patient with primary graft non-function. *J Hepatol*, 2003, 39, 649-653.

[178] Sussman, NL; Kelly, JH. Improved liver function following treatment with an extracorporeal liver assist device. *Artif Organs*, 1993, 17(1), 27-30.

[179] Xu, ASL; Luntz, TL; MacDonald, JM; Kubota, H; Hsu, E; London, RE; Reid, LM. Lineage Biology and Liver. In: Ed Lanza RP, Langer R, Vacanti J, editors. *Principles of Tissue Engineering, 2nd Edition*. San Diego, California, USA: Academic press; 2000; 559-598.

[180] Gerlach, J; Stoll, P; Schnoy, N; Neuhaus, P. Comparison of hollow fibre membranes for hepatocyte immobilisation in bioreactors. *Int J Art Org*, 1996,1 9, 610-616,

[181] Gerlach, J; Klöppel, K; Stoll, P; Vienken, J; Müller, Ch; Schauwecker, HH. Gas supply across membranes in bioreactors for hepatocyte culture. *Art Org*, 1990, 14, 328-333.

[182] Gerlach, J; Neuhaus, P. Culture model for primary hepatocytes. *In Vitro Cell Dev Biol*, 1994, 30A, 640-642.

[183] Gerlach, J; Brombacher, J; Courtney, JM; Neuhaus, P. Nonenzymatic versus enzymatic hepatocyte isolation from pig livers for larger scale investigations of liver cell perfusion systems. *Int J Art Org*, 1993, 16, 677-681.

[184] Gerlach, J; Brombacher, J; Klöppel, K; Smith, M; Schnoy, N; Neuhaus, P. Comparison of four methods for mass hepatocyte isolation from pig and human livers. *Transplantation*, 1994, 57, 1318-1322.

[185] Gerlach, J; Klöppel, K; Schön, MR; Brombacher, J; Courtney, JM; Unger, J; Neuhaus, P. Comparison of pig hepatocyte isolation using intraoperative perfusion without warm ischemia and isolation of cells from abattoir organs after warm ischemia. *Art Org*, 1993, 17, 950-953.

[186] Gerlach, J; Brombacher, J; Smith, M; Neuhaus, P. High yield hepatocyte isolation from pig livers for investigation of hybrid liver support systems: Influence of collagenase concentration and body weight. *J Surg Res*, 1996, 62, 85-89.

[187] Gerlach, J; Schauwecker, HH; Klöppel, K; Tauber, R; Müller, Ch; Bücherl, ES. Use of hepatocytes in adhesion and suspension cultures for liver support bioreactors. *Int J Art Org*, 1989, 12, 788-793.

[188] Gerlach, J; Klöppel, K; Müller, C; Schnoy, N; Smith, M; Neuhaus, P. Hepatocyte aggregate culture technique for bioreactors in hybrid liver support systems. *Int J Art Org,* 1993, 16/12, 843-846.

[189] Gerlach, J; Schnoy, N; Smith, M; Neuhaus, P. Hepatocyte culture between woven capillary networks - a microscopy study. *Art Org*, 1994, 18, 226-230.

[190] Gerlach, J. Long-term liver cell cultures in bioreactors and possible application for liver support. *Cell Biol Toxicol,* 1997, 13, 349-355.

[191] Zeilinger, K; Sauer, IM; Pless, G; Strobel, C; Rudzitis, J; Wang, A; Nussler, AK; Grebe, A; Mao, L; Auth, SH; Unger, J; Neuhaus, P; Gerlach, JC. Three-dimensional co-culture of primary human liver cells in bioreactors for in vitro drug studies: effects of the initial cell quality on the long-term maintenance of hepatocyte-specific functions. *Altern Lab Anim*, 2002, 30(5), 539-350.

[192] Gerlach, J; Ziemer, R; Neuhaus, P. Fulminant liver failure: relevance of extracorporeal hybrid liver support systems. *Int J Art Org*, 1996, 19, 7-13.

[193] Sauer, I; Gerlach, J. Modular extracorporeal liver support. Art Org 2002; 26:703-706

[194] Sauer I, Gerlach J. Concept of modular extracorporeal liver support for treatment of acute hepatic failure. *Metab Brain Dis*, 2002, 17 (4), 477-484.

[195] Gleisner, M; Bornemann, R; Stemerowicz, R; Meisler, M; Neuhaus, P; Gerlach, J. Immunisolation of hybrid liver support systems by semipermeable membranes. *Int J Artif Org*, 1997, 20, 644-649.

[196] Bornemann, R; Smith, MD; Gerlach, J. Consideration of potential immunological problems in the application of xenogenic hybrid liver support. *Int J Art Org*, 1996, 19, 655-663.

[197] Gerlach, J; Zeilinger, K; Sauer, IM; Mieder, T; Nauman, G; Grünwald, A; Pless, G; Holland, A; Mas, A; Vienken, J; Neuhaus P. Extracorporeal liver support: Porcine or human cell based systems? *Int J Artif Org*, 2002, 25 (10), 1013-1019.

[198] Evarts, RP; Nagy, P; Marsden, E; Thorgeirsson, SS. A precursor relationship exists between oval cells and hepatocytes in rat liver. *Carcinogenesis*, 1987, 8, 1737-1740.

[199] Mitaka, T; Sattler, GL; Putot, HC; Mochizuki, Y. Characteristics of small cell colonies developing in primary cultures of adult rat hepatocytes. *Virchows Archiv B Cell Pathol*, 1992, 62, 329-335.

[200] Mitaka, T; Sato, F; Mizoguchi, T; Yokono, T; Mochizuki, Y. Reconstruction of hepatic organoid by rat small hepatocytes and hepatic non-parenchymal cells. *Hepatology*, 1999, 29, 111-125.

[201] Dabeva, MD; Laconi, E; Oren, R; Petkov, PM; Hurston, E; Shafritz, DA. Liver regeneration and alpha-fetoprotein messenger RNA expression in the retrorsine model for hepatocyte transplantation. *Cancer Res.*, 1998, 58, 5825-5834.

[202] Omori, M; Evarts, RP; Omori, N; Hu, Z; Marsden, ER; Thorgeirsson SS. Expression of alpha-fetoprotein and stem cell factor/c-kit system in bile duct ligated young rats. *Hepatology,* 1997, 25(5) 1115-1122.

[203] Tournier, I; Legres, L; Schoevaert, D; Feldmann, G; Bernuau, D. Cellular analysis of alpha-fetoprotein gene activation during carbon tetrachloride and D-galactosamine-induced acute liver injury in rats. *Lab Invest*, 1988, 59(5), 657-665.

[204] Kubota, H; Reid, LM. Clonogenic hepatoblasts, common precursors for hepatocytic and biliary lineages, are lacking classical major histocompatibility complex class I antigen. *Proc Natl Acad Sci USA,* 2000, 97(22), 12132-12137.

[205] Crosby, HA; Kelly, DA; Strain, AJ. Human hepatic stem-like cells isolated using c-kit or CD34 can differentiate into biliary epithelium. *Gastroenterology,* 2001,120, 534-544.

[206] Baumann, U; Crosby, HA; Ramani, P; Kelly, DA; Strain, AJ. Expression of the stem cell factor receptor c-kit in normal and diseased pediatric liver: identification of a human hepatic progenitor cell? *Hepatology,* 1999, 30, 112-117.

[207] Fujio, K; Evarts, RP; Hu, Z; Marsden, ER; Thorgeirsson, SS. Expression of stem cell factor and its receptor, c-kit, during liver regeneration from putative stem cells in adult rat. *Lab Invest,* 1994, 70(4), 511-516.

[208] Fujio, K; Hu, Z; Evarts, RP; Marsden, ER; Niu, CH; Thorgeirsson, SS. Coexpression of stem cell factor and c-kit in embryonic and adult liver. *Exp Cell Res*, 1996, 224, 243-250.

[209] Omori, M; Omori, N; Evarts, RP; Teramoto, T; Thorgeirsson, SS. Coexpression of flt-3 ligand/flt-3 and SCF/c-kit signal transduction system in bile-duct-ligated SI and W mice. *Am J Pathol,* 1997, 150(4), 1179-1187.

[210] Matsusaka, S; Tsujimura, T; Toyosaka, A; Nakasho, K; Sugihara, A; Okamoto, E; Uematsu, K; Terada, N. Role of c-kit receptor tyrosine kinase in development of oval cells in the rat 2-acetylaminofluorene/partial hepatectomy model. *Hepatology,* 1999, 29(3), 670-676.

[211] Rao, MS; Yukawa, M; Omori, M; Thorgeirsson, SS; Reddy, JK. Expression of transcription factors and stem cell factor precedes hepatocyte differentiation in rat pancreas. *Gene Expr,* 1996, 6(1), 15-22.

[212] Omori, N; Omori, M; Evarts, RP; Teramoto, T; Miller, MJ; Hoang, TN; Thorgeirsson, SS. Partial cloning of rat CD34 cDNA and expression during stem cell-dependent liver regeneration in the adult rat. *Hepatology,* 1997, 26(3), 720-727.

[213] Saxena, R; Theise, ND; Crawford, JM. Microanatomy of the human liver-exploring the hidden interfaces. *Hepatology,* 1999, 30(6), 1339-1346.

[214] Lowes, KN; Brennan, BA; Yeoh, GC; Olynyk, JK. Oval cell numbers in human chronic liver diseases are directly related to disease severity. *Am J Pathol*, 1999, 154(2), 537-41.

[215] Yoon, J-H; Lee, H V-S; Lee, JS; Park, JB; Kim, CY. Development of non-transformed liver cell line with differentiated-hepatocyte and urea-synthetic functions: applicable for bioartificial liver. *Int J Artif Organs,* 1999, 18, 2127-2136.

[216] Kamiya, A; Kinoshita, T; Ito, Y; Matsui, T; Korikawa, Y; Senba, E; Nakashima, K; Taga, T; Yoshida, K; Kishsimoto, T; Miyajima, A. Fetal liver development requires a paracrine action of oncostatin M through the gp130 signal transducer. *EMBO J*, 1999, 18, 2127-2136.

[217] Kojima, N; Kinoshita, T; Kamiya, A; Nakaruma, K; Nakashima, K; Taga, T; Miyajima, A. Cell density-dependent regulation of hepatic development by a gp130-independent pathway. *Biochem Biophys Res Commun*, 2000, 277, 152-158.

[218] Sakai, Y; Jiang, J; Kojima, N; Kinoshita, T; Miyajima, A. enhanced *in vitro* maturation of fetal mouse liver cells with oncostatin M, nicotinamide and dimethylsulfoxide. *Cell Transplant*, 2002, 11, 435-441.

[219] Kinoshita, T; Sekiguchi, T; Xu, MJ; Ito, Y; Kamiya, A; Tsuji, K; Nakamura, T. Hepatic differentiation induced by oncostatin M attenuates fetal liver hematopoiesis. *Proc Natl Sci USA*, 1999, 96, 7265-7270.

[220] Alvaro, D; Gigliozzi, A; Attili, AF. Regulation and deregulation of cholangiocyte proliferation. Review. *J Hepatol*, 2000, 33, 333-340.

[221] Demetris, AJ; Seaberg, EC; Wennerberg, A; Ionellie, J; Michalopoulos, G. Ductular reaction after submassive necrosis in humans. Special emphasis on analysis of ductular hepatocytes. *Am J Pathol,* 1996, 149, 439-448.

[222] De Vos, R; Desmet, V. Ultrastructural characteristic of novel epithelial cell types identified in human pathologic liver specimens with chronic ductular reaction. *Am J Pathol*, 1992, 140, 1441-1450.

In: Trends in Stem Cell Research ISBN 1-59454-315-1
Editor: Erik V. Greer, pp. 101-128

Chapter IV

CTGF/Hcs24/CCN2 as a Regeneration Factor ("Regenerin") Acting on Various Types of Mesenchymal Cells

Masaharu Takigawa*
Department of Biochemistry and Molecular Dentistry, Okayama University Graduate School of Medicine and Dentistry, Okayama, Japan

Abstract

Connective tissue growth factor/hypertrophic chondrocyte-specific gene product 24 (CTGF/Hcs24/CCN2) is the second member of the CCN family of proteins and a multifunctional growth factor for fibroblasts, chondrocytes, osteoblasts and vascular endothelial cells. Depending on the type of cell with which it interacts, it promotes chomotaxis, migration, adhesion, proliferation, differentiation and/or extracellular matrix formation. Because the gene expression of CCN2/CTGF is very high in hypertrophic chondrocytes in the physiological state, a major physiological role for this factor has been suggested to be the promotion of endochondral ossification. However, its expression was also recently detected in various types of chondrocytes, osteoblasts and endothelial cells during embryonic development. The expression was also found to be up regulated in undifferentiated mesenchymal cells, osteoblasts, chondrocytes and endothelial cells during wound healing of bone, cartilage or skin, indicating the involvement of this factor in tissue regeneration. In fact, since CCN2/CTGF enhanced the proliferation and phenotypic maintenance of articular chondrocytes *in vitro*, without causing them to enter the hypertrophic state, recombinant CCN2/CTGF was administered *in vivo* to both rats of an experimental osteoarthritis model and to those with full-

* Masaharu Takigawa, D.D.S., Ph.D. Department of Biochemistry and Molecular Dentistry Okayama University Graduate School of Medicine and Dentistry, Okayama 700-8525, Japan. Tel.:+81-86-235-6645, Fax:+81-86-235-6649 E-mail: takigawa@md.okayama-u.ac.jp

thickness defects in their articular cartilage. The results showed the factor to be therapeutically useful for the regeneration and reconstruction of damaged cartilaginous tissues. Also, CCN2/CTGF enhanced the phenotypic expression of cartilage by mouse bone marrow-derived stromal cells, and induced chondrogenesis *in vitro*, indicating that CCN2/CTGF promotes at least chondrogenic differentiation of mesenchymal stem cells. Furthermore, CCN2/CTGF is known to be involved in fibrotic disorders, angiogenic diseases, and malignancies, and these pathological states may be due to uncontrolled overexpression of this growth factor. This review focuses on the role of CCN2/CTGF as a regeneration factor (regenerin) acting on various types of cells including mesenchymal stem cells.

Introduction

Connective tissue growth factor/hypertrophic chondrocyte-specific gene product 24 (CCN2/CTGF) is a member of the CCN protein family which is characterized by significant sequence homology and conservation of all 38 cysteine residues among its members (Figs. 1 and 2) [Bork, 1993; Brigstock, 1999; Lau and Lam, 1999; Perbal, 2004; Takigawa, 2003; Takigawa et al., 2003]. The members of this family originally consisted of CTGF(human)/Fisp12(mouse), Cyr61(mouse)/Cef10(chick), and Nov(chicken), and hence the family had been named "CCN" [Bork, 1993] (Figure 1); but recently new members such as ELM-1(murine)/WISP-1(human) [Hashimoto et al., 1998; Pennica et al., 1998], rCop-1(rat)/WISP-2(human)/CTGF-L(human/murine) [Kumar et al., 1999; Pennica et al., 1998] and WISP-3(human) [Pennica et al., 1998] have been added to this family [Bork, 1993; Brigstock, 1999; Lau and Lam, 1999; Perbal, 2004; Takigawa, 2003; Takigawa et al., 2003] (Figure 1). These proteins are characterized by 4 homologous modules, i.e., the IGFBP, von Willebrand factor type C repeat (VWC), thrombospondin type 1 repeat (TSP), and C-terminal (CT) modules (Figure 2). rCop-1(rat)/WISP-2(human)/CTGF-L(human/mouse), however, lacks the CT module [Kumar et al., 1999; Pennica et al., 1998] (Figure 1). In addition, a WISP-1 variant lacking the VWC module and a WISP-3 variant lacking both TSP and CT modules have recently been reported [Tanaka et al., 2003; Tanaka et al., 2001].

CTGF was originally discovered due to the cross-reactivity of a PDGF antiserum with a single polypeptide with a molecular weight of 38-kDa secreted by cultured human vein endothelial cells (HUVEC), and its cDNA was isolated from a HUVEC cDNA expression library with anti-PDGF and shown to encode a 349-amino acid protein [Bradham et al., 1991]. When CTGF was discovered, its partially purified protein was found to be both mitogenic and chemotactic for fibroblast-like cells *in vitro*; and so this protein was named "CTGF" [Bradham et al., 1991]. Before the discovery of CTGF, the cDNA of the mouse CTGF ortholog had been isolated as an immediate early gene by differential screening of a cDNA library from serum-stimulated NIH3T3 cells and named "fibroblast-inducible secreted protein-12" (fisp 12) [Ryseck et al., 1991]. The mouse ortholog was also isolated from TGF-β-stimulated mouse ARK-2B cells and named βIG-M2 [Brunner et al., 1991]. Fisp12/ βIG-M2 encodes 348 amino acids, contains 39 cysteine residues, and is 91 % homologous to CTGF. Independently of these studies, my research group cloned an mRNA highly expressed in hypertrophic chondrocytes from a human chondrocytic cell line, HCS-2/8 [Enomoto and Takigawa, 1992; Takigawa et al., 1991; Takigawa et al., 1989] and found that its gene product, which we named Hcs24, was identical with that of CTGF [Nakanishi et al., 1997; Takigawa, 2003; Takigawa et al., 2003]. We also found it to be a regulatory molecule involved in many stages of the process of endochondral ossification and angiogenesis.

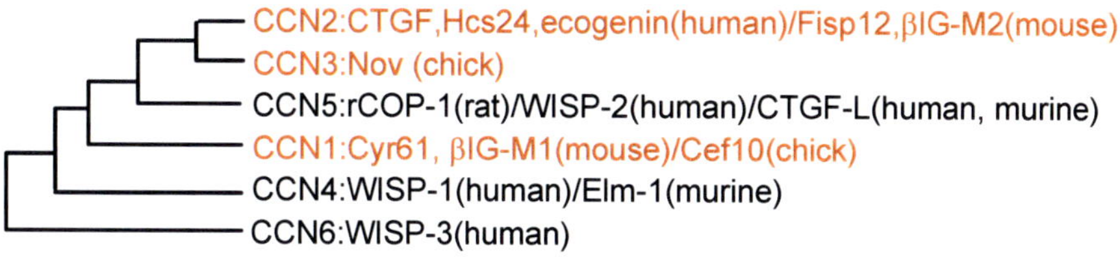

Figure 1. Nomenclature and dendrogram of CCN family members. Species described in parentheses show those in which the genes were originally found. Nomenclature for the CCN genes and proteins were proposed under the agreement of several major researchers of this field [Brigstock et al., 2003].

A unified nomenclature CCN1-6 for the CCN genes and proteins was recently proposed under the agreement of several major researchers of this field [Brigstock et al., 2003](Figure 1), and so CCN2/CTGF is used in this review. Detailed reviews of the CCN family can be found elsewhere [Moussad and Brigstock, 2000; Perbal, 2001; Takigawa, 2003]. This chapter describes the role of CCN2/CTGF as a regeneration factor (regenerin) acting on various types of mesenchymal cells including mesenchymal stem cells.

Gene Expression of CCN2/CTGF During Embryonic and Post Natal Development Suggests its Regenerin Action

As described above, CCN2/CTGF was found to be highly expressed in hypertrophic chondrocytes (Table 1). *In situ* hybridization using both whole-mount neonates (P0) and their longitudinal sections revealed the expression of CCN2/CTGF in hypertrophic chondrocytes and in cells in the zone of calcifying cartilage [Nakanishi et al., 1997; Nakanishi et al., 2000; Takigawa, 2003; Takigawa et al., 2003]. Whole-mount *in situ* hybridization of embryonic day 17(E17) mouse embryos showed that CCN2/CTGF was selectively expressed in the phalanges of the forelimbs [Yamaai et al., 2005]. Moreover, *in situ* hybridization using tissue sections of E17 and neonatal mice revealed gene expression of CCN2/CTGF in hypertrophic chondrocytes in the long bones, ribs, and vertebral columns, which cells were destined to form calcified tissues [Yamaai et al., 2005]. Even in 3-week-old mice, hypertrophic chondrocytes in the growth plate of long bones expressed CCN2/CTGF. Moreover, the expression of CCN2/CTGF was also observed in osteoblasts in the primary spongiosa, although its level was lower than that in the hypertrophic chondrocytes (Table 1). In addition, the expression of the factor was detected in both ameloblasts and odontoblasts in the developing tooth germ [Shimo et al., 2002; Yamaai et al., 2005] (Table1).

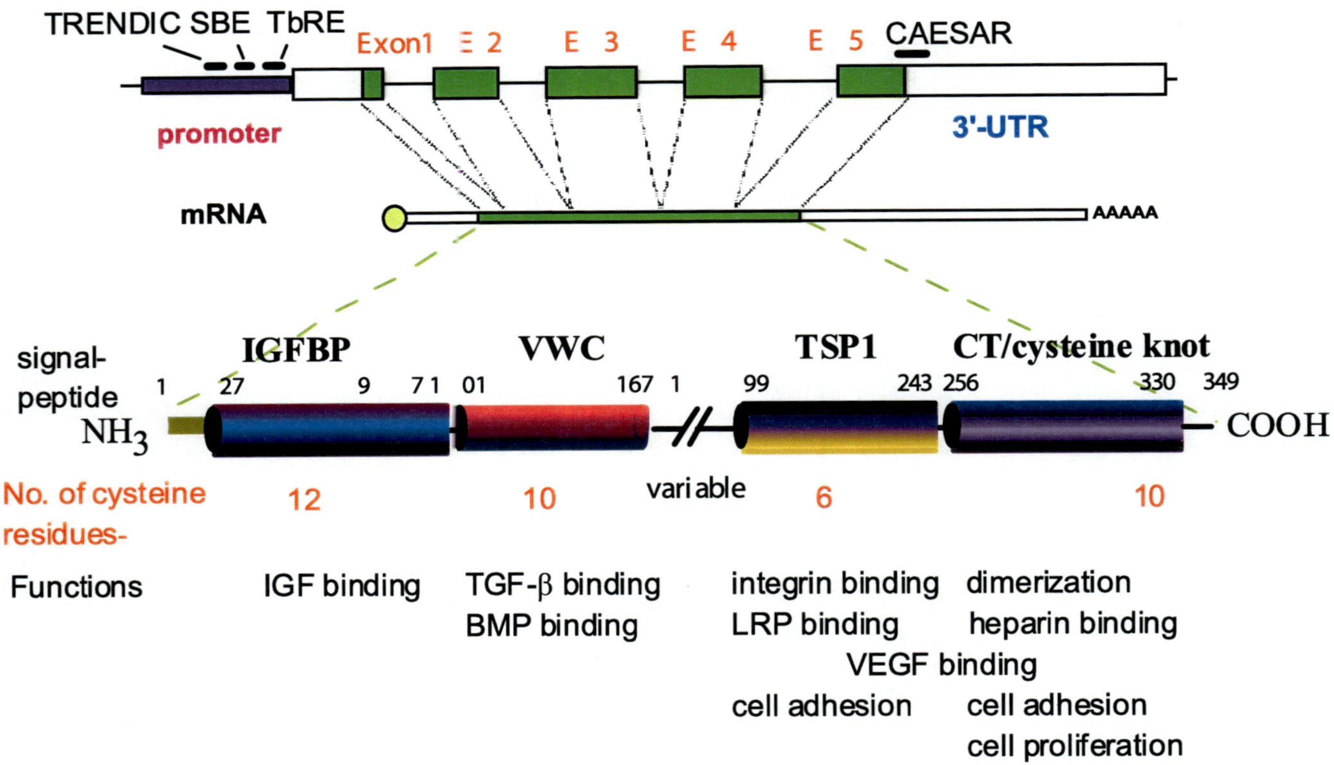

Figure 2. Modular structure and gene structure of CCN2/CTGF protein. The middle part of the figure illustrates the structure of the human CCN2/CTGF protein. IGFBP: IGF binding protein-like module, VWC: von Willebrand factor type C module, TSP1: thrombospondin type I repeat, CT: C-terminal module. The upper part of the figure illustrates the structure of the human *ctgf* gene. Exons are indicated by shading. Abbreviations: TRENDIC, transcriptional enhancer dominant in chondrocytes; SBE, Smad binding element; TbRE, transforming growth factor β responsive element; CAESAR, *cis*-acting element for structure-anchored repression; UTR, untranslated region. The mature *ccn2/ctgf* mRNA is also shown with a 5'-capped structure and polyadenyl tail. The lower part of the figure shows the function of each domain. BMP, bone morphogenetic protein; LRP, low density lipoprotein receptor-related protein; VEGF, vascular endothelial cell growth factor.

Endothelial cells in blood vessels in embryonic and neonatal mice also express the CCN2/CTGF gene (Table 1), although there was no detectable expression in normal human arteries [Igarashi et al., 1996; Oemar and Luscher, 1997; Oemar et al., 1997]. This difference may be due to the difference between rapidly growing embryos and adults rather than to a species difference; because actively migrating and proliferating endothelial cells, but not quiescent ones, express CCN2/CTGF [Shimo et al., 1998] (Table 1). A slight gene expression was also observed in brain and kidney, but the level was much lower than that in the hypertrophic zone of various cartilages [Takigawa et al., 2003] (Table 1).

Table 1. Gene expression of CCN2/CTGF

***In vivo**, Physiological state:*	
– hypertrophic chondrocytes	++
– proliferating and maturing chondrocytes in early embryo and during wound healing	+
– developing vascular endothelial cells	+
– endothelial cells in injured tissues	+
– osteoblasts & pre-osteoblasts in early embryo	+
– osteocytes stimulated by mechanical stress	+
– wounded periosteum	+
– stimulated undifferentiated mesenchymal cells	+
– periodontal cells stimulated by mechanical stress and wounding	+
– tooth germ	±
– neuron	+
– brain	±
– kidney	±
– skin, liver, spleen	–
***In vivo**, Pathological states in which CCN2/CTGF is induced:*	
– fibrosis (experimental skin fibrosis, scleroderma, systemic sclerosis, inflammatory bowel disease)	
– atherosclerosis (smooth muscle cells)	
– osteoarthritis	
– endothelial cells in tumors	
In culture	
In general: growing phase > quiescent phase	
chondrocytes, osteoblasts: bi-phasic, becoming high again in the highly differentiated state	

++: strong, +: moderate, ±: weak

In an *in vitro* study, CCN2/CTGF was highly expressed in rabbit growth cartilage cells in culture but was little expressed in cultured resting cartilage cells or osteoblastic cells [Nakanishi et al., 1997]. The expression of CCN2/CTGF increased during the continuous culture after confluence of rabbit growth cartilage cells and peaked in the hypertrophic phase [Nakanishi et al., 1997]. These findings agree with the results showing that its *in vivo* expression was predominant in hypertrophic chondrocytes in growth plate cartilage, suggesting a paracrine action of CCN2/CTGF on cells close to the hypertrophic zone of cartilage, which is a rapidly growing and remodeling tissue [Takigawa et al., 2003].

Expression and production of CCN2/CTGF by cultured cells is generally higher in the sparse, growing phase than in the confluent, quiescent phase [Shimo et al., 2001b]. For example, as described above, actively migrating and proliferating endothelial cells, but not quiescent ones, express CCN2/CTGF [Shimo et al., 1998]. Even chondrocytic HCS-2/8 cells express higher level of CCN2/CTGF in sparse culture than in confluent culture [Takigawa et al., 2003], but gene expression of CCN2/CTGF in chondrocytes goes up again after the cells reach confluence, at which time they differentiate to the hypertrophic state [Nakanishi et al., 1997]. A similar bi-phasic change was also found in osteoblasts in culture [Safadi et al., 2003] (Table 1).

As such, the high-level expression of CCN2/CTGF in rapidly growing and remodeling tissues in vivo and in activated cells in culture strongly suggest that CCN2/CTGF is an autocrine factor for tissue regeneration. Gene expression of CCN2/CTGF reported so far is summarized in Table 1.

cbfa1 is well known to be a transcription factor that directs mesenchymal stem cells to differentiate into osteoblasts [Komori et al., 1997]. Mice lacking *cbfa1 (cbfa1*-null mice) lack bone and so die just after birth [Komori et al., 1997]. However, *cbfa1* is also important in endochondral ossification [Enomoto et al., 2000; Inada, 1999]. The expression of *cbfa1* was detected from the zone of proliferating chondrocytes to the calcification zone [Enomoto et al., 2000; Inada, 1999]. Whole-mount *in situ* hybridization using E17 *cbfa1*-null mice revealed that the expression of CCN2/CTGF in the phalanges of the embryonic forepaws was completely undetectable [Yamaai et al., 2005]. Also by *in situ* hybridization using tissue sections, the expression of CCN2/CTGF in ribs and cervical vertebrae found in normal embryos was undetectable in *cbfa1*-null embryos [Yamaai et al., 2005]. These findings suggest that CCN2/CTGF may be a downstream regulator of *cbfa1* in chondrocytic differentiation and osteoblastic differentiation of mesenchymal stem cells during embryonic development.

Roles of CCN2/CTGF in Skeletal Regeneration

Bone is formed by 2 pathways. One is intramembranous bone formation; and another, endochondral ossification. In the process of intramembranous bone formation, mesenchymal stem cells differentiate to osteoblasts; and then these cells directly deposit bone matrix. Mesenchymal stem cells can also differentiate into chondrocytes.

These chondrocytes follow 2 distinct developmental pathways towards 2 distinct fates and functions. Chondrocytes that give rise to articular cartilage acquire their permanent phenotype to maintain normal joint function throughout life [Buckwalter, 1998]. In contrast, other chondrocytes, much more numerous, become organized into growth plates and undergo maturation, hypertrophic differentiation, calcification, and apoptosis, and are replaced by bone cells during the final stages of endochondral ossification [Cancedda, 1995; Hunziker, 1994].

Involvement of CCN2/CTGF in Endochondral Ossification

Endogenous overexpression of CCN2/CTGF in human chondrocytic HCS-2/8 cells, achieved by using recombinant adenoviruses that generated CCN2/CTGF sense RNA (mRNA), resulted in enhanced cellular proliferation and expression of aggrecan and type X collagen [Nakanishi et al., 2000]. Similarly, recombinant CCN2/CTGF (rCCN2/CTGF) protein (20 – 50 ng/ml) promoted the proliferation of HCS-2/8 cells and rabbit growth cartilage (RGC) cells in sparse and growing cultures [Nakanishi et al., 2000]. It also increased the proteoglycan synthesis and gene expressions of aggrecan and collagen type II, typical markers of chondrocyte maturation, in both types of cells in confluent cultures in which they were maturing. Furthermore, the rCCN2/CTGF effectively stimulated the gene expression of collagen type X, a marker of chondrocyte hypertrophy, in RGC cells in overconfluent culture in which the cells were in the prehypertrophic stage [Nakanishi et al., 2000]. Moreover, the rCCN2/CTGF stimulated alkaline phosphatase activity, a marker of calcification, and indeed induced matrix calcification of RGC cells in culture [Nakanishi et al., 2000]. Considering the high expression of CCN2/CTGF in hypertrophic chondrocytes, these results indicate that CCN2/CTGF produced by hypertrophic chondrocytes promotes the proliferation and differentiation of growth cartilage cells toward endochondral ossification (Figure 3).

At the final stage of endochondral ossification, calcifying cartilage is invaded by blood vessels, which leads to the recruitment of perivascular osteoblast-progenitor cells. Therefore, it is feasible that CCN2/CTGF acts on both osteoblasts and osteoblast progenitor cells and endothelial cells. In fact, CCN2/CTGF promoted migration, adhesion and proliferation of vascular endothelial cells and their tube formation in vitro as well as angiogenesis *in vivo* [Lau and Lam, 1999; Shimo et al., 2001a; Shimo et al., 1998; Shimo et al., 1999; Shimo et al., 2001b]. Moreover, rCCN2/CTGF promoted the proliferation of osteoblastic cell lines Saos-2 and MC3T3E1 [Nishida et al., 2000], and increased the mRNA expression of type I collagen, alkaline phosphatase, osteopontin and osteocalcin and the activity of alkaline phosphatase in these cells [Nishida et al., 2000]. It also stimulated collagen synthesis and the matrix mineralization in MC3T3-E1 cells [Nishida et al., 2000]. These findings taken together with the high expression level of CCN2/CTGF in hypertrophic chondrocytes in vivo indicate that CCN2/CTGF is a novel, paracrine regulator for the proliferation and differentiation of endothelial cells and osteoblasts in addition to chondrocytes during endochondral ossification. Therefore, we have proposed that bacause it acts on these types of cells and promotes the entire process of endochondral ossification, CCN2/CTGF should be called “ecogenin”(Figure 3).

The embryonic and neonatal growth of transgenic (Tg) mice that overproduced CCN2/CTGF under the control of the mouse type XI collagen promoter occurred normally, but such animals showed dwarfism within a few months of birth. X-ray analysis revealed that their bone density was decreased compared with that of normal mice, and the femora in the hindlimbs in particular showed an apparent low density [Nakanishi et al., 2001]. These results indicate that overexpression of CCN2/CTGF adversely affects certain steps of endochondral ossification, further supporting our hypothesis.

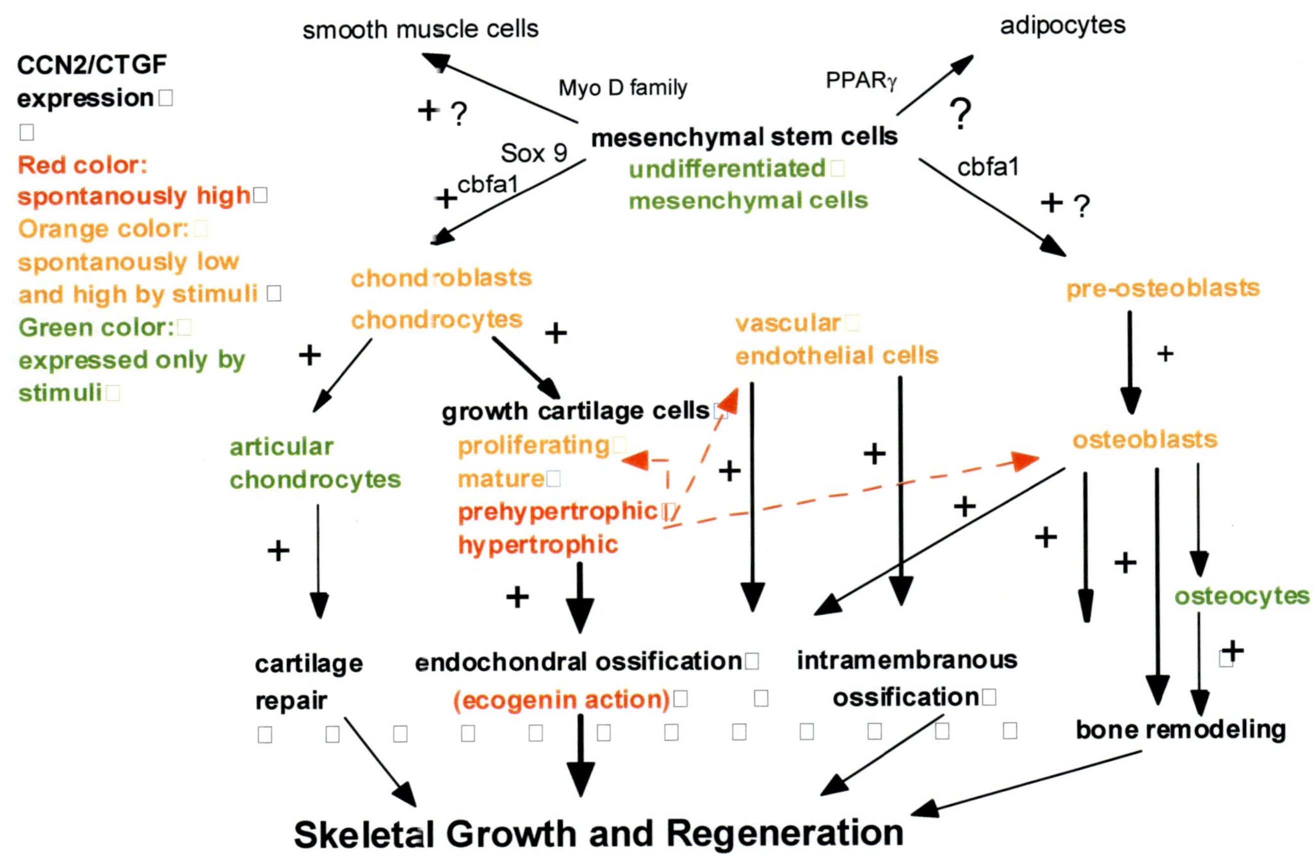

Figure 3. Possible roles of CTGF/CCN2 in skeletal development and regeneration. The level of CCN2/CTGF expression is shown in color. Red: spontaneously high, orange: spontaneously low and high by stimuli, green: expressed only by stimuli. + indicates the presence of direct evidence. Dotted arrows indicate ecogenin action: The arrows point to the target cells, and the root of arrows is next to CCN2/CTGF- producing cells. CCN2/CTGF released by the cells shown in orange and green may act in an autocrine/paracrine manner.

In addition, the stimulatory role of CCN2/CTGF in endochondral ossification shown by our *in vitro* and *in vivo* studies was confirmed by the results of gene knockout experiments in mice, which showed the *ctgf* deficient mice to display skeletal dysmorphisms due to impaired endochondral ossification [Ivkovic et al., 2003].

Possible role of CCN2/CTGF in Bone Regeneration

Wound healing is believed to occur by reproducing normal developmental processes. During fracture healing, both endochondral and intramembranous ossification processes occur. During fracture healing in a mouse rib model, the expression of CCN2/CTGF mRNA revealed by Northern blot analysis significantly increased on day 2 of fracture healing, reaching its peak on day 8, and then declined. CCN2/CTGF mRNA and protein were remarkably detected especially in hypertrophic chondrocytes and in proliferating chondrocytes in the regions of regenerating cartilage on days 8 and 14 after the fracture [Nakata et al., 2002], suggesting that proliferating chondrocytes can also express CCN2/CTGF in rapidly regenerating cartilage. CCN2/CTGF mRNA was also expressed in proliferating periosteal cells in the vicinity of the fracture sites on days 2 and 8, and in cells in the fibrous tissue around the callus on day 8 [Nakata et al., 2002] (Table 1). CCN2/CTGF was also detected in active osteoblasts in the regions of intramembranous ossification [Nakata et al., 2002] and in cells in fibrous tissue, in vascular endothelial cells in the callus, and in periosteal cells around the fracture sites, suggesting the involvement of CCN2/CTGF in regeneration of chondrocytes, osteoblasts and endothelial cells from precursor cells or mesenchymal stem cells.

CCN2/CTGF is also involved in alveolar bone regeneration after tooth extraction [Kanyama et al., 2003]. At the early healing stage after tooth extraction, CCN2/CTGF was expressed strongly in the residual periodontal ligament and in the endothelial cells and undifferentiated fibroblast-like cells, possibly mesenchymal stem cells, migrating into the granulation tissue at the bottom of the sockets (Table 1). At a later stage, osteoblast-like cells proliferated in the sockets with low CCN2/CTGF expression. These finding suggest that CCN2/CTGF produced by residual periodontal cells, endothelial cells and undifferentiated mesenchymal cells acts as an autocrine/paracrine factor in the sockets, causing the mesenchymal cells to differentiate into osteoblasts or the recruitment of osteoblasts from bone, which then replace the granulation tissue with bone.

To clarify the role of CCN2/CTGF in tissue regeneration of bone, we also investigated the localization and expression of CCN2/CTGF during distraction osteogenesis, which is a bone regeneration therapy to expand bone length [Kadota et al., 2003]. We performed osteotomy in the midshaft of the right femur and then started the distraction. After a 7-day lag phase, the distraction was done for 21 days (distraction phase) by using a small external fixator; and it was followed by a 7-day consolidation phase. Immunostaining showed the localization of CCN2/CTGF in various cells located in the bone-forming area around the osteotomy site. During the distraction phase, *in situ* hybridization showed that CCN2/CTGF mRNA was expressed not only in hypertrophic chondrocytes and osteoblasts but also in fibroblasts-like cells and mesenchymal cells in sites of endochondral ossification, and not only in osteoblasts but also in pre-osteoblasts and fibroblasts-like cells in sites of intramembranous ossification. RT-PCR revealed a higher level of CCN2/CTGF mRNA expression in the distracted group than in the non-distracted group. These results suggest that CCN2/CTGF is involved in not only endochondral but also intramembranous ossification and plays important roles in regeneration of bone during distraction osteogenesis.

The role of CCN2/CTGF in bone regeneration via intramembranous ossification was also reported, in which *in vivo* delivery of rCCN2/CTGF into the femoral marrow cavity induced osteogenesis in an animal model [Safadi et al., 2003].

In addition to that in mammalians, gene expression of CCN2/CTGF was also found during regeneration of the newt limb [Cash et al., 1998]. The gene was expressed in hypertrophic chondrocytes in the restored cartilaginous skeleton.

Possible role of CCN2/CTGF in Bone Remodeling

As mentioned above, CCN2/CTGF mRNA is minimally expressed in osteoblasts when compared with its expression level in hypertrophic chondrocytes. However, even in normal rats, osteoblasts and osteocytes in alveolar bone adjacent to the periodontal ligament express a moderate level of CCN2/CTGF [Yamashiro et al., 2001]. This may be due to a high level of bone-forming activity of these cells in this area because of physiological tooth movement (Figure 3; Table 1). In fact, little or no expression was observed in osteocytes embedded deeper in the alveolar bone [Yamashiro et al., 2001]. CCN2/CTGF was also shown to be expressed in normal long bones during the period of growth or modeling [Safadi et al., 2003]. Moreover, CCN2/CTGF was over-expressed in osteopetrotic (*op*) rats, which animals exhibit defects in osteoclast function, severe high bone density, and up-regulation of bone matrix and mineralization-related genes [Xu et al., 2000]. This finding is consistent with the *in vitro* data shown above and suggests some role for this factor in normal skeletal modeling/remodeling. The dramatic over-expression in the *op* mutant skeleton may be secondary to the uncoupling of bone resorption and bone formation, resulting in dysregulation of osteoblast gene expression and function [Xu et al., 2000].

Interestingly, experimental tooth movement increased the expression of CCN2/CTGF mRNA in osteocytes and osteoblasts around the periodontal ligament, and intense expression of CCN2/CTGF extended to osteocytes situated deep in alveolar bone matrix apart from periodontal ligament (Figure 3; Table 1). These findings suggest that osteoblasts activated by certain stimuli also express CCN2/CTGF and that the expressed CCN2/CTGF may play some roles in bone formation as well as in bone remodeling.

Role of CCN2/CTGF in Regeneration of Articular Cartilage

Like rabbit growth cartilage (RGC) cells, rabbit articular cartilage (RAC) cells transfected with recombinant adenoviruses generating mRNA for CCN2/CTGF synthesized more proteoglycan than the control cells [Nishida et al., 2002]. Also, rabbit RAC cells treated with recombinant CCN2/CTGF (rCCN2/CTGF) (20 – 50 ng/ml) showed increased synthesis of DNA and proteoglycans. Although the rCCN2/CTGF stimulated the gene expression of type II collagen and aggrecan core protein, which are markers of chondrocyte maturation, in both RGC and RAC cells, the gene expression of type X collagen, a marker of hypertrophic chondrocytes, was stimulated by rCCN2/CTGF in RGC cells, but not in RAC cells [Nishida et al., 2002]. Oppositely, gene expression of tenascin-C, a marker of articular chondrocytes, was stimulated by rCCN2/CTGF in RAC cells, but not in RGC cells [Nishida et al., 2002]. Moreover, rCCN2/CTGF effectively increased both alkaline phosphatase (ALPase) activity and matrix calcification of RGC cells, but not those of RAC cells [Nishida et al., 2002]. These results indicate that CCN2/CTGF promotes the proliferation and differentiation of articular chondrocytes, but does not promote their hypertrophy or calcification [Nishida et al., 2002]. The difference between the effects of CCN2/CTGF on RGC and RAC would be dependent on the difference in the characteristics of these different types of chondrocytes.

Another molecule of the CCN family, CCN1/Cyr61, was also shown to act on chondrocytes but seems to be related to early chondrogenesis [Wong et al., 1997]. The effective concentration ranges of its recombinant protein was 0.3-5 μg/ml, one much higher than that of CCN2/CTGF (20-50 ng/ml) [Nakanishi et al., 2000].

CCN2/CTGF is normally not expressed in articular cartilage cells of adult rodents, although it's weak expression is observed in growing, young animals [Nawachi et al., 2002]. However, even in adult rodents, quantitative real-time reverse-transcription polymerase chain reaction (RT-PCR) assays showed a significant increase in the level of CTGF/CCN2 mRNA in the monoiodoacetic acid (MIA)-induced OA model [Nishida et al., 2004] (Figure 3). Immunohistochemical analysis and *in situ* hybridization revealed that the clustered chondrocytes, in which clustering indicates an attempt to repair the damaged cartilage, produced CTGF/CCN2. Therefore, CTGF/CCN2 was suspected to play critical roles in cartilage regeneration. In fact, rCTGF/CCN2 incorporated in gelatin hydrogel (rCTGF/CCN2-hydrogel) and injected as a single dose into the joint cavity of MIA-induced OA model rats repaired their articular cartilage to the extent that it became histologically similar to normal articular cartilage [Nishida et al., 2004]. When rCTGF/CCN2-hydrogel-collagen complex was implanted into the defects on the surface of articular cartilage *in situ*, new cartilage filled the defect 4 weeks postoperatively, whereas only soft tissue repair occurred when the PBS-hydrogel-collagen was implanted [Nishida et al., 2004]. These findings suggest the utility of CTGF/CCN2 in the regeneration of articular cartilage.

Angiogenic Activity of CCN2/CTGF May be Applied for Regeneration of Vascular System

Although CCN2/CTGF had been originally identified in the conditioned medium from HUVEC [Bradham et al., 1991], no detectable *in vivo* expression was found in normal adult human arteries [Igarashi et al., 1996; Oemar and Luscher, 1997; Oemar et al., 1997]. However, the presence of CCN2/CTGF protein and/or mRNA was detected in endothelial cells during human embryonic development [Surveyor et al., 1998]. This difference may be due to the difference between rapidly growing embryos and adults; because actively migrating and proliferating endothelial cells in culture, but not quiescent ones, were shown to express CCN2/CTGF [Shimo et al., 1998; Takigawa, 2003; Yamaai et al., 1999]. However, the expression of CCN2/CTGF mRNA in endothelial cells even in mouse embryos was lower than that in hypertrophic chondrocytes [Takigawa, 2003; Yamaai et al., 1999]. With respect to pathogenic processes, endothelial cells of atherosclerotic blood vessels expressed CCN2/CTGF [Oemar and Luscher, 1997; Oemar et al., 1997]. Moreover, xenografts of MDA231 human breast cancer cells induced invasion of endothelial cells that could be strongly stained with anti-CTGF antibody [Shimo et al., 1998]. Therefore, CCN2/CTGF is thought to be expressed in activated, but not quiescent, endothelial cells *in vivo*. (Table 1; Figure 3)

Since suppression of endogenous expression of CCN2/CTGF in cultured endothelial cells by the addition of antisense oligomer or by transfection with expression vectors that generated its antisense RNA markedly inhibited the abilities of the cells to proliferate and migrate, CCN2/CTGF should be involved in the proliferation and migration of the cells *via* an autocrine pathway [Shimo et al., 1998; Takigawa, 2003]. In fact, rCCN2/CTGF (20–50 ng/ml) promoted the adhesion, proliferation, and migration of vascular endothelial cells and induced tube formation by them [Shimo et al., 1999; Takigawa, 2003] and recombinant mouse CCN2 (Fisp12) also stimulated endothelial cell adhesion and migration through integrin $\alpha_v\beta_3$ and promoted endothelial cell survival [Babic et al., 1999]. These findings indicate that endothelial cells are a target for CCN2/CTGF. *In vivo*, application of rCCN2/CTGF to the chicken chorioallantoic membrane (CAM) resulted in a gross angiogenic response [Shimo et al., 1999; Takigawa, 2003] and also induced neovascularization in rat corneal micropocket implants [Babic et al., 1999]. Moreover, impaired angiogenesis at the final stage of endochondral ossification was also observed in CCN2/CTGF deficient mice. In addition to these physiological phenomena, human breast cancer (MDA231) cells, which express a high level of CCN2/CTGF mRNA and secrete much CCN2/CTGF protein, showed a high level of neovascularization when xenografted in nude mice; whereas human squamous cell carcinoma (A431) cells, which express a low level of CCN2/CTGF mRNA and secrete little CCN2/CTGF protein, showed a low level of neovascularization [Shimo et al., 2001b]. Furthermore, CCN2/CTGF also increased the expression by MDA231 cells of a number of metalloproteinases, which are involved in endothelial cell invasion [Kondo et al., 2002]. These findings indicate that CCN2/CTGF is a novel, potent autocrine/paracrine angiogenesis factor that functions in multi-stages in this process.

Similar angiogenic activity was reported for CCN1/Cyr61 [Lau and Lam, 1999]. Moreover, CCN1/Cry61-deficient mice were shown to be embryonic lethal due to failure of angiogenesis during placental development [Mo et al., 2002].

In contrast, Inoki et al. [Inoki et al., 2002] reported recently that recombinant CCN2/CTGF and the thrombospondin type-1 repeat domain of CCN2/CTGF inhibited vascular endothelial cell growth factor (VEGF)165-induced angiogenesis by forming a complex with $VEGF_{165}$ (Figure 2), although they also showed that their recombinant CCN2/CTGF itself had a weak angiogenic activity. Since the angiogenic activity of CCN2/CTGF has been confirmed by several groups, the relation between CCN2/CTGF and VEGF should be investigated more in detail.

The expression of CCN2/CTGF in endothelial cells was induced by wounding of monolayer cultures [Shimo et al., 1998]. Moreover, the expression was induced by other angiogenic factors such as VEGF, basic fibroblast growth factor (bFGF), and CCN2/CTGF itself [Shimo et al., 1998]. Hypoxic induction of angiogenesis by human breast cancer cells (MDA-231) can be ascribed at least in part to CCN2/CTGF [Shimo et al., 2001a; Shimo et al., 2001b]. These findings suggest that upregulation of CCN2/CTGF by angiogenic stimuli is a common phenomenon in endothelial cells. CCN2/CTGF might be a common autocrine mediator of all angiogenesis factors. If this hypothesis proves to be correct, CCN2/CTGF would be very important therapeutic target for the regeneration of vascular systems in addition to the treatment of angiogeneic diseases.

Possible Role of CCN2/CTGF in Regeneration of Other Tissues

CCN2/CTGF is expressed in fibroblasts or fibroblast-like cells during normal development [Kireeva et al., 1997; Surveyor and Brigstock, 1999; Surveyor et al., 1998] and wound healing of skin [Igarashi et al., 1993] and cornea [Blalock et al., 2003]. During the wound healing process in most tissues, fibroblasts activated to proliferate and migrate are the major player in this event. Although it is not clear whether these undifferentiated types of fibroblasts are derived from quiescent fibroblasts or from mesenchymal stem cells or from both, in any case they fill the defects. CCN2/CTGF is reported to be both mitogenic and chemotactic for fibroblast-like cells [Brigstock et al., 1997; Frazier et al., 1996] (Figure 2) and its mitogenic action is synergized by other growth factors and heparin [Brigstock et al., 1997; Frazier et al., 1996; Kireeva et al., 1997]. It was also reported that CCN2/CTGF directly stimulated adhesion of fibroblasts and that this adhesion-promoting effect is thought to be mediated by integrins [Chen et al., 2001; Kireeva et al., 1997] (Figure 2). CCN2/CTGF also stimulated the production of extracellular matrix components and their receptors such as collagens [Castro, 2003; Dammeier et al., 1998b; Duncan et al., 1999; Frazier et al., 1996; Murphy et al., 1999; Riser and Cortes, 2001; Riser et al., 2000], fibronectin [Frazier et al., 1996; Riser and Cortes, 2001; Riser et al., 2000; Twigg et al., 2002] and integrin α_5 [Frazier et al., 1996]. Considering these biological actions of CCN2/CTGF, the roles of CCN2/CTGF in wound healing may be summarized as follows: First of all, fibroblast-derived CCN2/CTGF stimulates migration of fibroblasts in the injured tissue. Then, it mediates the adhesion of these cells to settle them at the defect site and initiates tissue regeneration by increasing the proliferation of the cells to provide an adequate number of them. Finally, CCN2/CTGF promotes the deposition of ECM components to repair the tissue defect. So, all the activities of CCN2/CTGF described above contribute to the regeneration of the injured tissues. In addition, as mentioned in the previous section, CCN2/CTGF is capable of inducing angiogenesis, which is essential for most tissues. Therefore, it is clear that CCN2/CTGF is a key player involved in the wound-healing cascade.

CCN2/CTGF is also suggested to be involved in repair of the intestinal mucosa [Dammeier et al., 1998b] and in *in vitro* wound healing of renal epithelium cells [Pawar et al., 1995]. Therefore, CCN2/CTGF may be important for regeneration of not only connective tissues but also other tissues as well, although the latter have a little been investigated.

Platelet-Derived CCN2/CTGF may be a Trigger for Tissue Regeneration

In many tissues/organs, blood clotting is the first step of wound healing after injury. Since CCN2/CTGF is induced and produced by undifferentiated mesenchymal cells, probably including mesenchymal stem cells, which migrate into the defect, it may play the role as an autocrine/paracrine regeneration factor throughout the process of wound healing. However, at the initial step, CCN2/CTGF may be supplied by blood clots, because CCN2/CTGF was recently found to be abundant in platelets and released from activated platelets [Cicha et al., 2004; Kubota et al., 2004]. The CCN2/CTGF released from platelets may be important for the initiation of proliferation, migration and adhesion of undifferentiated mesenchymal cells, maybe including stem cells to regenerate wounded tissues.

Persistent Overexpression of CCN2/CTGF in Non-Skeletal Connective Tissues Causes the Pathological State Fibrosis

CCN2/CTGF is present and/or overexpressed in fibrotic skin disorders such as systemic and localized sclerosis, keloid and scar tissues, and in fibrotic lesions of various organs and tissues including the lung, kidney, liver, cardiovascular system, pancreas, lens, bowel, and gingiva, and the stroma of various tumors [Takigawa, 2003]. These findings strongly suggest the pathological significance of CCN2/CTGF in these fibrotic disorders [Leask et al., 2002]. In this regard, CCN2/CTGF has been suggested to be a downstream mediator of TGF-β which is a well-known growth factor involved in fibrosis [Grotendorst, 1997]. It is well known that the mouse ortholog of CTGF, βIG-M2, was originally isolated from mouse ARK-2B cells stimulated by TGF-β [Brunner et al., 1991] TGF-β directly activated transcription of the CCN2/CTGF gene in human skin fibroblasts through a TGF-β response element [Grotendorst et al., 1996]. Further, CCN2/CTGF mRNA was strongly expressed in the fibroblasts in mouse skin at the site of a TGF-β injection [Shinozaki et al., 1997], and TGF-β was also shown to induce CCN2/CTGF expression in both cardiac fibroblasts and cardiac myocytes [Chen et al., 2000], in rabbit corneal fibroblasts [Folger et al., 2001], and in cultured human lung fibroblasts [Kucich et al., 2001]. TGF-β1-treated scleroderma fibroblasts produced greater amounts of CCN2/CTGF, than did the factor-treated normal fibroblasts [Kikuchi et al., 1995]. However, it was also reported that persistent fibrosis requires TGF-β as an induction factor and CCN2/CTGF as a maintenance factor [Mori et al., 1999]. Maintenance of fibrosis by autocrine overexpression of CCN2/CTGF was also reported by another group [Shi-wen et al., 2000]. Therefore, although it is unclear whether CCN2/CTGF is a downstream mediator of TGF-β in the pathogenesis of fibrosis, persistent overexpression of CCN2/CTGF should be considered to be involved in the pathogenesis of fibrosis (Figure 4). The involvement of CCN2/CTGF in fibrosis is in line with the fact that CCN2/CTGF is highly expressed during the wound healing process. If up-regulated expression of CCN2/CTGF is appropriately terminated, normal wound healing would occur (Figure 4). However, if the overexpression continues, it would cause fibrosis and scaring. No disease in skeletal tissues with continuous overexpression of CCN2/CTGF has been reported, although there was a report that *ctgf* transgenic mice in which *ctgf* was forced to be overexpressed in cartilage have skeletal disorders [Nakanishi et al., 2001]. Presumably, unless this gene is dramatically overexpressed, nothing would happen in these tissues in which the high level of extracellular matrix (ECM) synthesis occurs in the normal state.

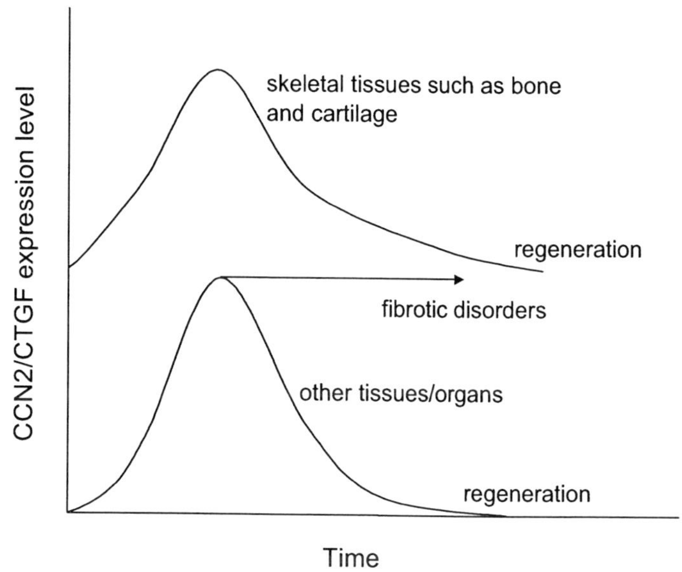

Figure 4. CTGF/CCN2 as a regeneration factor (regenerin). CTGF/CCN2 is induced after wounding in various tissues and regeneration occurs to repair the defect in the tissues. In the case of skeletal tissues, the induced CTGF/CCN2 gradually decreases in level as the wound is repaired. This process is similar to that of normal development. So far, there is no report that continuous overexpression causes any skeletal disorder, except that *ctgf* transgenic mice have skeletal disorders. In contrast, in tissues other than skeletal tissues, down-regulation of the induced CTGF/CCN2 is essential for proper regeneration. If overexpression of CTGF/CCN2 continues, it may cause fibrotic disorders. From this point of view, the true role of CTGF/CCN2 is to regenerate these tissues, so should be called "regenerin", but its proper regulation is important for preventing fibrotic disorders.

Controlling the Action of CCN2/CTGF and its Gene Expression is Critical for Proper Regeneration of Various Organs

Almost all organs except avascular tissues such as cartilage comprise a mixture of cell types, and regeneration thus requires the coordinated growth of various tissue components. As described above, CCN2/CTGF is an important regeneration factor at least for the connective tissues. Therefore, if, due to elevated expression of CCN2/CTGF, the regeneration rate of connective tissue frame work becomes higher than that of the parenchymal cells in organs such as liver, kidney, and lung, disturbance of the regeneration of parenchymal cell may occur, as seen in liver cirrhosis. CCN2/CTGF is a multi-functional growth factor that potentiates either growth or differentiation of several types of mesenchymal cells, but its expression *in vivo* is normally low. However, the expression in certain tissues is upregulated during development and cytodifferentiation and is induced in certain pathological states. Therefore, it is very important to understand the molecular regulatory mechanism of action and gene expression of CCN2/CTGF and then to control them for establishment of a successful strategy for tissue regeneration.

Mechanism of Action of CCN2/CTGF

Several receptors or receptor like-molecules for CCN2/CTGF have been reported. Classical binding studies using iodinated rCCN2/CTGF showed both a high- and a low-affinity binding sites to be present on chondrocytic HCS-2/8 cells [Nishida et al., 1998; Takigawa et al., 2003] and osteoblasts [Nishida et al., 2000; Takigawa et al., 2003]. Also, cross-linking studies using chondrocytic cells, osteoblastic cells and endothelial cells revealed the formation of ^{125}I-CCN2/CTGF-receptor complexes with an apparent molecular weight of 280 kDa [Nishida et al., 2000; Nishida et al., 1998; Takigawa et al., 2003]. This receptor on HCS-2/8 cells was phosphorylated by stimulation of the cells with CCN2/CTGF, suggesting it to be a signal-transducing receptor [Nishida et al., 1998]. Although it was shown that the CCN2/CTGF-induced proliferation of chondrocytes was mediated by ERK and that the CCN2/CTGF-stimulated proteoglycan synthesis in chondrocytes was mediated by p38 MAPK [Yosimichi et al., 2001], a link between the receptor and these MAP kinases is unclear and should be clarified in the future.

On the other hand, a binding protein for CCN2/CTGF with a mass of approximately 620 kDa from bone marrow stroma cell line BMS2 was purified and identified as a low-density lipoprotein receptor-related protein/α_2-macroglobulin receptor (LRP) [Segarini et al., 2001]. This would suggest the role of CCN2/CTGF in mesenchymal stem cells, but it is unclear that the receptor is in fact a signal-transducing receptor for CCN2/CTGF.

Concerning the adhesion-promoting activity of CCN2/CTGF toward various types of cells, this activity is now well documented to be mediated by cell-surface integrins [Babic et al., 1999; Jedsadayanmata et al., 1999; Lau and Lam, 1999] such as $\alpha_V\beta_3$ on endothelial cells [Babic et al., 1999; Lau and Lam, 1999] and $\alpha_{IIb}\beta_3$ on activated platelets [Babic et al., 1999; Jedsadayanmata et al., 1999; Lau and Lam, 1999]. However, there is no evidence that other activities such as the promotion of proliferation and differentiation are mediated by integrins.

Because of their unique modular structure, CCN proteins are generally highly interactive with other molecules (Figure 2). In fact, CCN2/CTGF is known to be an ECM-associated growth factor, and so may exert its action in a matricrine manner. In addition to the receptor candidates described above, perlecan has been suggested to act as a co-receptor and to play an indispensable role for the paracrine action of CCN2/CTGF in cartilage [Nishida et al., 2003]. In this regard, it is noteworthy that malformation of the growth plate was observed in perlecan gene deficient mice [Arikawa-Hirasawa et al., 1999].

In addition to binding to ECM components, CCN2/CTGF was reported to bind to growth factors as well. As indicated by its name, the IGFBP module binds IGFs, although its affinity is lower than that of IGFBPs (Figure 2). This interaction might modulate IGF action. Moreover, rCCN2/CTGF and the thrombospondin type-1 repeat domain of CCN2/CTGF each form a complex with $VEGF_{165}$ and modulate VEGF action [Inoki et al., 2002] (Figure 2). Surprisingly, CCN2/CTGF was described as an antagonist of VEGF, although the angiogenic effects of CCN2/CTGF were also affirmed. Furthermore, CCN2/CTGF was recently shown to directly bind BMP-4 and TGF-β through its cysteine-rich domain and to be able to antagonize BMP-4 activity by preventing the binding of BMP-4 to its receptors but to have the opposite effect, i.e., enhancement of receptor binding, on TGF-β [Abreu et al., 2002] (Figure 2).

Extracellular Molecules that Regulate Gene Expression of CCN2/CTGF

Because TGF-β has been shown to be highly expressed in the hypertrophic region of cartilage tissue [Joyce et al., 1990] and to induce the expression of CCN2/CTGF in chondrocytes [Nakanishi et al., 1997], the predominant expression of CCN2/CTGF mRNA in the hypertrophic chondrocytes in their physiological state may be due to the high level of TGF-β [Takigawa et al., 2003]. It is well known that TGF-β is a potent inducer of matrix formation and that cartilage contains many matrix components such as collagen and proteoglycans. Therefore, the upregulated expression of CCN2/CTGF induced by TGF-β would be a naturally occurring phenomenon in cartilage, although its overexpression in non-skeletal tissues could cause fibrosis. Similarly, BMP-2, which is involved in bone formation, induced CCN2/CTGF gene expression in the chondrocytic cell line HCS-2/8 [Nakanishi et al., 1997] and the highest immunoreactivity for BMP-2-7 was detected in the hypertrophic and calcifying zone chondrocytes of the growth plate [Anderson, 2000]. So, it is also feasible that the high-level expression of CCN2/CTGF in hypertrophic chondrocytes is at least partially due to high levels of BMPs and that CCN2/CTGF mediates the action of BMP on bone formation, including endochondral ossification.

In addition to TGF-β, various other growth factors such as PDGF, EGF, and FGF induced gene expression of CCN2/CTGF in fibroblasts, although the effect was less than that of TGF-β [Grotendorst et al., 1996; Igarashi et al., 1993; Ryseck et al., 1991]. Moreover, bFGF and VEGF induced production of CCN2/CTGF in bovine aorta endothelial cells [Shimo et al., 2001b]. CCN2/CTGF was also autoinduced by CCN2/CTGF itself in rat mesangial cells [Riser et al., 2000] and in bovine endothelial cells [Shimo et al., 2001b]. Besides growth factors, dexamethasone caused a large increase in CCN2/CTGF expression in Balb/c 3T3 cells [Dammeier et al., 1998a] and chondrocytic HCS-2/8 cells [Kubota et al., 2003].

As a physical factor, mechanical stress induced CCN2/CTGF gene expression in osteoblasts and osteocytes [Yamashiro et al., 2001], chondrocytes [Wong et al., 2003] and fibroblasts [Schild and Trueb, 2002]. Moreover, hypoxic stress increased both *ccn2/ctgf* gene expression level and CCN2/CTGF secretion by tumor cells [Kondo et al., 2002; Shimo et al., 2001a]. The observed increase in the steady-state mRNA level by hypoxia was accounted for not by a transcriptional activation of the *ccn2/ctgf* promoter, but by a post-transcriptional alteration of the stability of *ccn2/ctgf* mRNA. Hypoxic induction of CCN2/CTGF was also reported to occur in primary cultures of bovine chondrocytes [93].

On the other hand, hepatocyte growth factor (HGF) counteracted TGF-β1 by attenuating CTGF induction, and prevented renal fibrogenesis [Inoue et al., 2003]. Several inhibitors of the gene expression of CCN2/CTGF have been reported. Examples include cyclic AMP and agents that elevate its level [Duncan et al., 1999; Kothapalli et al., 1998] such as forskolin [Kothapalli et al., 1998] and choleratoxin [Kothapalli et al., 1998], prostaglandin E_2 [Ricupero et al., 1999], and tumor necrosis factor-α [Ricupero et al., 1999].

Based on our better understanding of the mechanisms regulating CCN2/CTGF gene expression, the development of positive regulatory molecules should be therapeutically useful for tissue regeneration; whereas that of negative regulatory one may be therapeutically useful for fibrotic disorders.

Genetic Elements Controlling Gene Expression of CCN2/CTGF

To date, several *cis*-elements for the transcription control of *ccn2/ctgf* have been described (Figure 2). The most classical one is the TGF-β response element (TbRE) which was found in a study on fibroblastic cells [Grotendorst et al., 1996]. Although the TbRE plays certain roles in the regulation of *ccn2/ctgf* gene expression [Eguchi et al., 2001], a more critical *cis*-regulatory element, designated TRENDIC, which stands for a *tr*anscriptional *en*hancer *d*ominant *in* *c*hondrocytes, was found in the region upstream of TbRE but near it [Eguchi et al., 2002](Figure 2). Between these 2 enhancers, a DNA segment with a Smad binding consensus sequence (SBE) can be found. Certain nuclear proteins bind specifically to SBE, TbRE and TRENDIC (Figure 2). Therefore, mutual interaction between a binding counterpart of one of these elements and that of the adjacent element may well occur, which would enable complex regulation of gene expression. Moreover, a tandem repeat of the TEF1- binding consensus sequence in the promoter area was recently reported [Leask et al., 2003]. Interestingly, this element was found to mediate the *ccn2/ctgf* gene induction by TGF-β, despite its distinct identity from TbRE. Identification of their binding counterparts - in other words, specific transcription factors - would give us critical insight into the complex interplay of these genetic drivers of CCN2/CTGF production.

It is widely recognized that the 3'-untranslated region (3'-UTR) of various mRNAs contain critical genetic elements that determine the fate of their own mRNAs. Since *ccn2/ctgf* mRNA is characterized by its long 3'-UTR, 3'-UTRs of this mRNA from different vertebrates including humans were linked to the 3'-end of a reporter mRNA; and their effects on gene expression were then investigated. As a result, all of them remarkably repressed the reporter gene expression at similar levels [Kondo et al., 2000; Kubota et al., 1999; Mukudai et al., 2003], suggesting crucial roles of the 3'-UTR for the regulation of *ccn2/ctgf* gene expression among vertebrates.

The strong repressive function of the 3'-UTR appears to be mediated by the cooperation of several distinct *cis*-elements. One of these *cis*-acting elements in the 3'-UTR was initially discovered in human *ccn2/ctgf* [Kondo et al., 2000; Kubota et al., 1999, 2000; Mukudai et al., 2003] and named *c*is-*a*cting *e*lement of *s*tructure-*a*nchored *r*epression (CAESAR, Figure 2).

This *cis*-acting element is located at the junction of the coding region and 3'-UTR. Because the repressive effect was found to be realized without a change in the steady-state mRNA level, this element is an RNA regulatory element acting at post-transcriptional stages.

Thus, agents that stimulate CCN2/CTGF production, action or signaling pathways could lead to the development of new therapeutic approaches for tissue regeneration whereas agents that inhibit them could lead to the development of new therapeutic approaches for the control of fibrotic disorders.

Inducing Effect of CTGF/CCN2 on Chondrogenic Differentiation of Mouse Bone Marrow-Derived Mesenchymal Stem Cells

To investigate whether rCTGF/CCN2 could stimulate chondrogenic cell differentiation or not, we cultured bone marrow-derived mesenchymal stem cells (MSCs) for 14 days in α-MEM with or without 30 or 50 ng/ml of rCTGF/CCN2 [Nishida et al., 2004]. The cultures were thereafter processed for histochemical detection of cartilage matrix by using toludine blue staining. Some cartilage matrix was present in control cultures and in 30 ng/ml rCTGF/CCN2-treated cultures of MSCs, but a much larger amount of it was detected in 50 ng/ml rCTGF/CCN2-treated ones. In addition, to clarify the role of CTGF/CCN2 in the chondrocytic differentiation of MSCs, we used Northern blot analysis to investigate the effect of rCTGF/CCN2 on the mRNA expression of type II collagen and aggrecan in cultured MSCs. Chondrogenesis, as represented by the expression of type II collagen and aggrecan mRNAs, was clearly induced in MSCs treated with 50 ng/ml rCTGF/CCN2. These findings suggest that CTGF/CCN2 induce chondrogenic differentiation of mouse bone marrow-derived MSCs.

It is well-known that MSCs can differentiate into osteoblasts, myoblasts and adipocytes in addition to chondrocytes. The findings described in the section about bone regeneration indicate a possible role of CCN2/CTGF in differentiation of MSCs into osteoblasts. However, so far there is no report on the role of CCN2/CTGF in differentiation of MSCs into adipocytes or myoblasts. Concerning myogenic differentiation, inhibition of CCN2/CTGF expression resulted in decreased differentiation of human rhabdomyosarcoma cells [Croci et al., 2004]. Therefore, it would be interested to investigate whether CCN2/CTGF determines the direction of differentiation of MSCs or promotes the differentiation of MSCs of which direction had been already commited.

Concluding Remarks

In this chapter, I described that CCN2/CTGF is a multifunctional growth/differentiation factor for fibroblasts, chondrocytes, osteoblasts, and vascular endothelial cells, as well as for mesenchymal stem cells. Depending on the type of cell with which it interacts, it promotes chemotaxis, migration, adhesion, proliferation, differentiation and/or extracellular matrix formation. These cell/tissue specific functions may be due to interaction with other molecules such as growth factors and ECM components. Because gene expression of CCN2/CTGF is very high in hypertrophic chondrocytes in the physiological state, a major physiological role for this factor, which may be referred to as "ecogenin", has been suggested to be the promotion of endochondral ossification. However, accumulating data that show its gene expression in various types of cells including undifferentiated mesenchymal cells during embryonic development and during would healing of bone, cartilage, and skin indicate the involvement of this factor in tissue regeneration. In fact, recombinant CCN2/CTGF regenerated damaged cartilage by acting on chondrocytes and bone marrow-derived mesenchymal stem cells. In the case of organs/tissues composed of parenchymal cells and a connective tissue framework, the balance between regeneration of parenchymal cells and that of connective tissue framework is important for proper regeneration of the organs/tissues. Although CCN2/CTGF is known to be involved in fibrotic disorders and angiogenic diseases, these phenomena are characterized by overregeneration of connective tissue rather than that of parenchymal cells in the organs/tissues. For proper regeneration of these organs/tissues, regeneration of the connective tissue framework is indispensable but should be controlled properly. From this point of view, CCN2/CTGF should be thought of as a regeneration factor, which may be called "regenerin", rather than as a pathological factor even in these organs/tissues. Thus, delivery of CCN2/CTGF itself and the controlling of its action and gene expression would be therapeutically beneficial for promotion of proper controlled tissue regeneration as well as for preventing pathological changes.

References

Abreu, J. G., Ketpura, N. I., Reversade, B., & De Robertis, E. M. (2002) Connective-tissue growth factor (CTGF) modulates cell signalling by BMP and TGF-β. *Nat Cell Biol, 4*, 599-604.

Anderson, H. C., Hodges, P. T., Aguilera, X. M., Missana, L., & Moylan, P. E. (2000) Bone morphogenetic protein (BMP) localization in developing human and rat growth plate, metaphysis, epiphysis, and articular cartilage. *J Histochem Cytochem, 48*, 1493-1502.

Arikawa-Hirasawa, E., Watanabe, H., Takami, H., Hassell, J., & Yamada, Y. (1999) Perlecan is essential for cartilage and cephalic development. *Nat. Genet., 23*, 354-358.

Babic, A. M., Chen, C. C., & Lau, L. F. (1999) Fisp12/mouse connective tissue growth factor mediates endothelial cell adhesion and migration through integrin $\alpha_v\beta_3$, promotes endothelial cell survival, and induces angiogenesis in vivo. *Mol Cell Biol, 19*, 2958-2966.

Blalock, T. D., Duncan, M. R., Varela, J. C., Goldstein, M. H., Tuli, S. S., Grotendorst, G.R., & Schultz, G. S. (2003) Connective tissue growth factor expression and action in human corneal fibroblast cultures and rat corneas after photorefractive keratectomy. *Invest Ophthalmol Vis Sci, 44*, 1879-1887.

Bork, P. (1993) The modular architecture of a new family of growth regulators related to connective tissue growth factor. *FEBS Lett, 327*, 125-130.

Bradham, D. M., Igarashi, A, Potter, R. L., & Grotendorst, G. R. (1991) Connective tissue growth factor: a cysteine-rich mitogen secreted by human vascular endothelial cells is related to the SRC-induced immediate early gene product CEF-10. *J Cell Biol, 114*, 1285-1294.

Brigstock, D. R. (1999) The connective tissue growth factor/cysteine-rich 61/nephroblastoma overexpressed (CCN) family. *Endocr Rev, 20*, 189-206.

Brigstock, D. R., Goldschmeding, R., Katsube, K. I., Lam, S. C., Lau, L.F., Lyons, K., Naus, C., Perbal, B., Riser, B., Takigawa, M., & Yeger, H. (2003) Proposal for a unified CCN nomenclature. *Mol Pathol, 56*, 127-128.

Brigstock, D. R., Steffen, C. L., Kim, G. Y., Vegunta, R. K., Diehl, J. R., & Harding, P. A. (1997) Purification and characterization of novel heparin-binding growth factors in uterine secretory fluids. Identification as heparin-regulated Mr 10,000 forms of connective tissue growth factor. *J Biol Chem, 272*, 20275-20282.

Brunner, A., Chinn, J., Neubauer, M., & Purchio, A. F. (1991) Identification of a gene family regulated by transforming growth factor β. *DNA Cell Biol., 10*, 293-300.

Buckwalter, J. A., & Mankin, H. J. (1998) Articular cartilage: tissue design and chondrocyte-matrix interactions. *Instr. Course Lect., 47*, 477-486.

Cancedda, R., Cancedda, F. D., & Castagnola, P. (1995) Chondrocyte differentiation. *Int. Rev. Cyt., 159*, 265-358.

Cash, D. E., Gates, P. B., Imokawa, Y., & Brockes, J. P. (1998) Identification of newt connective tissue growth factor as a target of retinoid regulation in limb blastemal cells. *Gene, 222*, 119-124.

Castro, F. P., Jr. (2003) Stingers, cervical cord neurapraxia, and stenosis. *Clin Sports Med, 22*, 483-492.

Chen, C. C., Chen, N., & Lau, L. F. (2001) The angiogenic factors Cyr61 and connective tissue growth factor induce adhesive signaling in primary human skin fibroblasts. *J Biol Chem, 276*, 10443-10452.

Chen, M. M., Lam, A., Abraham, J. A., Schreiner, G.F., & Joly, A. H. (2000) CTGF expression is induced by TGF-β in cardiac fibroblasts and cardiac myocytes: a potential role in heart fibrosis. *J Mol Cell Cardiol, 32*, 1805-1819.

Cicha, I., Garlichs, C. D., Daniel, W. G., & Goppelt-Struebe, M. (2004) Activated human platelets release connective tissue growth factor. *Thromb Haemost, 91*, 755-760.

Croci, S., Landuzzi, L., Astolfi, A., Nicoletti, G., Rosolen, A., Sartori, F., Follo, M. Y., Oliver, N., De Giovanni, C., Nanni, P., & Lollini, P. L. (2004) Inhibition of connective tissue growth factor (CTGF/CCN2) expression decreases the survival and myogenic differentiation of human rhabdomyosarcoma cells. *Cancer Res, 64*, 1730-1736.

Dammeier, J., Beer, H. D., Brauchle, M., & Werner, S. (1998a) Dexamethasone is a novel potent inducer of connective tissue growth factor expression. Implications for glucocorticoid therapy. *J Biol Chem, 273*, 18185-18190.

Dammeier, J., Brauchle, M., Falk, W., Grotendorst, G. R., & Werner, S. (1998b) Connective tissue growth factor: a novel regulator of mucosal repair and fibrosis in inflammatory bowel disease? *Int J Biochem Cell Biol, 30*, 909-922.

Duncan, M. R., Frazier, K. S., Abramson, S., Williams, S., Klapper, H., Huang, X., & Grotendorst, G. R. (1999) Connective tissue growth factor mediates transforming growth factor β-induced collagen synthesis: down-regulation by cAMP. *Faseb J, 13*, 1774-1786.

Eguchi, T., Kubota, S., Kondo, S., Kuboki, T., Yatani, H., & Takigawa, M. (2002) A novel cis-element that enhances connective tissue growth factor gene expression in chondrocytic cells. *Biochem Biophys Res Commun, 295*, 445-451.

Eguchi, T., Kubota, S., Kondo, S., Shimo, T., Hattori, T., Nakanishi, T., Kuboki, T., Yatani, H., & Takigawa, M. (2001) Regulatory mechanism of human connective tissue growth factor (CTGF/Hcs24) gene expression in a human chondrocytic cell line, HCS-2/8. *J Biochem (Tokyo), 130*, 79-87.

Enomoto, H., Enomoto-Iwamoto, M., Iwamoto, M., Nomura, S., Himeno, M., Kitamura, Y., Kishimoto, T., & Komori, T. (2000) Cbfal is a positive regulatory factor in chondrocyte maturation. *J Biol Chem, 275*, 8695-8702.

Enomoto, M., & Takigawa, M. (1992) Regulation of tumor derived and immortalized chondrocytes. In M. Adolphe (Ed.), *Biological Regulation of the Chondrocytes* (pp. 321-338). Boca Raton, CRC Press.

Folger, P.A., Zekaria, D., Grotendorst, G., & Masur, S. K. (2001) Transforming growth factor-β-stimulated connective tissue growth factor expression during corneal myofibroblast differentiation. *Invest Ophthalmol Vis Sci, 42*, 2534-2541.

Frazier, K., Williams, S., Kothapalli, D., Klapper, H., & Grotendorst, G. R. (1996) Stimulation of fibroblast cell growth, matrix production, and granulation tissue formation by connective tissue growth factor. *J Invest Dermatol, 107*, 404-411.

Grotendorst, G. R. (1997) Connective tissue growth factor: a mediator of TGF-β action on fibroblasts. *Cytokine Growth Factor Rev, 8*, 171-179.

Grotendorst, G. R., Okochi, H., & Hayashi, N. (1996) A novel transforming growth factor β response element controls the expression of the connective tissue growth factor gene. *Cell Growth Differ, 7*, 469-480.

Hashimoto, Y., Shindo-Okada, N., Tani, M., Nagamachi, Y., Takeuchi, K., Shiroishi, T., Toma, H., & Yokota, J. (1998) Expression of the Elml gene, a novel gene of the CCN (connective tissue growth factor, Cyr61/Cef10, and neuroblastoma overexpressed gene) family, suppresses In vivo tumor growth and metastasis of K-1735 murine melanoma cells. *J Exp Med, 187*, 289-296.

Hunziker, E. B. (1994) Mechanism of longitudinal bone growth and its regulation by growth plate chondrocytes. *Microsc. Res. Tech., 28*, 505-519.

Igarashi, A., Nashiro, K., Kikuchi, K., Sato, S., Ihn, H., Fujimoto, M., Grotendorst, G. R., & Takehara, K. (1996) Connective tissue growth factor gene expression in tissue sections from localized scleroderma, keloid, and other fibrotic skin disorders. *J Invest Dermatol, 106*, 729-733.

Igarashi, A., Okochi, H., Bradham, D. M., & Grotendorst, G. R. (1993) Regulation of connective tissue growth factor gene expression in human skin fibroblasts and during wound repair. *Mol Biol Cell, 4*, 637-645.

Inada, M., Yasui, T., Nomura, S., Miyake, S., Deguchi, K., Himeno, M., Sato, M., Yamagiwa, H., Kimura, T., Yasui, N., Ochi, T., Endo, N., Kitamura, Y., Kishimoto, T., & Komori, T. (1999) Maturational disturbance of chondrocytes in Cbfa1-deficient mice. *Dev Dynam, 214*, 279-290.

Inoki, I., Shiomi, T., Hashimoto, G., Enomoto, H., Nakamura, H., Makino, K., Ikeda, E., Takata, S., Kobayashi, K., & Okada, Y. (2002) Connective tissue growth factor binds vascular endothelial growth factor (VEGF) and inhibits VEGF-induced angiogenesis. *Faseb J, 16*, 219-221.

Inoue, T., Okada, H., Kobayashi, T., Watanabe, Y., Kanno, Y., Kopp, J. B., Nishida, T., Takigawa, M., Ueno, M., Nakamura, T., & Suzuki, H. (2003) Hepatocyte growth factor counteracts transforming growth factor-β1, through attenuation of connective tissue growth factor induction, and prevents renal fibrogenesis in 5/6 nephrectomized mice. *Faseb J, 17*, 268-270.

Ivkovic, S., Yoon, B. S., Popoff, S. N., Safadi, F. F., Libuda, D. E., Stephenson, R. C., Daluiski, A., & Lyons, K. M. (2003) Connective tissue growth factor coordinates chondrogenesis and angiogenesis during skeletal development. *Development, 130*, 2779-2791.

Jedsadayanmata, A., Chen, C. C., Kireeva, M. L., Lau, L. F., & Lam, S. C. (1999) Activation-dependent adhesion of human platelets to Cyr61 and Fisp12/mouse connective tissue growth factor is mediated through integrin $\alpha II_b \beta_3$. *J Biol Chem, 274*, 24321-24327.

Joyce, M. E., Jingushi, S., & Bolander, M. E. (1990) Transforming growth factor-β in the regulation of fracture repair. *Orthop. Clin. North. Am., 21*, 199-209.

Kadota, H., Nakanishi, T., Asaumi, K., Yamaai, T., Nakata, E., Mitani, S., Aoki, K., Aiga, A., Inoue, H., & Takigawa, M. (2003) Expression of connective tissue growth factor/hypertrophic chondrocyte-specific gene product 24 (CTGF/Hcs24/CCN2) during distraction osteogenesis. *J Bone Miner Metab, 22,* 293-302.

Kanyama, M., Kuboki, T., Akiyama, K., Nawachi, K., Miyauchi, F. M., Yatani, H., Kubota, S., Nakanishi, T., & Takigawa, M. (2003) Connective tissue growth factor expressed in rat alveolar bone regeneration sites after tooth extraction. *Arch Oral Biol, 48*, 723-730.

Kikuchi, K., Kadono, T., Ihn, H., Sato, S., Igarashi, A., Nakagawa, H., Tamaki, K., & Takehara, K. (1995) Growth regulation in scleroderma fibroblasts: increased response to transforming growth factor-β1. *J Invest Dermatol, 105*, 128-132.

Kireeva, M. L., Latinkic, B. V., Kolesnikova, T. V., Chen, C. C., Yang, G. P., Abler, A. S., & Lau, L. F. (1997) Cyr61 and Fisp12 are both ECM-associated signaling molecules: activities, metabolism, and localization during development. *Exp Cell Res, 233*, 63-77.

Komori, T., Yagi, H., Nomura, S., Yamaguchi, A., Sasaki, K., Deguchi, K., Shimizu, Y., Bronson, R. T., Gao, Y.-H., Inada, M., Sato, M., Okamoto, R., Kitamura, Y., Yoshiki, S., & Kishimoto, T. (1997) Target disruption of Cbfa1 results in a complete lack of bone formation owing to maturational arrest of osteoblasts. *Cell, 89*, 755-764.

Kondo, S., Kubota, S., Eguchi, T., Hattori, T., Nakanishi, T., Sugahara, T., & Takigawa, M. (2000) Characterization of a mouse *ctgf* 3'-UTR segment that mediates repressive regulation of gene expression. *Biochem Biophys Res Commun, 278*, 119-124.

Kondo, S., Kubota, S., Shimo, T., Nishida, T., Yosimichi, G., Eguchi, T., Sugahara, T., & Takigawa, M. (2002) Connective tissue growth factor increased by hypoxia may initiate angiogenesis in collaboration with matrix metalloproteinases. *Carcinogenesis, 23*, 769-776.

Kothapalli, D., Hayashi, N., & Grotendorst, G. R. (1998) Inhibition of TGF-β-stimulated CTGF gene expression and anchorage-independent growth by cAMP identifies a CTGF-dependent restriction point in the cell cycle. *Faseb J, 12*, 1151-1161.

Kubota, S., Hattori, T., Nakanishi, T., & Takigawa, M. (1999) Involvement of *cis*-acting repressive element(s) in the 3'-untranslated region of human connective tissue growth factor gene. *FEBS Lett, 450*, 84-88.

Kubota, S., Kawata, K., Yanagita, T., Doi, H., Kitoh, T., Takigawa, M. (2004) Abundant retention and release of connective tissue growth factor (CTGF/CCN2) by platelets, which promotes wound healing and tissue regeneration. *J Biochem (Tokyo), 136,* 279-282.

Kubota, S., Kondo, S., Eguchi, T., Hattori, T., Nakanishi, T., Pomerantz, R. J., Takigawa, M. (2000) Identification of an RNA element that confers post-transcriptional repression of connective tissue growth factor/hypertrophic chondrocyte specific 24 *(ctgf/hcs24)* gene: similarities to retroviral RNA-protein interactions. *Oncogene 19*, 4773-4786.

Kubota, S., Moritani, N. H., Kawaki, H., Mimura, H., Minato, M., & Takigawa, M. (2003) Transcriptional induction of connective tissue growth factor/hypertrophic chondrocyte-specific 24 gene by dexamethasone in human chondrocytic cells. *Bone, 33*, 694-702.

Kucich, U., Rosenbloom, J. C., Herrick, D. J., Abrams, W. R., Hamilton, A. D., Sebti, S. M., & Rosenbloom, J. (2001) Signaling events required for transforming growth factor-β stimulation of connective tissue growth factor expression by cultured human lung fibroblasts. *Arch Biochem Biophys, 395*, 103-112.

Kumar, S., Hand, A. T., Connor, J. R., Dodds, R. A., Ryan, P. J., Trill, J. J., Fisher, S. M., Nuttall, M. E., Lipshutz, D. B., Zou, C., Hwang, S. M., Votta, B. J., James, I. E., Rieman, D. J., Gowen, M., & Lee, J. C. (1999) Identification and cloning of a connective tissue growth factor-like cDNA from human osteoblasts encoding a novel regulator of osteoblast functions. *J Biol Chem, 274*, 17123-17131.

Lau, L. F., & Lam, S. C. (1999) The CCN family of angiogenic regulators: the integrin connection. *Exp Cell Res, 248*, 44-57.

Leask, A., Holmes, A., & Abraham, D. J. (2002) Connective tissue growth factor: a new and important player in the pathogenesis of fibrosis. *Curr Rheumatol Rep, 4*, 136-142.

Leask, A., Holmes, A., Black, C. M., & Abraham, D. J. (2003) Connective tissue growth factor gene regulation. Requirements for its induction by transforming growth factor-β2 in fibroblasts. *J Biol Chem, 278*, 13008-13015.

Mo, F. E., Muntean, A. G., Chen, C. C., Stolz, D. B., Watkins, S. C., & Lau, L. F. (2002) CYR61 (CCN1) is essential for placental development and vascular integrity. *Mol Cell Biol, 22*, 8709-8720.

Mori, T., Kawara, S., Shinozaki, M., Hayashi, N., Kakinuma, T., Igarashi, A., Takigawa, M., Nakanishi, T., & Takehara, K. (1999) Role and interaction of connective tissue growth factor with transforming growth factor-β in persistent fibrosis: A mouse fibrosis model. *J Cell Physiol, 181*, 153-159.

Moussad, E. E., & Brigstock, D. R. (2000) Connective tissue growth factor: what's in a name? *Mol Genet Metab, 71*, 276-292.

Mukudai, Y., Kubota, S., & Takigawa, M. (2003) Conserved repressive regulation of connective tissue growth factor/hypertrophic chondrocyte-specific gene 24 (*ctgf/hcs24*) enabled by different elements and factors among vertebrate species. *Biol Chem, 384*, 1-9.

Murphy, M., Godson, C., Cannon, S., Kato, S., Mackenzie, H. S., Martin, F., & Brady, H. R. (1999) Suppression subtractive hybridization identifies high glucose levels as a stimulus for expression of connective tissue growth factor and other genes in human mesangial cells. *J Biol Chem, 274*, 5830-5834.

Nakanishi, T., Kimura, Y., Tamura, T., Ichikawa, H., Yamaai, Y., Sugimoto, T., & Takigawa, M. (1997) Cloning of a mRNA preferentially expressed in chondrocytes by differential display-PCR from a human chondrocytic cell line that is identical with connective tissue growth factor (CTGF) mRNA. *Biochem Biophys Res Commun, 234*, 206-210.

Nakanishi, T., Nishida, T., Shimo, T., Kobayashi, K., Kubo, T., Tamatani, T., Tezuka, K., & Takigawa, M. (2000) Effects of CTGF/Hcs24, a product of a hypertrophic chondrocyte-specific gene, on the proliferation and differentiation of chondrocytes in culture. *Endocrinology, 141*, 264-273.

Nakanishi, T., Yamaai, T., Asano, M., Nawachi, K., Suzuki, M., Sugimoto, T., & Takigawa, M. (2001) Overexpression of connective tissue growth factor/hypertrophic chondrocyte-specific gene product 24 decreases bone density in adult mice and induces dwarfism. *Biochem Biophys Res Commun, 281*, 678-681.

Nakata, E., Nakanishi, T., Kawai, A., Asaumi, K., Yamaai, T., Asano, M., Nishida, T., Mitani, S., Inoue, H., & Takigawa, M. (2002) Expression of connective tissue growth factor/hypertrophic chondrocyte-specific gene product 24 (CTGF/Hcs24) during fracture healing. *Bone, 31*, 441-447.

Nawachi, K., Inoue, M., Kubota, S., Nishida, T., Yosimichi, G., Nakanishi, T., Kanyama, M., Kuboki, T., Yatani, H., Yamaai, T., & Takigawa, M. (2002) Tyrosine kinase-type receptor ErbB4 in chondrocytes: interaction with connective tissue growth factor and distribution in cartilage. *FEBS Lett, 528*, 109-113.

Nishida, T., Kubota, S., Fukunaga, T., Kondo, S., Yosimichi, G., Nakanishi, T., Takano-Yamamoto, T., & Takigawa, M. (2003) CTGF/Hcs24, hypertrophic chondrocyte-specific gene product, interacts with perlecan in regulating the proliferation and differentiation of chondrocytes. *J Cell Physiol, 196*, 265-275.

Nishida, T., Kubota, S., Kojima, S., Kuboki, T., Nakao, K., Kushibiki, T., Tabata, Y., & Takigawa, M. (2004) Regeneration of defects in articular cartilage in rat knee joints by CCN2 (connective tissue growth factor). *J Bone Miner Res, 19,* 1308-1319.

Nishida, T, Kubota, S., Nakanishi, T., Kuboki, T., Yosimichi, G., Kondo, S., & Takigawa, M. (2002) CTGF/Hcs24, a hypertrophic chondrocyte-specific gene product, stimulates proliferation and differentiation, but not hypertrophy of cultured articular chondrocytes. *J Cell Physiol, 192*, 55-63.

Nishida, T, Nakanishi, T, Asano, M, Shimo, T, & Takigawa, M (2000) Effects of CTGF/Hcs24, a hypertrophic chondrocyte-specific gene product, on the proliferation and differentiation of osteoblastic cells *in vitro*. *J Cell Physiol, 184*, 197-206.

Nishida, T., Nakanishi, T., Shimo, T., Asano, M., Hattori, T., Tamatani, T., Tezuka, K., & Takigawa, M. (1998) Demonstration of receptors specific for connective tissue growth factor on a human chondrocytic cell line (HCS-2/8). *Biochem Biophys Res Commun, 247*, 905-909.

Oemar, B. S., & Luscher, T. F. (1997) Connective tissue growth factor. Friend or foe? *Arterioscler Thromb Vasc Biol, 17*, 1483-1489.

Oemar, B. S., Werner, A., Garnier, J. M., Do, D. D., Godoy, N., Nauck, M., Marz, W., Rupp, J., Pech, M., & Luscher, T. F. (1997) Human connective tissue growth factor is expressed in advanced atherosclerotic lesions. *Circulation, 95*, 831-839.

Pawar, S., Kartha, S., & Toback, F. G. (1995) Differential gene expression in migrating renal epithelial cells after wounding. *J Cell Physiol, 165*, 556-565.

Pennica, D., Swanson, T. A., Welsh, J. W., Roy, M. A., Lawrence, D. A., Lee, J., Brush, J., Taneyhill, L. A., Deuel, B., Lew, M., Watanabe, C., Cohen, R. L., Melhem, M. F., Finley, G. G., Quirke, P., Goddard, A. D., Hillan, K. J., Gurney, A. L., Botstein, D., & Levine, A. J. (1998) WISP genes are members of the connective tissue growth factor family that are up-regulated in Wnt-1-transformed cells and aberrantly expressed in human colon tumors. *Proc Natl Acad Sci U S A, 95*, 14717-14722.

Perbal, B. (2001) NOV (nephroblastoma overexpressed) and the CCN family of genes: structural and functional issues. *Mol Pathol, 54*, 57-79.

Perbal, B. (2004) CCN proteins: multifunctional signalling regulators. *Lancet, 363*, 62-64.

Ricupero, D. A., Rishikof, D. C., Kuang, P. P., Poliks, C. F., & Goldstein, R. H. (1999) Regulation of connective tissue growth factor expression by prostaglandin E_2. *Am J Physiol, 277*, L1165-1171.

Riser, B. L., & Cortes, P. (2001) Connective tissue growth factor and its regulation: a new element in diabetic glomerulosclerosis. *Ren Fail, 23*, 459-470.

Riser, B. L., Denichilo, M., Cortes, P., Baker, C., Grondin, J. M., Yee, J., & Narins, R. G. (2000) Regulation of connective tissue growth factor activity in cultured rat mesangial cells and its expression in experimental diabetic glomerulosclerosis. *J Am Soc Nephrol, 11*, 25-38.

Ryseck, R. P., Macdonald-Bravo, H., Mattei, M. G., & Bravo, R. (1991) Structure, mapping, and expression of *fisp-12*, a growth factor-inducible gene encoding a secreted cysteine-rich protein. *Cell Growth Differ, 2*, 225-233.

Safadi, F. F., Xu, J., Smock, S. L., Kanaan, R. A., Selim, A. H., Odgren, P. R., Marks, S. C., Jr., Owen, T. A., & Popoff, S. N. (2003) Expression of connective tissue growth factor in bone: its role in osteoblast proliferation and differentiation *in vitro* and bone formation *in vivo*. *J Cell Physiol, 196*, 51-62.

Schild, C., & Trueb, B. (2002) Mechanical stress is required for high-level expression of connective tissue growth factor. *Exp Cell Res, 274*, 83-91.

Segarini, P. R., Nesbitt, J. E., Li, D., Hays, L. G., Yates, J. R. 3rd, & Carmichael, D. F. (2001) The low density lipoprotein receptor-related protein/α_2-macroglobulin receptor is a receptor for connective tissue growth factor. *J Biol Chem, 276*, 40659-40667.

Shi-wen, X., Pennington, D., Holmes, A., Leask, A., Bradham, D., Beauchamp, J. R., Fonseca, C., du Bois, R. M., Martin, G. R., Black, C. M., & Abraham, D. J. (2000) Autocrine overexpression of CTGF maintains fibrosis: RDA analysis of fibrosis genes in systemic sclerosis. *Exp Cell Res, 259*, 213-224.

Shimo, T., Kubota, S., Kondo, S., Nakanishi, T., Sasaki, A., Mese, H., Matsumura, T., & Takigawa, M. (2001a) Connective tissue growth factor as a major angiogenic agent that is induced by hypoxia in a human breast cancer cell line. *Cancer Lett, 174*, 57-64.

Shimo, T., Nakanishi, T., Kimura, Y., Nishida, T., Ishizeki, K., Matsumura, T., & Takigawa, M. (1998) Inhibition of endogenous expression of connective tissue growth factor by its antisense oligonucleotide and antisense RNA suppresses proliferation and migration of vascular endothelial cells. *J Biochem (Tokyo), 124*, 130-140.

Shimo, T., Nakanishi, T., Nishida, T., Asano, M., Kanyama, M., Kuboki, T., Tamatani, T., Tezuka, K., Takemura, M., Matsumura, T., & Takigawa, M. (1999) Connective tissue growth factor induces the proliferation, migration, and tube formation of vascular endothelial cells *in vitro*, and angiogenesis *in vivo*. *J Biochem (Tokyo), 126*, 137-145.

Shimo, T., Nakanishi, T., Nishida, T., Asano, M., Sasaki, A., Kanyama, M., Kuboki, T., Matsumura, T., & Takigawa, M. (2001b) Involvement of CTGF, a hypertrophic chondrocyte-specific gene product, in tumor angiogenesis. *Oncology, 61*, 315-322.

Shimo, T., Wu, C., Billings, P. C., Piddington, R., Rosenbloom, J., Pacifici, M., & Koyama, E. (2002) Expression, gene regulation, and roles of Fisp12/CTGF in developing tooth germs. *Dev Dyn, 224*, 267-278.

Shinozaki, M., Kawara, S., Hayashi, N., Kakinuma, T., Igarashi, A., & Takehara, K. (1997) Induction of subcutaneous tissue fibrosis in newborn mice by transforming growth factor β--simultaneous application with basic fibroblast growth factor causes persistent fibrosis. *Biochem Biophys Res Commun, 240*, 292-297.

Surveyor, G. A., & Brigstock, D. R. (1999) Immunohistochemical localization of connective tissue growth factor (CTGF) in the mouse embryo between days 7.5 and 14.5 of gestation. *Growth Factors, 17*, 115-124.

Surveyor, G. A., Wilson, A. K., & Brigstock, D. R. (1998) Localization of connective tissue growth factor during the period of embryo implantation in the mouse. *Biol Reprod, 59*, 1207-1213.

Takigawa, M. (2003) CTGF/Hcs24 as a multifunctional growth factor for fibroblasts, chondrocytes and vascular endothelial cells. *Drug News Perspect, 16*, 11-21.

Takigawa, M., Nakanishi, T., Kubota, S., & Nishida, T. (2003) Role of CTGF/HCS24/ecogenin in skeletal growth control. *J Cell Physiol, 194*, 256-266.

Takigawa, M., Pan, H.-O., Kinoshita, A., Tajima, K., & Takano, Y. (1991) Establishment from a human chondrosarcoma of a new immortal cell line with high tumorigenicity *in vivo*, which is able to form proteoglycan-rich cartilage-like nodules and to respond to insulin *in vitro*. *Int. J. Cancer, 48*, 717-725.

Takigawa, M., Tajima, K., Pan, H.-O., Enomoto, M., Kinoshita, A., Suzuki, F., Takano, Y., & Mori, Y. (1989) Establishment of a clonal human chondrosarcoma cell line with cartilage phenotypes. *Cancer Res., 49*, 3996-4002.

Tanaka, S., Sugimachi, K., Kameyama, T., Maehara, S., Shirabe, K., Shimada, M., Wands, J. R., & Maehara, Y. (2003) Human WISP1v, a member of the CCN family, is associated with invasive cholangiocarcinoma. *Hepatology, 37*, 1122-1129.

Tanaka, S., Sugimachi, K., Saeki, H., Kinoshita, J., Ohga, T., Shimada, M., & Maehara, Y. (2001) A novel variant of WISP1 lacking a Von Willebrand type C module overexpressed in scirrhous gastric carcinoma. *Oncogene, 20*, 5525-5532.

Twigg, S. M., Joly, A. H., Chen, M. M., Tsubaki, J., Kim, H. S., Hwa, V., Oh, Y., & Rosenfeld, R. G. (2002) Connective tissue growth factor/IGF-binding protein-related protein-2 is a mediator in the induction of fibronectin by advanced glycosylation end-products in human dermal fibroblasts. *Endocrinology, 143*, 1260-1269.

Wong, M., Kireeva, M. L., Kolesnikova, T. V., & Lau, L. F. (1997) Cyr61, product of a growth factor-inducible immediate-early gene, regulates chondrogenesis in mouse limb bud mesenchymal cells. *Dev. Biol. 192,* 492-508.

Xu, J., Smock, S. L., Safadi, F. F., Rosenzweig, A. B., Odgren, P. R., Marks, S. C., Jr., Owen, T. A., & Popoff, S. N. (2000) Cloning the full-length cDNA for rat connective tissue growth factor: implications for skeletal development. *J Cell Biochem, 77*, 103-115.

Yamaai, T., Asano, M., Nakanishi, T., Nishida, T., Yosimichi, G., Hattori, T., Sugimoto, T., & Takigawa, M. (1999) Gene expression pattern of cartilage-derived growth factor CTGF/Hcs24 during mouse development. *Jap. J. Oral Biol., 41,* 464. (in Japanese)

Yamaai, T., Nakanishi, T., Asano, M., Nawachi K., Yosimichi, G., Ohyama, K., Komori, T., Sugimoto, T., & Takigawa, M. (2005) Gene expression pattern of cartilage-derived growth factor CTGF/Hcs24 during mouse development. *J. Bone Miner Metab., in press.*

Yamashiro, T., Fukunaga, T., Kobashi, N., Kamioka, H., Nakanishi, T., Takigawa, M., & Takano-Yamamoto, T. (2001) Mechanical stimulation induces CTGF expression in rat osteocytes. *J Dent Res, 80*, 461-465.

Yosimichi, G., Nakanishi, T., Nishida, T., Hattori, T., Takano-Yamamoto, T., & Takigawa, M. (2001) CTGF/Hcs24 induces chondrocyte differentiation through a p38 mitogen-activated protein kinase (p38MAPK), and proliferation through a p44/42 MAPK/extracellular-signal regulated kinase (ERK). *Eur J Biochem, 268*, 6058-6065.

Acknowledgments

This work was supported in part by grants from the program Grants-in-Aid for Scientific Research (S) and Exploratory Research of the Ministry of Education, Culture, Sports, Science and Technology of Japan and Japan Society for the Promotion of Science. The author wish to thank Drs. Satoshi Kubota, Takashi Nishida, Yoshiki Mukudai, Tomoichiro Yamaai and other faculty members and graduate students who collaborate with me at Okayama University Graduate School of Medicine and Dentistry and Dental School. I also give thanks to Ms. Yuki Nonami for technical and secretarial assistance.

In: Trends in Stem Cell Research
Editor: Erik V. Greer, pp. 129-146

ISBN 1-59454-315-1

Chapter V

Peripheral Blood Progenitor Cell Collection in Small Children

Julián Sevilla*, Miguel Angel Díaz, Sandra Fernández-Plaza, Marta González-Vicent and Luis Madero

Hospital Infantil Universitario Niño Jesús, Madrid, Spain

Abstract

Peripheral blood progenitor cells (PBPC) are increasingly used as a source of stem cells for either autologous or allogeneic hematopoietic transplantation. Technical problems are derived from the size of the patient or donor in the pediatric setting. Thereby, vascular access, anticoagulation and extracorporeal volume need special consideration in children. Several groups have published over the last years their experience in mobilization and transplantation in small children. In this paper our own experience and that of other groups is reviewed. Mobilization with cytokines alone (in donors and patients) or in combination with chemotherapy (patients) has been reported to be efficient and well tolerated by these children. Several variables are related to the collection in donors and patients. This issue is discussed in this paper based on published studies. PBPC collection is a safe procedure as reported in the literature, when performed by experienced apheresis teams.

Key words: Leukapheresis, LVL, children, G-CSF, mobilization.

* Hospital Infantil Universitario Niño Jesús. Avd Menéndez Pelayo 65. Madrid. 28009. Spain. FAX number: +34 915035902 e-mail: jsevilla.hnjs@salud.madrid.org

Introduction

Leukapheresis for peripheral blood progenitor cells (PBPC) collection is currently the usual procedure to obtain stem cells to restore hematopoiesis after high dose chemotherapy. Faster granulocyte and platelet engraftment lead most trasplant centers to collect PBPC instead of bone marrow stem cells. Moreover, leukapheresis is becoming easier with the experience of the apheresis teams and new software facilities, and avoid general anesthesia and an operating theatre. All these reasons diminish the economical resources, complications and secondary effects related to the procedure. Some technical aspects such as vascular access, anticoagulation and volume of extracorporeal circulation are important, and required special considerations, in small children. In allogeneic transplantation is also essential to determinate the patient/donor size for evaluating the collected dose of PBPC in small donors. In this article we review some technical and clinical aspects related to PBPC collection in pediatric population.

Technical Aspects

The same separators used for adults are used in pediatric leukapheresis with certain procedural or disposable modifications [1, 2, 3, 4, 5]. Factors that must be evaluated are extracorporeal circuit volume, blood flow rates, type of anticoagulant and vascular access. Other important issue to be considered is the emotional stress for them and their relatives.

Extracorporeal Circuit Volume

One of the main limiting factors for leukapheresis in children is the patient whole blood volume [1,2,3,6]. Volume shifts depend on the machine used. Changes in blood volume caused by discontinuous flow machines are poorly tolerated in children. Therefore these apheresis machines are limited to larger children. Recently Koristek *et al* have described a discontinuous manual technique for PBPC collection in low weight children (<10 kg) [7]. It consists on the collection of 8 ml/kg of total blood (10% of total blood volume), and immediately injected into a transfer bag and centrifuged. After it, the buffy-coat is pooled in a collection bag, and the remaining plasma and red blood cells are reinfused. Then a subsequent cycle starts. With this protocol, the authors obtained sufficient number of PBPC for hematopoietic rescue after high dose chemotherapy, and higher dose of stem cells than with autologous whole blood transfusions (an option developed by the UKC CSG Baby Brain Protocol)[7].

However, most centers used continuous flow cell separators in small children, since the maintained flow of these separators avoids large volume shifts during the procedure.

As the extracorporeal volume may exceed up to 40% of a child blood volume (for instance a child of 10 kg undergoing a leukapheresis in a machine using a blood warmer), is essential, in some cases, to consider the child total blood volume and his/her hematocrit to perform the priming of the circuit with red blood cells. Initially it was related to body weight

[3] either 25 kg [8] or 20 kg [9, 10]. Other authors prefer to prime always if the extracorporeal volume is higher than 15% [1, 2] or 25% [6] of child total blood volume or, if the patient is anemic. Recently Orbach *et al* [11] have proposed to prime the extracorporeal line, in children weighing less than 15 kg, only with high molecular weight hydroxyethyl starch or 4% albumin. Transfusion of red blood cells was only performed before the procedure if hemoglobin was less than 12 gr/dl. For these authors, it may facilitate apheresis but it does not avoid transfusion of red blood cells in 62.5% [15/24] of children. Two patients also required red blood cell priming of the circuit. This approach might avoid some unnecessary transfusions and the risk associated with it. Therefore, it may lead to decrease the long term complications associated with the procedure without worsen the results of the leukapheresis.

Blood Flow Rates

It is important in pediatric leukapheresis the use of apheresis machines that allowed low flow rates. It is due to the small vascular access and the citrate toxicity (latter discussed).

Type of Anticoagulation

Anticoagulant Citrate Dextrose formula A (ACD-A) is the most widely used anticoagulant in apheresis. In children, it has a clear secondary effect, hypocalcemia. It is due to its calcium binding effect, and the lower hepatic metabolism of the citrate in small children. Minor clinical symptoms of hypocalcemia in infants and small children may be understimated or recognized leading to severe toxicity. Therefore the use of citrate as the only anticoagulant requires careful attention to any clinical symptom of hypocalcemia (abdominal pain, dysphoria, cardiac rhythm disturbances or hypotension). In some cases ionized calcium levels must be monitorized during the procedure. [1, 3, 6].

There have been to answers to this problem. Some authors prefer to increased the ionized calcium level either by bolus administration of calcium during the procedure [1, 3, 11, 12, 13] or by continuous infusion [3, 14]. Other groups have used either heparin alone or a combination of heparin with ACD at higher rate (30/1) to avoid, or at least to decrease, the dose of citrate infused during the collection [1, 3, 6, 8, 10].

Vascular Access

Although peripheral veins can be used in adolescents and older children, most pediatric patients require a double lumen central venous catheter for the collection [1, 3, 6]. These catheters should be sufficiently stiff to avoid collapse under the negative draw pressure of the cell separator, and include port offset to minimize recirculation. The majority of apheresis teams use one subcutaneous tunneled, double-lumen central venous catheter [3]. However other groups have preferred different approaches. Some of them use either one central venous line [11, 12], or an arterial line for collection [15], and return to a peripheral vein. And others use a double lumen central venous catheter [16]. In our experience, the latter approach is appropriate for the collection in children, if this access is later used for transplantation [10, 17].

Gorlin *et al* suggested an algorithm for venous access selection in pediatric population [8]. The smallest children should undergo the procedure with a 10 Fr Hickman double-lumen catheter or a 7-9 Fr single-lumen draw line and a 7 Fr double lumen return [8]. Our team usually performs these procedures through a 7 Fr double-lumen catheter placed in femoral vein.

Psycho-Social Aspects

PBPC collection by leukapheresis may be stressful for children and his or her relatives. The procedure must be clearly explained to the family by nurses and physicians. It relaxes them avoiding some grade of stress on the child. It is also important for the child to be housed in a familiar environment, and previously meet the nursing staff. The use of some familiar toys or games or, the presence of a friend, or brother or sister, usually makes children to feel more comfortable. All these measures may help the child to deal with his or her fears and anxieties.

PBPC Mobilization

Harvesting the maximum dose of $CD34^+$ cells is necessary after PBPC collection in order to shorten the granulocyte and platelet engraftment, or facilitate subsequent graft manipulation. In adults, a minimum dose of 2-2.5 $x10^6$ $CD34^+$ cells/kg have been suggested for hematopoietic rescue after high dose chemotherapy based on granulocyte and platelet engraftment analysis [18, 19, 20]. In children, our group proposed a minimum dose of 2 $x10^6$ $CD34^+$ cells/kg after early and long-term engraftment evaluation [21], but similarly to adults, a dose of 5 $x10^6$ $CD34^+$ cells/kg is considered optimum [19, 21]. However, in order to achieve these doses, an adequate mobilization schedule and leukapheresis efficacy is necessary. Several mobilization regimens have been proposed for it based on chemotherapy, cytokines or a combination of both.

Initially, children underwent leukapheresis on the recovery phase of neutropenia after tumor specific chemotherapy [22, 23]. The addition of G-CSF to chemotherapy for priming was later performed with encouraging results [12, 24, 25]. It has been reported to result in higher number of $CD34^+$ cells, but it has several disadvantages such as a longer mobilization period, with several toxicities and, in some cases febrile neutropenia. Moreover, optimal timing for collection is not easily scheduled. All these issues make the collection difficult in some apheresis units. The use of G-CSF alone (5-16 μg/kg/d for 5-8 days) for mobilization came later and may avoid these problems [26, 27].

Although theoretically the simultaneous used of other cytokines could add some priming effect. The addition of GM-CSF to a regular G-CSF schedule neither enhanced the stem cell yield nor the hematological recovery in our series [10, 28]. These results were similar to those reported in a randomized trial in adult patients [29].

However, some studies have suggested a dose-response to G-CSF for priming in different groups of patients [30, 31, 32, 33]. Weaver et al performed a randomized study in adults finding that breast cancer patients receiving 10 μg/kg/d had a 3.3-fold increase in the probability of not achieving the optimum dose of $CD34^+$ cells/kg as compared to patients receiving 20-40 μg/kg/d [34]. However, Halle et al obtained similar $CD34^+$ cell yield between pediatric patients receiving 10 or 20 μg/kg/d [35]. These authors concluded that mobilization with G-CSF at the highest dose could minimize the duration of priming, but would not reduce the number of leukapheresis [35]. A prospective randomized trial in children performed by Kawano et al showed a dose-response effect for G-CSF in a range of 75 to 400 $\mu g/m^2$ independently of chemotherapy [36]. Kröger et al recommended to divide the total daily dose based on their study in adult breast cancer patients comparing 2x5μg/kg versus 1x10 μg/kg [37]. Our data, recently published, are also coincident with the dose response effect and the total daily dose division [38]. In that study we found that children mobilized with 12 μg/kg/12h for four days had better collections than a prior series mobilized with 12 μg/kg/d [28, 38]. The tolerance of this schema was good, with mild bone pain or myalgia in 23% of patients [38]. Similar results have been found with this schedule by other study in children [39]. We have also reviewed our own experience exclusively in patients weighting less than 20 kg. In this study the $CD34^+$ cell count in peripheral blood before the procedure, a variable considered to be related to the efficiency of G-CSF priming when the mobilization is performed on patients on steady-state hematopoiesis, was also higher in those children mobilized with higher doses of G-CSF (24 μg/kg/day) than those mobilized with lower doses (12 μg/kg/day) (86 vs 40 median $CD34^+$ cell/μl respectively; $p<0.05$) [10].

There have been several retrospective studies performed to defined risk factors related to poor mobilization of PBPC [19, 40]. The number of chemotherapy cycles (up to six cycles), the use of alkylating agents, prior radiotherapy, type of malignancy, age, and bone marrow involvement by the tumor have been described as risk factors for defficient priming [19, 40]. Melphalan, carmustine, mechlorethamine, lomustine, procarbazine, platinum compounds and chlorambucil hinder the collection of PBPC more than other agents [40, 41]. Drake et al [42] described a chemotherapy scoring system that was later validated by Clark and Brammer [41]. Recently, we analized the variables related to better collection in small children in our unit [10]. Several variables were associated to higher $CD34^+$ cell count/μl in peripheral blood before apheresis (as above mention a variable related to the efficiency of priming with G-CSF alone). The association with the priming schedule has been previously discussed. Those children mobilized with higher and divided doses, had higher $CD34^+$ cell count/μl in peripheral blood before apheresis, than those mobilized with lower doses. Other variables are listed in table 1. Patients previously treated with more than 6 cycles of chemotherapy ($p=0.004$) and radiotherapy ($p=0.008$) had lower $CD34^+$ cell counts than children not previously treated. Children diagnosed with solid tumors have a median $CD34^+$ cell count/μl higher than patients diagnosed with hematologic malignancies (80 vs 32), although this latter association may be related to the treatment previously received instead of the disease, since several patients with solid tumors had never received chemotherapy. In this series, age was also related with the quality of priming in univariate and multivariate analysis. Younger patients had better counts than older ones. This variable was also found to influence the yield in a previous study [38].

Table 1. Factors related to mobilization in children

Variable	Better	Worst
Age	Younger	Older
Diagnosis	Solid tumor	Hematologic malignancy
Prior chemotherapy	None or less than 6 cycles	More than 6 cycles
Prior radiotherapy	No	Yes
Priming schedule	Higher and daily divided doses (12 μg/kg/12 hours for four days)	12 μg/kg/24 hours for four days

Pediatric Donors

Leukapheresis is also substituting bone marrow harvest for stem cell colletion for allogeneic hematopoietic transplant. In most adult donors PBPC collection is performed after priming with G-CSF at a dose of 10-12 μg/kg per day [43, 44]. Although mobilization of pediatric donors for PBPC harvest was not recommended for ethical reasons by a European panel, this procedure is increasingly used [45]. Some series have been published regarding this issue [9, 35, 46-51]. Most pediatric groups follow the same protocol for PBPC mobilization proposed for adult donors (10 μg/kg per day for 4-5 days) although others have used a daily divided dose such as 6 μg/kg twice daily [46].

The dose effect of G-CSF has been also studied in healthy donors. Waller et al used a regimen consisting of G-CSF 12 μg/kg twice daily with a reliably high numbers of PBPC in adult donors mobilized with this regimen compared to the regular schedule (10 μg/kg per day), although secondary effects were observed more frequently in the population mobilized with the highest dose [52]. Tanaka et al also found dose-effect of G-CSF in healthy volunteers [53]. Interestingly, these authors reported the release of less mature progenitor cells ($CD34^+/33^-$, or $CD34^+$/HLA-DR^- cells) when higher doses of G-CSF (100 $\mu g/m^2$ vs 200 $\mu g/m^2$) were used [53]. Arbona et al have reported better yields in healthy donors (mainly adults) with regimens with higher and divided doses for priming, without significant increased in the toxicities [54]. In the only prospective randomized study dividing the daily dose, Kröger et al have observed that $CD34^+$ yield is higher in adult donors in the divided dose group requiring fewer apheresis, without increasing toxicities or cost [55]. Similar results were obtained by Lee et al in healthy adult donors and children with non-malignant diseases [56]. However these results are not consistent in other adult series [57].

De la Rubia et al has recently reported a similar pattern of $CD34^+$ cell mobilization in patients younger and older than 18 years [50]. Increased age was related in their study with a more frequent requirement of repeated apheresis to achieve the required $CD34^+$ cell dose

[50]. In contrast, Kawano et al found no difference between pediatric and adult donors for the collection after G-CSF priming [9]. A recent retrospective study of the Spanish National Donor Registry focused on factors predicting peripheral blood progenitor cell collection from pediatric donors [51]. In this paper several variables were studied in 105 aphereses performed in 76 healthy children. Median age was 10 year old (range 1-18). Their median body weight was 42 kg (range 9-89). Variables were study in association with the amount of $CD34^+$ cells collected, related either to donor or recipient body weigh. Younger age, large volume leukapheresis and the first apheresis day versus subsequent ones, were those related in a multivariate analysis with better collections [51]. In all cases the secondary effects of mobilization were mild and managed with minor analgesics, as reported in part in a prior study by this group [50, 51].

Secondary effects related to G-CSF mobilization have higher incidence in older than younger patients. Murata et al described acceptable short-term toxicities with G-CSF mobilization, with headache being more frequent in donors younger than 35 year old in their retrospective analysis of adverse events in 94 donors [58]. However, for most authors, and in our own experience, G-CSF seems to have fewer secondary effects on children donors [9, 48, 50, 58, 59]. Several rare adverse events have been reported following G-CSF priming in normal donors such as splenic rupture, acute iritis, acute gouty arthritis and anaphylactoid reaction [60]. The only rare adverse event reported in pediatric donors was observed in our institution. He was a 10 month old boy who developed a transient cutaneous papular, erythematous eruption consistent with an activated lymphocyte skin infiltration in the biopsy [61].

Glycosylated G-CSF (lenograstim) has been also used for mobilization at doses between 3-10 μg/kg/d. It provides effective donor mobilization of PBPC, but has neither diminished secondary effects nor improved $CD34^+$ cell harvest [62, 63, 64]. However, lenograstim has been reported to mobilize PBPC more efficiently than the nonglycosylated form when used at the same dose (10 μg/kg per day) [65].

Apheresis of the Smallest Children

There is scanty published experience on pediatric leukapheresis for PBPC collection. As above discussed the restriction in the use of leukapheresis for PBPC collection in children are the difficulty of adequate vascular access, volume shifts and the risk of metabolic problems. Lasky et al [22] and Takaue et al [23] first published their experience on aphereses for PBPC collection in the last 80´s. Some papers have reported the experience of other groups in this field over the next years [12, 13, 15, 25, 26, 46, 66-71]. Although some low weight patients have been included in different pediatric papers, the experience on the smallest children have been summarized in some studies. Theses series are heterogeneous since different previous treatments, diagnosis, mobilization schedules and cell separators were used. All these reasons and the inter-patient variability hinder any critical comparison (table 2). Demeocq et al [12] reported their experience with 20 children weighing less than 25 kg (median 14.5 kg). They had been diagnosed with different malignant diseases and underwent leukapheresis with a continuos flow blood separator (Cobe Spectra). Mobilization had been performed with chemotherapy or chemotherapy plus G-CSF. Either peripheral veins for access and return, or a central catheter for vascular access with return through peripheral veins or other central catheter were used for vascular access. The extracorporeal line had been primed in 19 patients with red blood cells. Acid-citrate-dextrose (ACD-A) had been used as an anticoagulant in a ratio of 1:14, and calcium gluconate had been administered in bolus during the procedure. Median blood flow rate was 13.85 ml/min (ranged 7 to 22). Apheresis had been of 2-3 hour duration, and every patient had undergone a mean of 3.3 procedures (range 1-4). The procedures had been performed under mild sedation, with no serious complications, except for a platelet loss of 36%. If platelet count fell below 50 x 10^9/l the autologous platelet-rich plasma was returned to the patient (12). Takaue et al later reported their series of 38 children weighing less than 20 kg (15). A total of 81 aphereses had been performed to collect PBPCs with a continuos flow blood separator (CS-3000; Fenwall Laboratory, Dearfield, IL), mainly after chemotherapy mobilization. The extracorporeal line had been primed with red blood cells. ACD was used as the anticoagulant in a ratio 1:6.7. The median blood volume processed per kg patient body weight was 270 ml. And, the median blood flow rate was 30 ml/min (range 17-46). Therefore, aphereses were of less than 3 hour duration. The mean number of apheresis per patient was 2.1. Median CD34$^+$ cell yield was 15x10^6/kg (range 2-107). Even through an arterial access, no significant problems were reported related to the puncture site and vascular access. Profound cytopenia was reported in all cases. Thrombopenia leaded to platelet-rich plasma reinfusion. Two cases of transient and reversible hemodynamic secondary adverse events (cyanosis and hypotension) were reported [15].

Table 2. Main characteristics of published series on PBPC collection in small children

Year	Author	Higher weight	Mobilization	Cell separator	Antico-agulation	Main adverse event
1994	Demeocq et al [12]	25 kg	Chemotherapy Chemotherapy +G-CSF	Cobe – Spectra	ACD-A	Thrombopenia
1995	Takaue et al [15]	20 kg	Chemotherapy	CS-3000	ACD-A	Hypotension Thrombopenia
1996	Nussbaumer et al [69]	10 kg	Chemotherapy +G-CSF	CS-3000	ACD-A	Thrombopenia
1996	Gorlin et al [8]	25 kg	Chemotherapy +G-CSF	Cobe-Spectra	ACD-A + heparin	Related to vascular access
1997	Urban et al [70]	20 kg	Chemotherapy +G-CSF	CS-3000 plus	ACD-A	Thrombopenia
1998	Kanold et al [13]	15 kg	Chemotherapy +G-CSF Chemotherapy +GM-CSF	Cobe-Spectra	ACD	Thrombopenia
1998	Díaz et al [17]	25 kg	G-CSF	Cobe-Spectra	ACD	Thrombopenia
2003	Sevilla et al [10]	20 kg	G-CSF or G-CSF and GM-CSF	Cobe-Spectra	ACD alone or ACD-A + heparin	Hemodynamic and thrombopenia

Nussbaumer et al to perform the collection in three children of less than 10 kg body weight diagnosed with neuroblastoma with a CS 3000 Plus blood cell separator (Baxter Healthcare Fenwall Division, Deerfield, IL, USA) [69]. Leukapheresis were performed after chemotherapy and G-CSF priming. A total of 15 procedures were performed. Sedation was not necessary in any patient. The extracorporeal line of the cell separator had been primed with a mixture of two units of red blood cells and 500 ml of 5% albumin adjusted to the patient hematocrit using normal saline, and ACD had been used as anticoagulation at a 1:10 ratio. Calcium was not administered during the procedure. A median of 4 collections was needed per patient (range: 4-7) to achieved the target mononuclear cell dose ($\geq 4x10^8$/kg). The blood flow rate was 20 ml/min. A mean of 1617 ml were processed per patient and the mean duration of apheresis was 88.5 minutes. No adverse events were observed related to vascular access, citrate toxicity or hemodynamic complications. Platelet depletion was the main secondary effect of the procedure in this series [69].

Chemotherapy plus G-CSF was used by Gorlin et al for mobilization in 14 further patients weighing less than 25 kg [8]. Afterwards PBPC collection were performed using a COBE Spectra (COBE BCT Inc. Lakewood, CO), primed with red blood cells [8]. Anticoagulation was performed with ACD and heparin to reduce citrate toxicity. It allowed the ratio to be increased to 25-30:1, decreasing the total dose of citrate infused. The mean number of procedures per patient was 6, and a mean of 561 ml/kg was processed. Blood flow rates ranged from 20-50 ml/min. Median $CD34^+$ cell/kg per collection was 4.5 x 10^6. Even after a mean of 6 leukaphereses, only 80% of patients achieved 2.5 x 10^6 $CD34^+$ cell/kg in this series. These patients suffered several central venous catheter related complications such as occlusion, placement problems, and even pneumothorax. But the only complication observed during the procedure was one single case of citrate toxicity developed by the smallest patient [8].

Another series of children weighing less than 20 kg was published by Urban et al in 1997 [70]. Mobilization with G-CSF + chemotherapy was used in 7 patients. Nine children underwent the collection using a continuos flow blood separator (Fenwall CS-3000 plus). In this series the extracorporeal line ha been primed with a mixture of washed red blood cells, albumin and normal saline. ACD-A had been used as anticoagulant (ratio 10:1). Children had received calcium prophylaxis prior to the procedure and the autologous platelet-rich plasma had been automatically reinfused. A median of 4 aphereses was performed in every patient. The median blood flow rate was 25 ml/min (range: 22-29). A median duration of 112 minutes was employed for each procedure. A median of 10.34 (0.45-62.7) x 10^6 $CD34^+$ cell/kg was collected after 4 apheresis [70]. No complications were described related to central access insertion (subclavian and jugularis). No adverse events related to citrate toxicity were observed, and the only side effect reported was platelet depletion (median platelet loss 23%).

A retrospective analysis of 109 leukaphereses in 30 children was published by Kanold et al in 1997 [13]. In this study this group reviewed their experience that had been previously partially reported [12]. PBPC harvest had been performed using a continuos flow blood separator (Cobe Spectra). Different mobilization regimens had been used mainly dependent on different chemotherapies and G-CSF or GM-CSF. Based on their own experience, they introduced several modifications in their protocol. Initially, priming had been performed only in patients with a body weight lower than 15 kg and, finally, they used a more restrictive criteria, priming only those patients with a blood volume lower than 1 liter and/or a hemoglobin level lower than 10 g/dl. Most procedures were performed through a central line, although it was placed at diagnosis for treatment. ACD was used as anticoagulant with a median ratio of 13:1 (range: 7-26). To decrease metabolic problems related to hypocalcemia, calcium was administered systematically before and during the procedure and the ACD/blood ratio was used at the minimally active ratio. Median blood flow rate was 13 ml/min (range: 7-39). Median single procedure duration was 150 min (range 39-255), and a median of 1.75 blood volumes (range: 0.4-4.6) was processed. A median of 2 leukapheresis was performed in every patient. The median $CD34^+$ cell yield in a series of leukapheresis sessions [1-4] was $5.9x10^6$ $CD34^+$ cell/kg. No signs of citrate toxicity or hemodynamic side effects were reported. Platelet count diminished after each procedure with a median 37.6% loss (range 13.4-82.5), but no platelet transfusion was needed. In this paper the authors also discussed their experience with large volume leukapheresis (LVL) sessions (more than 3 total blood volumes processed). Tolerance was excellent and none of the children suffered adverse events. In six patients, a single LVL was enough to achieve the target $CD34^+$ cell dose for transplantation. Platelet count dropped significantly, median 37.5% (33.9-54.9). Finally, the authors also tried to clarify the factors that may affect the PBPC collection [13]. They could not find an association between patient's age, diagnosis, duration of preleukapheresis treatment and mobilization regimens and $CD34^+$ cell yield.

As mentioned above, we have extended our previous published data in a recent study [10, 17]. Eighty-nine large volume leukapheresis, performed in sixty six children weighing less than 25 kg were analyzed in this paper [10].The cell separator was primed with red blood cells. Initially ACD was used as anticoagulant [17], but since June 1997 a mixture of 500 ml ACD and 5000 iu of preservative free heparin was used for anticoagulation. It allowed to decrease the total dose of citrate. In this series 46 procedures were performed with ACD and 43 with ACD plus heparin. on 32 patients to 88 LVL procedures performed on patients weighing less than 25 kg[10]. Vasovagal events (either pallor and/or diaphoresis, mild hypotension or tachycardia) were the most frequent adverse events related with leukapheresis. However, only three cases of systolic hypotension were recorded as severe adverse events during the procedure, and these patients rapidly respond to plasma volume expanders. Pain or hematoma at a venipuncture site was diagnosed in four patients, and three patients suffered minor bleeding after immediate removal of central venous catheters (femoral) that did not require treatment. Median platelet drop was 52%, and it was only related to the total blood volumes processed (p=0.0001). No serious complications were related to vascular access, and no procedure needed to be interrupted as the result of physical difficulty. Regarding variables affecting the $CD34^+$ cell yield, in our previous series we found a trend, but not significant correlation, between the duration of previous chemotherapy and the $CD34^+$ cell yield [17]. In the recent study, with higher number of children, we have finally related the yield, not only to receiving more than 6 cycles of chemotherapy (p=0.02), but also to prior radiotherapy (p=0.01) [10]. It was also related to the underlying disease, with those patients diagnosed with hematologyc malignancy collecting lower amount of PBPC than those diagnosed with solid tumors [10]. Finally, the variable closer related to the yield was the $CD34^+$ cel count in peripheral blood before the procedure, as described in several previous studies in adults and children [6, 10, 19, 68].

Large Volume Leukapheresis and Recruitment Phenomenon

There are two tendencies to avoid secondary effects of PBPC collection in children. Some teams prefer to perform several short apheresis in order to reduce the metabolic and hemodynamic complications related to each procedure, although it is widely recognized that apheresis performed on the second or third day collects a lower number of PBPC [71, 72]. Others, and this our tendency, prefer to perform only one procedure, processing as much blood as possible. Thereby, the patient is only at risk once. LVL has been reported in several adult patients to collect higher number of $CD34^+$ cells than considered to be in peripheral blood before the procedure [73-75], it due to recruitment. But recruitment has not been defined similarly by all the authors. Some authors considered recruitment as a time or blood volume dependent phenomenon leading to a $CD34^+$ cells release into peripheral blood after the beginning of apheresis [72, 75, 76, 78]. Some kind of PBPC mobilization by the procedure itself has been proposed, but it has not been demonstrated [72, 75, 77, 78]. Studies focus on $CD34^+$ cell kinetics during leukapheresis are not consistent and they do not clarify this phenomenon. While some investigators have reported a decrease in $CD34^+$ cells in peripheral blood after processing three or four blood volumes [76, 77, 9, 80], others could not demonstrate this decrease [72, 75, 78, 81]. We describe recruitment merely as collecting more PBPCs than those previously estimated (based on the $CD34^+$ cell count and the patient's blood volume) to be in peripheral blood at the beginning of the leukapheresis. Following this definition, most procedures (72%) had recruitment in our series with LVL [10]. A mean of 1.2-fold greater $CD34^+$ cells was collected than were present in peripheral blood before the procedure. The only variable clearly affecting the recruitment was the total blood volume processed ($p=0.0005$). Based on this result we continue performing LVL. Moreover, 95% of children weighing less than 20 kg achieved the target $CD34^+$ cell dose by LVL after an adequate mobilization regimen [10]. In older patients only 83.6% achieved the target dose with the same schedule, as previously discussed, age is an important variable for PBPC collection [38].

Conclusions

Leukaphereses in children may be performed with any of the reported continuous blood cell separator with no sedation. Several variables have been related to the yield in children, similar to those described in adults. Adverse events are mainly related either to vascular access or to metabolic or hemodynamic changes. Some technical modifications are used in children to decrease these complications. Anyway, taking all these complications together, they are not higher than those adverse events or toxicities reported after general anesthesia and bone marrow harvesting.

References

[1] Gorlin JB. Therapeutic plasma exchange and cytapheresis in pediatric patients. *Transfusion Science* (1999) 21: 21-36.

[2] Galacki DM. An overview of therapeutic apheresis in pediatrics. *Journal of Clinical Apheresis* (1997) 12: 1-3.

[3] Marson P, MG Petris and G de Silvestro. Collection of peripheral blood stem cells in pediatric patients: a concise review on technical aspects. *Bone Marrow Transplant* (1998) 22, Suppl. 5: S7-S11.

[4] Bambi F, Faulkner LB, Azzari C, Gelli AMG, Tamburini A, Tintori V et al. Pediatric peripheral blood progenitor cell collection: Haemonetics MCS 3P versus COBE Spectra versus Fresenius AS104. *Transfusion* (1998) 38: 70-74.

[5] Moog R, Müller N. Peripheral blood progenitor cell colelctions in children. *Transfusion* (1999) 39: 334-335.

[6] Díaz MA, J Kanold, MG Vicent, P Halle, L Madero and F Deméocq. Using peripheral blood progenitor cells (PBPC) for transplantation in pediatric patients: a state-of-the-art review. *Bone Marrow Transplant* (2000) 26: 1291-1298.

[7] Koristek Z, Sterba J, Havranova D and Mayer J. Technique for PBSC harvesting in children of weight under 10 kg. *Bone Marrow Transplant* (2002) 29: 57-61.

[8] Gorlin JB, D Humphreys, P Kent, D Galacki, SV Kevy, S Grupp, L Diller, H Weinstein, H Grier and R Shamberger. Pediatric large volume peripheral blood progenitor cell collections from patients under 25kg: A primer. *J Clin Apheresis* (1996) 11: 195-203.

[9] Kawano Y, Y Takaue, T Watanabe, T Abe, Y Okamoto, A Iwai, T Iwai, A Watanabe, E Ito, A Makimoto, R Nakagawa, H Watanabe, J Sato, K Suenaga, H Suzuya, T Ohnishi, S Kanamaru, M Kaneko and Y Kuroda. Efficacy of the mobilization of peripheral blood stem cells by granulocyte colony-stimulating factor in pediatric donors. *Cancer Research* (1999) 59:3321-3324.

[10] Sevilla J, González-Vicent M, Madero L, García-Sánchez F and Díaz MA. Large volume leukapheresis in small children: safety profile and variables affecting peripheral blood progenitor cell collection. *Bone Marrow Transplant* (2003) 31: 263-267.

[11] Orbach D, Hojjat-Assari S, Doz F, Pacquement H, Guillaume A, Mathiot C et al. Peripheral blood stem cell collection in 24 low-weight infants: experience of a single centre. *Bone Marrow Transplant* (2003) 31: 171-174.

[12] Deméocq F, J Kanold, J Chassagne, MJ Bezou, P Lutz, L deLumley, I Philip, JP Vannier, G Margueritte, JP Lamagnére, MC Baranzelli, A Lenat, H Carla and G Malpuech. Successful blood stem cell collection and transplant in children weighing less than 25 kg. *Bone Marrow Transplant* (1994) 13: 43-50.

[13] Kanold J, P Halle, C Rapatel, M Berger, P Gembara, L deLumley, JP Vannier, P Lutz, D Plantaz, G Malpuech and F Deméocq. Safe and efficient peripheral blood stem cell collection in the smallest of children. *Ther Apher* (1997) 2: 49-57.

[14] Bolan CD and Leitman SF. Management of anticoagulation-associated toxicity during large-volume leukapheresis of peripheral blood stem cell donors. *Blood* (2002) 99: 1878.

[15] Takaue Y, Kawano Y, Abe T, Okamoto Y, Suzue T, Shimizu T, Saito S, Sato J, Makimoto A, Nakagawa R, Watanabe T, Ito M and Kuroda Y. Collection and transplantation of peripheral blood stem cells in very small children weighing 20 kg or less. *Blood* (1995) 86: 372-380.

[16] Shariatmadar S, Noto TA. Femoral vascular access for large volume collection of peripheral blood progenitor cells. *Journal of Clinical Apheresis* (1998) 13:99-102.

[17] Diaz MA, A Alegre, A Benito, M Villa and L Madero. Peripheral blood progenitor cell collection by large-volume leukapheresis in low-weight children. *J Hematother* (1998) 7: 63-68.

[18] Bender JG, LB To, S Williams and LS Schwartzberg. Defining a therapeutic dose of peripheral blood stem cells. *J Hematother* (1992) 4: 329-341.

[19] Bensinger W, F Appelbaum, S Rowley, R Storb, J Sandres, K Lilleby, T Gooley, T Demirer, K Schiffman, C Weaver, R Clift, T Chauncey, J Klarnet, P Montgomery, S Petersdorf, P Weiden, R Witherspoon and CD Buckner. Factors that influence collection and engraftment of autologous peripheral-blood stem cells. *J Clin Oncol* (1995) 13: 2547-2555.

[20] Weaver CH, J Potz, J Redmond, K Tauer, LS Schwartzberg, P Kaywin, R Drapkin, B Grant, P Unger, C Allen, K Longin, B Zhen, B Hazelton and CD Buckner. Engraftment and outcomes of patients receiving myeloablative therapy followed by autologous peripheral blood stem cells with a low $CD34^+$ cell content. *Bone Marrow Transplant* (1997) 19: 1103-1110.

[21] Diaz MA, MG Vicent, F Garcia-Sanchez, JL Vicario and L Madero. Long-term hematopoietic engraftment after autologous peripheral blood progenitor cell transplantation in pediatric patients: effects of the CD34+ cell dose. *Vox Sang* (2000) 79: 145-150.

[22] Lasky LC, B Bostrom, J Smith, TJ Moss and NKC Ramsay. Clinical collection and use of peripheral blood stem cells in pediatric patients. *Transplantation* (1989) 47: 613-616.

[23] Takaue Y, T Watanabe, Y Kawano, T Koyama, T Abe, T Suzue, J Satoh, T Shimokawa, T Ninomiya, M Kosaka, E Shimizu, T Ogura and Y Kuroda. Isolation and storage of peripheral blood hematopoietic stem cells for autotransplantation into children with cancer. *Blood* (1989) 74: 1245-1251.

[24] Fukuda M, S Kojima, K Matsumoto and T Matsuyama. Autotransplantation of peripheral blood stem cells mobilized by chemotherapy and recombinant human granulocyte colony-stimulating factor in childhood neuroblastoma and non-Hodgkin's lymphoma. *Br J Haematol* (1992) 80: 327-331.

[25] Leibundgut K, A von Rohr, K Brülhart, A Hirt, E Ischi, C Jeanneret, J Muff, A Ridolfi-Lüthy, HP Wagner and A Tobler. The number of circulating CD34+ blood cells predicts the colony-forming capacity of leukapheresis products in children. *Bone Marrow Transplant* (1995) 15: 25-31.

[26] Kanold J, CH Rapatel, M Berger, J Chassagne, P Lutz, L deLumley, D Plantaz, JP Vannier, G Malpuech and F Deméocq. Use of G-CSF to mobilize peripheral blood stem cells for collection from children. *Br J Haematol* (1994) 88: 633-635.

[27] Diaz MA, M Villa, A Alegre, ML Lamana, A de la Vega, A Granda and L Madero. Collection and transplantation of peripheral blood progenitor cells mobilized by G-CSF alone in children with malignancies. *Br J Haematol* (1996) 94:148 - 154.

[28] Madero L, M González-Vicent, J Molina, R Madero, V Quintero and MA Díaz. Use of concurrent G-CSF + GM-CSF vs G-CSF alone for mobilization of peripheral blood stem cells in children with malignant disease. *Bone Marrow Transplant* (2000) 26: 365-369.

[29] Spitzer G, D Adkins, M Mathews, W Velasquez, C Bowers, F Dunphy, N Kronmueller, R Niemeyer, W McIntyre and P Petruska. Randomized comparison of G-CSF + GM-CSF vs G-CSF alone for mobilization of peripheral blood stem cells: effects on hematopoietic recovery after high-dose chemotherapy. *Bone Marrow Transplant* (1997) 20: 921-930.

[30] Zeller W, K Gutensohn, M Stockschläder, J Dierlamm, N Kröger, G Koehne, K Hummel, H Kabisch, HJ Weh, P Kühnl, DK Hossfeld and AR Zander. Increase of mobilized CD34-positive peripheral blood progenitor cells in patients with Hodgkin´s disease, non-Hodgkin´s lymphoma, and cancer of the testis. *Bone Marrow Transplant* (1996) 17: 709-713.

[31] Nademanee A, I Sniecinski, GM Schmidt, AC Dagis, MR O´Donnell, DS Snyder, PM Parker, AS Stein, EP Smith, A Molina, DE Stepan, G Somlo, KA Margolin, D Woo, JC Niland and S Forman. High-dose therapy followed by autologous peripheral-blood stem-cell transplantation for patients with Hodgkin´s disease and non-Hodgkin´s lymphoma using unprimed and granulocyte colony-stimulating factor-mobilized peripheral-blood stem cells. *J Clin Oncol* (1994) 12: 2176-2186.

[32] Somlo G, I Sniecinski, T Odom-Maryon, B Nowicki, W Chow, V Hamasaki, L Leong, K Margolin, R Morgan, J Raschko, S Shibata, M Tetef, A Molina, RJ Berenson, SJ Forman and JH Doroshow. Effect of CD34+ selection and various schedules of stem cell reinfusion and granulocyte colony-stimulating factor priming on hematopoietic recovery after high-dose chemotherapy for breast cancer. *Blood* (1997) 89: 1521-1528.

[33] Lie AKW, CH Hui, T Rawling, PG Dyson, D Thorp, J Benic, CM Rawling, I Toogood, N Horvath, PJ Simmons and LB To. Granulocyte colony-stimulating factor (G-CSF) dose-dependent efficacy in peripheral blood stem cell mobilization in patients who had failed initial mobilization with chemotherapy and G-CSF. *Bone Marrow Transplant* (1998) 22: 853-857.

[34] Weaver CH, R Birch, FA Greco, L Schwartzberg, B McAneny, M Moore, D Oviatt, J Redmond, C George, T Alberico, P Johnson and CD Buckner. Mobilization and harvesting of peripheral blood stem cells: randomized evaluations of different doses of filgrastim. *Br J Haematol* (1998) 100: 338-347.

[35] Halle P, J Kanold, C Rapatel, N Boiret, M Berger, JL Stephan, E Albuisson, O Tournilhac, J Bonhomme and F Deméocq. Granulocyte colony-stimulating factor alone at 20 μg/kg vs 10 μg/kg for peripheral blood stem cell mobilization in children. *Pediatr Transplantation* (2000) 4: 285- 288.

[36] Kawano Y, T Watanabe and Y Takaue. Mobilization/harvest and transplantation with blood stem cells, manipulated or unmanipulated. *Pediatr Transplantation* (1999) 3, suppl. 1: 65-71.

[37] Kröger N, W Zeller, HT Hassan, W Krüger, K Gutensohn, C Löliger and AR Zander. Stem cell mobilization with G-CSF alone in breast cancer patients: higher progenitor cell yield by delivering divided doses (2x5 μg/kg) compared to a single dose (1x10 μg /kg). *Bone Marrow Transplant* (1999) 23: 125-129.

[38] Sevilla J, González-Vicent M, Madero L, García-Sanchez F, Diaz MA. Granulocyte colony-stimulating factor alone at 12 μg/kg twice a day for 4 days for peripheral blood progenitor cell priming in pediatric patients. *Bone Marrow Transplant* (2002) 30: 417-420.

[39] Pérez-Dueñas B, Alcorta I, Estella J, Rives S, Toll T, Tuset E. Safety and efficacy of high-dose G-CSF (24 μg/kg) alone for PBSC mobilization in children *Bone Marrow Transplant* (2002) 30: 987-988.

[40] Stiff PJ. Management strategies for the hard-to-mobilize patient. *Bone Marrow Transplant* (1999) 23, suppl, 2: S29-S33.

[41] Clark RE and CG Brammer. Previous treatment predicts the efficiency of blood progenitor cell mobilisation: validation of a chemotherapy scoring system. *Bone Marrow Transplant* (1998) 22: 859-863.

[42] Drake M, L Ranaghan, TC Morris, L Nolan, ZR desai, AE Irvine, A Jordan, K Magill and S Price. Analysis of the effect of prior therapy on progenitor cell yield: use of a chemotherapy scoring system. *Br J Haematol* (1997) 98: 745-749.

[43] Anderlini P, M Körbling, D Dale, A Gratwohl, N Schmitz, D Stroncek, C Howe, S Leitman, M Horowitz, E Gluckman, S Rowley, D Przepiorka and R Champlin. Allogeneic blood stem cell transplantation: considerations for donors. *Blood* (1997) 90: 903-908.

[44] Grigg AP, AW Roberts, H Raunow, S Houghton, JE Layton, AW Boyd, KM McGrath and D Maher. Optimizing dose and scheduling of filgrastim (Granulocyte colony-stimulating factor) for mobilization and collection of peripheral blood progenitor cells in normal volunteers. *Blood* (1995) 86: 4437-4445.

[45] Schaison G, OB Eden, G Henze, WA Kamps, F Locatelli, J Ninane, J Ortega, P Riikonen and HP Wagner. Recommendations on the use of colony-stimulating factors in children: conclusions of a European panel. *Eur J Pediatr* (1998) 157: 955-966.

[46] Körbling M, KW Chan, p Anderlini, D Seong, A durett, A Langlinais, D Przepiorka, J Gajewski, P Miller, J Sundberg, P Alilaen, P Bojko, N Mirza, D Claxton, K van Besien, I Khouri, B Andersson, R Mehra and R Champlin. Allogeneic peripheral blood stem cell transplantation using normal patient-related pediatric donors. *Bone Marrow Transplant* (1996) 18: 885-890.

[47] Li CK, PM Yuen, KW Chik, MMK Sing, K Li, KS Tsang, A Wong, TF Leung and H Lai. Allogeneic peripheral blood stem cell transplant in children. *Med Pediatr Oncol* (1998) 30:147-151.

[48] González M, A Benito, MA Díaz, L Madero. Peripheral blood progenitor cell (PBPC) collection by large-volume leukapheresis from pediatric donors. *Bone Marrow Transplant* (1999) 23: 631-633.

[49] Pahys J, V Fisher, M Carneval, R Yomtovian, R Sarode and M Nieder. Successful large volume leukapheresis on a small infant allogeneic donor. *Bone Marrow Transplant* (2000) 26: 339-341.

[50] De la Rubia J, MA Díaz, A Verdeguer, MJ Pascual, c Arbona, R Arrieta, S Brunet, J Bargay, C Martínez, I Espigado, D Serrano, A Alegre, F de Arriba, J de la Serna, C Zamora, L Benlloch and MA Sanz. Donor age-related differences in PBPC mobilization with rHuG-CSF. *Transfusion* (2001) 41: 201-205.

[51] Diaz MA, Sevilla J, de la Rubia J, Verdeguer A, Espigado I, G-Vicent M et al. Factors predicting peripheral blood progenitor cell collection from pediatric donors for allogeneic transplantation. *Haematologica* (2003) 88: 919-922.

[52] Waller CF, H Bertz, MK Wenger, S Fetscher, M Hardung, M Engelhardt, D Behringer, W Lange, R Mertelsmann and J Finke. Mobilization of peripheral blood progenitor cells for allogeneic transplantation: efficacy and toxicity of a high-dose rhG-CSF regimen. *Bone Marrow Transplant* (1996) 18: 279-283.

[53] Tanaka R, T Matsudaira, J Aizawa, y Ebihara, K Muraoka, K Tsuji, K Ikebuchi, K Kodama, F Takaku and T Nakahata. Characterization of peripheral blood progenitor cells (PBPC) mobilized by filgrastim (rHuG-CSF) in normal volunteers: dose-effect relationship for filgrastim with the character of mobilized PBPC. *Br J Haematol* (1996) 92: 795-803.

[54] Arbona C, F Prosper, I Benet, F Mena, C Solano and J Garcia-Conde. Comparison between once a day vs twice a day G-CSF for mobilization of peripheral blood progenitor cells (PBPC) in normal donors for allogeneic PBPC transplantation. *Bone Marrow Transplant* (1998) 22: 39-45.

[55] Kröger N, H Renges, W Krüger, K Gutensohn, C Löliger, I Carrero, L Cortes and AR Zander. A randomized comparison of once versus twice daily recombinant human granulocyte colony-stimulating factor (filgrastim) for stem cell mobilization in healthy donors for allogeneic transplantation. *Br J Haematol* (2000) 111: 761-765.

[56] Lee V, CK Li, MMK Shing, KW Chik, K Li, KS Tsang, DC Zhao, DH Lai, A Wong and PMP Yuen. Single vs twice daily G-CSF dose for peripheral blood stem cells harvest in normal donors and children with non-malignant diseases. *Bone Marrow Transplant* (2000) 25: 931-935.

[57] Anderlini P, M Donato, MJ Lauppe, YO Huh, TG Martin, KW Chan, RE Champlin and M Körbling. A comparative study of once-daily versus twice-daily filgrastim administration for the mobilization and collection of $CD34^+$ peripheral blood progenitor cells in normal donors. *Br J Haematol* (2000) 109: 770-772.

[58] Murata M, M Harada, S Kato, S Takahashi, H Ogawa, S Okamoto, S Tsuchiya, H Sakamaki, Y Akiyama and Y Kodera. Peripheral blood stem cell mobilization and apheresis: analysis of adverse events in 94 normal donors. *Bone Marrow Transplant* (1999) 24: 1065-1071.

[59] Watanabe T, Y Kawano, A Watanabe and Y Takaue. Autologous and allogeneic transplantation with peripheral blood $CD34^+$ cells: a pediatric experience. *Haematologica* (1999) 84: 167-176.

[60] Gutierrez-Delgado F and W Bensinger. Safety of granulocyte colony-stimulating factor in normal donors. *Curr Opin Hematol* (2001) 8: 155-160.

[61] Torrelo A, L Madero, IG Mediero, A Zambrano (2000). A cutaneous eruption from G-CSF in a healthy donor. *Pediatric Dermatology* 17: 205-207.

[62] Watts MJ, I Addison, SG Long, S Hartley, S Warrington, M Boyce and DC Linch. Crossover study of the haematological effects and pharmacokinetics of glycosylated and non-glycosylated G-CSF in healthy volunteers. *Br J Haematol* (1997) 98: 474-479.

[63] Blaise D, E Jourdan, M Michallet, JP Jouet, JM Boiron, G Michel, C Faucher, N Fégueux, MP Schuller, N Badri, C Chabannon and D Maraninchi. Mobilisation of healthy donors with lenograstim and transplantation of HLA-genoidentical blood progenitors in 54 patients with hematological malignancies: a pilot study. *Bone Marrow Transplant* (1998) 22: 1153-1158.

[64] Höglund M, B Smedmyr, B Simonsson, T Tötterman and M Bengtsson. Dose-dependent mobilisation of haematopoietic progenitor cells in healthy volunteers receiving glycosylated rHuG-CSF. *Bone Marrow Transplant* (1996) 18: 19-27.

[65] Hoglund M, B Smedmyr, M Bengtsson, TH Totterman, V Cour-Chabernaud, A Yver and B Simonsson. Mobilization of CD34+ cells by glycosylated and nonglycosylated G-CSF in healthy volunteers—a comparative study. *Eur J Haematol* (1997) 59: 177-183.

[66] Lasky LC, SB Fox, J Smith and B Bostrom. Collection and use of peripheral blood stem cells in very small children. *Bone Marrow Transplant* (1991) 7: 281-284.

[67] Alegre A, MA Díaz, L Madero, A Granda, A de la Vega, M Villa, MJ Fernández-Villalta, JJ Gil-Fernández and JM Fernández-Rañada. Large-volume leukapheresis for peripheral blood stem cell collection in children: a simplified single-apheresis approach. *Bone Marrow Transplant.* (1996) 17, 923-927.

[68] Díaz MA, A Alegre, M Villa, A Granda, A de la Vega, M Ramirez, D Ruano, A Gonzalez, JM Merino and L Madero. Pediatric experience with autologous peripheral blood progenitor cell transplantation: influence of $CD34^+$ cell dose in engraftment kinetics. *Bone Marrow Transplant* (1996) 18: 699-703.

[69] Nussbaumer W, Schönitzer, T Trieb, FM Fink, K Maurer-Dengg, P Höcker, A Wagner, H Schwaighofer, D Nachbaur and D Niederwieser. Peripheral blood stem cell (PBSC) collection in extremely low-weight infants. *Bone Marrow Transplant* (1996) 18: 15-17.

[70] Urban C, W Schwinger, M Benesch, H Lackner, R Kerbl, R Gilli, U Pätzold and S Burdach. Feasibility of peripheral blood stem cell (PBSC) and peripheral blood mononuclear cell (PBMNC) separation in children with a body weight below 20 kg. *Med Ped Oncol* (1997) 29: 115-120.

[71] Kajiume T, S Yoshimi, K Kobayashi and N Kataoka. The relation between the number of PBSC and number of leukapheresis in children. *J Hematother* (1999) 8: 185-188.

[72] Abe T, A Makimoto, Y Kawano, Y Okamoto, T Kajiume, R Nakagawa, H Watanabe, J Sato, T Watanabe, Y Kuroda and Y Takaue. Intra-apheresis recruitment of blood progenitor cells in children. *Transfusion* (1998) 38: 944-950.

[73] Malachowski ME, RL Comenzo, CD Hillyer, KO Tiegerman and EM Berkman. Large-volume leukapheresis for peripheral blood stem cell collection in patients with hematologic malignancies. *Transfusion* (1992) 32: 732-735.

[74] Hillyer CD, DA Lackey III, KK Hart, LL Stempora, RA Bray, JG Bender and AD Donnenberg. $CD34^+$ progenitors and colony-forming units-granulocyte macrophage

are recruited during large-volume leukapheresis and concentrated by counterflow centrifugal elutriation. *Transfusion* (1993) 33: 316-321.

[75] Smolowiccz AG, K Villman, G Berlin and U Tidefelt. Kinetics of peripheral blood stem cell harvest during a single apheresis. *Transfusion* (1999) 39: 403-409.

[76] Cassens U, PH Momkvist, M Zuehlsdorf,, M Mohr, J Kienast, WE Berdel and W Sibrowski. Kinetics of standardized large volume leukapheresis (LVL) in patients do not show recruitment phenomenon of peripheral blood progenitor cells (PBPC). *Bone Marrow Transplant* (2001) 28: 13-20.

[77] Gorlin JB, EC Vamvakas, E Cooke, D Galacki, R Geha, D Humphreys and P Kent. Large volume leukapheresis in pediatric patients: processing more blood diminishes the apparent magnitude of intra-apheresis recruitment. *Transfusion* (1996) 36: 879-885.

[78] Cull G, J Ivey, P Chase, R Picciuto, R Herrmann and P Cannell. Collection and recruitment of CD34+ cells during large-volume leukapheresis. *J Hematother* (1997) 6: 309-314.

[79] Passos-Coelho JL, MA Machado, P Lúcio, F Leal-da-Costa, MR Gomes-Silva and A Parreira. Large-volume leukaphereses may be more efficient than standard-volume leukaphereses for collection of peripheral blood progenitor cells. *J Hematother* (1997) 6: 465-474.

[80] Cassens U, P Ostkamp-Ostermann, N van der Werf, H Garritsen, H Ostermann and W Sibrowski. Volume-dependent collection of peripheral blood progenitor cells during large-volume leukapheresis for patients with solid tumours and haematological malignancies. *Transfus Med* (1999) 9: 311-320.

[81] Yamaguchi E, K Yamato and Y Miyata. Kinetics of peripheral blood stem cell collection in large-volume leukapheresis for pediatric patients undergoing chemotherapy and adult patients before chemotherapy. *J Hematother Stem Cell Res* (2000) 9: 565-572.

In: Trends in Stem Cell Research
Editor: Erik V. Greer, pp. 147-167

ISBN 1-59454-315-1

Chapter VI

Neural Crest Stem Cells and Organogenesis

Hidetoshi Yamazaki [1,2*]***, Miya Yoshino***[2] ***and Shin-Ichi Hayashi***[2]

[1]Division of Regenerative Medicine and Therapeutics, Department of Genetic Medicine and Regenerative Therapeutics, Institute of Regenerative Medicine and Biofunction, Tottori University Graduate School of Medical Science, and the [2]Division of Immunology, Department of Molecular and Cellular Biology, School of Life Science, Faculty of Medicine, Tottori University, Yonago, Tottori, Japan

Abstract

The neural crest (NC) is a multipotent cell population generated from the tip of the neural fold. NC cells differentiate into neurons, glial cells, osteoblasts, odontoblasts and melanocytes, and participate in organogenesis of the cranio-face, parathyroid, heart and thymus. The thymus is a vertebrate-specific organ related to the adaptive immune system. Since the failure of NC cells results in immunodeficiency, NC cells should be concerned with thymic organogenesis. The differentiation of NC cells has been thought to be regulated by environmental factors during the migration of these cells into peripheral tissues, and in accord with this, multipotent NC-derived cells can rarely be detected in the peripheral tissues. Recently, we found that NC-derived cells with the potential to differentiate into neurons, glial cells and melanocytes are present in the developing thymus, and that these cells are rarely detected at late embryonic stages. Multipotent NC-derived cells were also found to be present in the developing teeth. These findings suggest that multipotent NC-derived cells are present in peripheral tissues and may

[*] Hidetoshi Yamazaki, DDS, Ph.D. Division of Regenerative Medicine and Therapeutics, Department of Genetic Medicine and Regenerative Therapeutics, Institute of Regenerative Medicine and Biofunction, Tottori University Graduate School of Medical Science, 86 Nishi-Machi, 683-8503, Yonago, Tottori, Japan. Tel: +81-859-34-8270, Fax: +81-859-34-8272 E-mail: yamazaki@grape.med.tottori-u.ac.jp

participate in organogenesis. We discuss here whether NC cells may contribute to the organogenesis of lymphoid organs other than thymus, and of the central nervous system.

Introduction

Neural crest (NC) cells are generated from the tip of dorsal neural tube at about embryonic day 9.0 (E9.0) in mice [1-4]. They give rise to multiple lineage cells, including peripheral nerve cells, osteoblasts, chondrocytes, odontoblasts and melanocytes [1-4] (Fig. 1A). Migrating NC cells or their progeny cells participate in the organogenesis of the cranium, including the face and jaw, and of the thyroid, heart, and thymus [1,5-9]. Since NC cells first appeared phylogenetically in vertebrates, these organs are only present in vertebrates [1,10,11] (Fig. 1B). The failure of NC function results in immunodeficiencies and craniofacial abnormalities [12-15]. Especially, the adaptive immunity system associated with the evolutionary emergence of lymphocytes and the jaw with teeth making it possible to crush and consume hard foods are deeply associated with NC cells. The acquisition of NC cells might thus have given vertebrates survival advantages.

In this review, we address the relationship between the NC cells and thymic and tooth development. We further discuss the possibilities that NC cells also participate in the organogenesis of lymphoid organs other than the thymus, and in the organogenesis of the central nervous system.

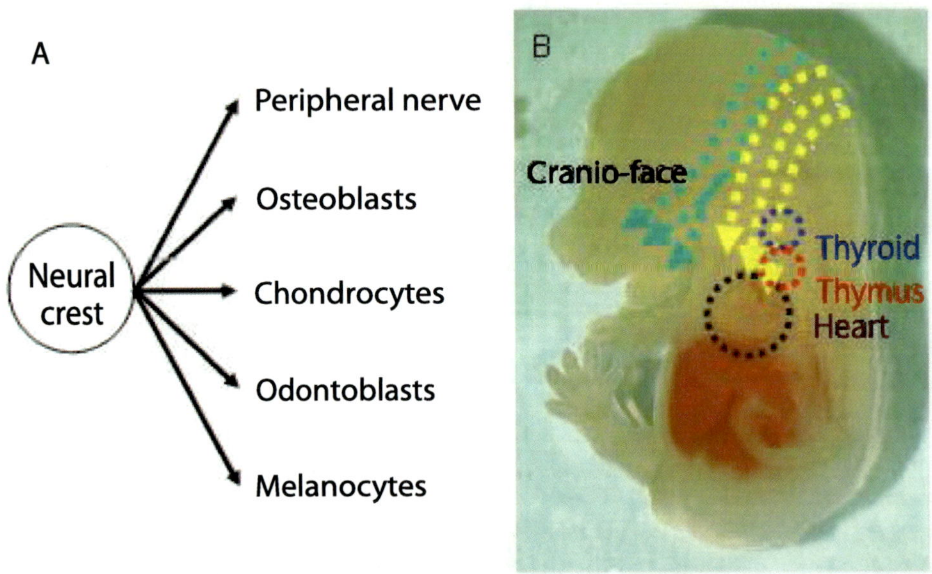

Figure 1. The potential of neural crest (NC) cells to differentiate, and their contribution to organogenesis (A) NC cells differentiate into various cell lineages: peripheral nerve cells, osteoblasts and chondrocytes in the cranium and face region, odontoblasts in the teeth, and melanocytes in the skin. (B) NC cells or NC-derived cells participate in the organogenesis in the cranio-face, thyroid, thymus and heart at the embryonic stage. This figure shows E14.5 day-old mice.

Organogenesis of the Thymus

The defense mechanisms in vertebrates, in contrast to those in invertebrates, involve not only innate immunity but also adaptive immunity [16,17]. Adaptive immunity is antigen-specific, and carried out by T and B lymphocytes, which have been acquired only in vertebrates [15,16]. T lymphocytes differentiate in the thymus, and emigrate to peripheral lymphoid organs such as the lymph nodes and spleen [18-20]. Therefore, a lack of the thymus results in the deficiency of T lymphocytes [12-15].

It is thought that cells derived from all three germ layers, endoderm, ectoderm and mesoderm, and NC cells, so called fourth germ participate in the organization of the thymus [21] (Fig. 2A). The thymic epithelium is derived from the endoderm and/or ectoderm, and the thymic mesenchyme from the mesoderm and/or NC cells [21] Thymic organogenesis is divided temporally into three stages (Fig. 2B): early (E9.5-E11), middle (E11.5-E15), and late (E15.5- birth) organogenesis stages [22].

In the early organogenesis stage, the initial positioning induction of thymic organogenesis, and subsequently outgrowth of the thymic rudiments, occurs. Interactions between epithelium and mesenchyme in the absence of hematopoietic cells are critical in this stage [23,24]. The thymic organization is initiated by interactions between the endoderm in the third pharyngeal pouch and NC-derived mesenchyme from the third and fourth pharyngeal arches, although it is still not clear which of these, the endoderm or NC-derived cells, determines the onset of thymic organogensis (Fig. 2A). The thymic organogenesis is first visible at approximately E10.5. At this stage, several transcriptional factors, *Hoxa3, Pax1, Pax9,* and *Tbx1,* and growth factors such as *RA, Shh* and *FGF8* that are expressed on the endoderm in the third pharyngeal pouch, are thought to play an important role in the onset of the thymic organization [25-31]. NC-derived cells that express *Hoxa3*, *Tbx1, Pax3, Pdgfrα,* and *Etra* participate in the thymic organization. Subsequently, *Eya1* and *Six1*, important genes for the fate determination of the thymic epithelium, are expressed on early thymic rudiments [32-34]. They induce *FoxNI* expression specifically in the thymic epithelium [36,37]. Nude mice carrying a mutation of *FoxNI* show the loss of T lymphocytes because of the abnormality of the thymic epithelium [38,39].

The middle organogenesis stage is primarily concerned with the patterning and initial epithelial differentiation in thymic rudiments. At this stage, hematopoietic cells immigrate there [24,40,41]. This process requires interactions between epithelium-mesenchyme [42-44]. These interactions result in the generation of regional differences within the thymic rudiments such as the cortical and medullary epithelia [45]. Previously, the cortical and medullary epithelia were thought to be derived from the ectoderm and endoderm, respectively [21]. However, recent reports propose that all thymic epithelial cells both in the cortex and medulla arise from common progenitor cells originating from the endoderm [7,46,47]. At this stage, since thymic rudiments are not yet vascularized, hematopoietic cells may cross the condensing mesenchymal layer and enter the thymic rudiment by unknown mechanisms. It is possible that the epithelium or mesenchyme may secrete factor(s) promoting the immigration of hematopoietic cells. Because the differentiation and proliferation of thymic epithelial cells are known to require signals from other cells in the developing organs such as T lymphocytes, the interaction between epithelium and/or mesenchyme and hematopoietic cells is also critical for thymic organogenesis [18-20].

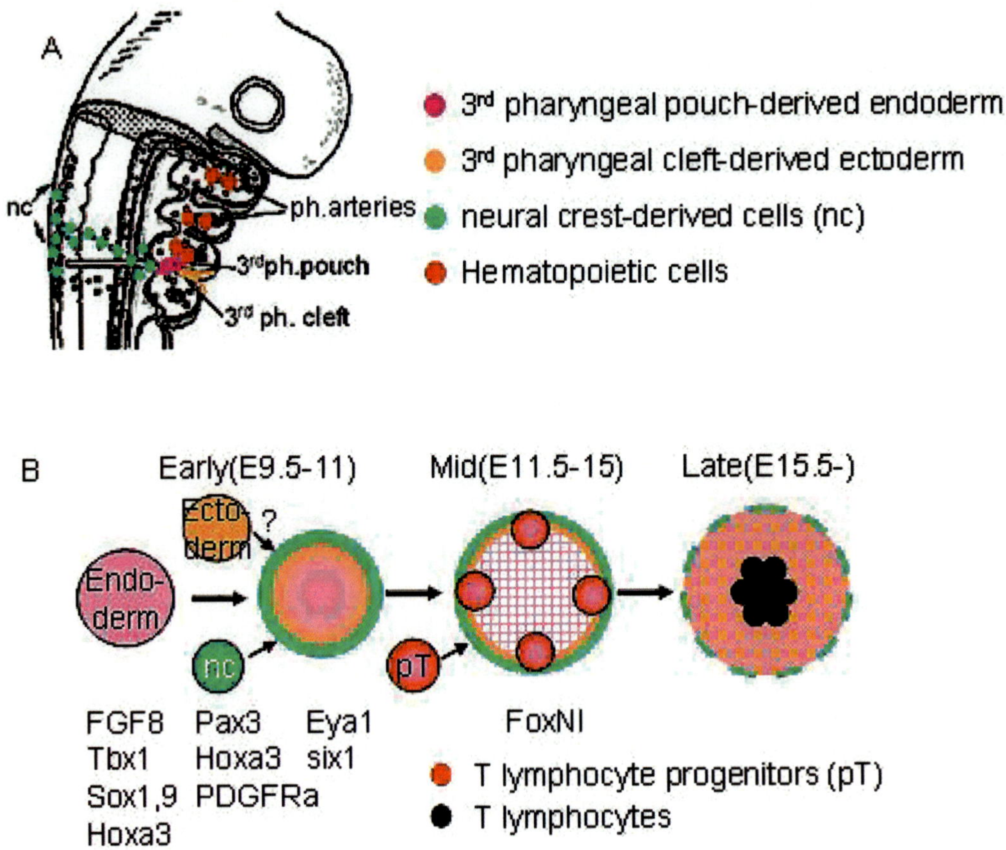

Figure 2. Murine thymic organogenesis. This diagram depicts the earliest stages of thymus organogenesis, up to the differentiation of T lymphocytes. (A) Four distinct cell lineages, namely 3rd pharyngeal pouch-derived endodermal cells, 3rd pharyngeal cleft-derived ectodermal cells, NC-derived mesenchymal cells that have migrated into the 3rd pharyngeal arch, and mesoderm-derived mesenchymal cells (not shown here) and hematopoietic cells, participate in thymic organogenesis. (B) Thymic organogenesis is divided into three stages. In the early stage (E9.5-11), endoderm-derived epithelial cells give rise to the thymic rudiment. NC-derived mesenchymal cells that have migrated into the 3rd pharyngeal arches and/or ectoderm-derived epithelial cells participate in the thymic rudiment. These interactions between mesenchymal and epithelial cells define the character of the thymus. Transcription factors known to play a role in thymus organogenesis are indicated under the cell type where they are expressed. In the middle stage (E11.5-15), hematopoietic cells (T lymphocyte progenitors: pT) invade into the thymic rudiment through mesenchymal cells around the thymus, because blood vessels have not been formed in the thymus. Then, interactions between epithelial cells and T lymphocyte progenitors allow undifferentiated thymic epithelial cells to differentiate into thymic epithelial cells that support T lymphocyte differentiation. Especially, the acquisition of the potential to generate thymic epithelial cells is indicated by the expression of *FoxNI.* In the late stage (E15.5-), thymic epithelium differentiates into cortex and medullar epithelia, which support the differentiation of $CD4^+CD8^+$ T lymphocytes into $CD4^+$ or $CD8^+$ mature T lymphocytes.

In the late developmental stage, the selection of the repertoire of T lymphocyte antigen receptors occurs accompanied by the differentiation from $CD4^{-}CD8^{-}$ double negative (DN) T lymphocytes to $CD4^{+}CD8^{+}$ double positive (DP) T lymphocytes, and subsequently the positive and negative selection of self-antigens occur [18-20,48]. This stage continues through the first postnatal week in mice. Thymic epithelial cells acquire full functional competence, and the thymic microenvironment for T lymphocyte development is accomplished. Then, $CD4^{+}$ or $CD8^{+}$ single positive (SP) mature T lymphocytes that have finished undergoing positive and negative selection move to the medulla from the cortex in the thymus, and emigrate to peripheral lymphoid organs [49].

Avian NC Cells in the Developing Thymus

Although the participation of avian NC cells in thymic organogenesis has been reported by Le Douarin [1,7,50], the presence of NC-derived cells in the postnatal thymus has not been clearly demonstrated. The thymus is composed of three major lineages of cells, hematopoietic cells including T lymphocytes, epithelial cells that support T lymphocyte differentiation, and mesenchymal cells surrounding the thymus [18-20]. Experiments using chick and quail inter-species chimera indicate that NC-derived cells differentiate into interlobular and outer-thymic mesenchymal cells, but not into hematopoietic cells or thymic epithelial cells [7]. Furthermore, ablation of NC cells by the microcautery of neural folds induces thymic abnormalities that reduce both mesenchymal and epithelial cells [51,52]. NC-derived cells differentiate into mesenchymal cells around the thymus, and might regulate the differentiation and proliferation of thymic epithelial cells.

Presence of Murine NC-Derived Cells in the Developing Thymus

Although there have been a few reports showing that NC-derived cells might be present in the developing murine thymus, the temporal and spatial features of the presence of NC-derived cells in the thymus has not been defined in detail [7,8]. The potential of the NC-derived cells to differentiate has not been reported yet. To clarify the period of the presence of NC-derived cells in the thymus, it is important to trace NC-derived cells through embryonic stages. However, no molecule or antibody specific for murine NC lineage cells is known yet, although an antibody against HNK-1 is specific for avian NC cells [53]. Recently, to trace NC-derived cells, transgenic mice carrying reporter genes under the control of the promoter regions of the *Pax3* or *Wnt1* gene or human tissue plasminogen activator have been used [8,54,55,56]. NC cells are classified into cranial and trunk NC cells [1,2]. Cranial NC cells are thought to contribute mainly to the cranio-face, thymus and heart, and in particular, the cells that contribute in and around the heart are called cardiac NC cells. NC cells participating in other regions including the dorsal root ganglion and skin are called trunk NC cells. The use of these control regions enabled us to detect cranial NC cells or their derived cells, respectively. We chased NC-derived cells in the developing thymus using *P0-cre* mice and *Rosa26R* mice, which allowed us to trace not only cranial NC-derived cells but also trunk NC-derived cells as LacZ-positive cells [9,55,57]. The results obtained with these mice indicate that NC-derived cells are present in the fetal thymus at E14.5, and are mainly located around the thymus (Fig. 3). The distribution suggests that NC-derived cells might correspond to mesenchymal cells around the thymus [8,9,52].

Next, the period when NC-derived cells are present in the developing thymus was assessed. Interestingly, NC-derived cells are detected in the thymus at E13.5 and E14.5, but rarely detected beyond E16.5 (data not shown) [9]. These results suggest that NC-derived cells play important roles in thymic organogenesis in the early and middle stages. It is not clear why NC-derived cells are reduced in the thymus at late embryonic stages. In fact, it has been reported that NC-derived cells are rarely detected in the postnatal thymus [8].

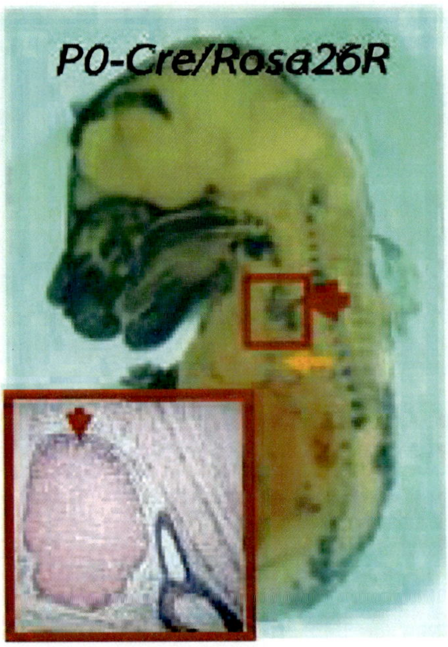

Figure 3. Presence of NC-derived cells in the fetal thymus. Cells that expressed the *P0* gene were detected by staining of E14.5-day-old *P0-Cre/Rosa26R*-double *Tg* mice with LacZ. LacZ^{+} cells were present in the thymus (red arrow) as well as face and heart (yellow arrow). No LacZ^{+} cells were present in any tissues of age-matched *Rosa26R-Tg* mice (data not shown). The section of the E14.5 day-old *P0-cre/Rosa26R* mice showed that NC-derived cells are mainly located around the thymus (higher magnification of the region in the box; arrowhead).

Roles of Mesenchymal Cells in the Developing Thymus

T lymphocyte progenitors immigrate into the thymic primordium at the middle stage (Fig. 2B). In this process, mutual interactions of T lymphocyte progenitors, thymic epithelial cells, and mesenchymal cells are important. For example, chemoattractants for T lymphocyte progenitors (Fractakline and TECK) are expressed on thymic epithelial cells [41,58]. Furthermore, mesenchymal cells surrounding the thymus express adhesion molecules and extra-cellular matrix (ECM), ligands for integrins on T lymphocyte progenitors [59-62].

In the thymus, T lymphocyte progenitors proliferate and differentiate into mature T lymphocytes [18-20]. The differentiation of both T lymphocytes and thymic epithelial cells is important in these processes because thymic epithelial cells express IL-7 and Notch ligands, and regulate the differentiation of T lymphocyte progenitors [63-65]. Wnt family proteins are expressed on T lymphocyte progenitors [36,66] and are known to regulate the specific expression of FoxNI on thymic epithelial cells [36].

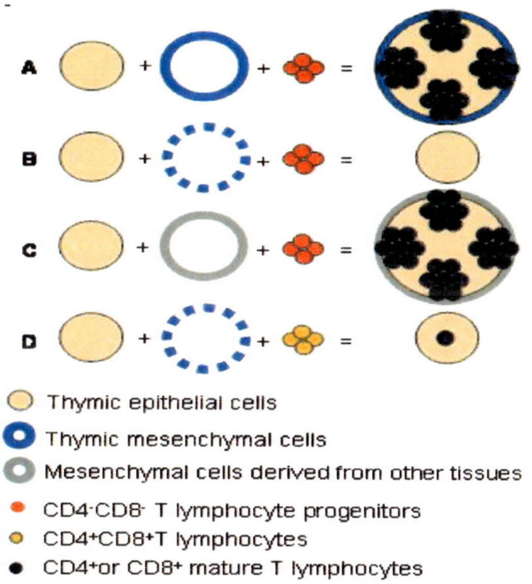

Figure 4. Analysis of roles of thymic mesenchyme in developing thymus using thymic organ culture E12.5 thymic rudiments are treated with dGuo to remove the hematopoietic cells but not stromal cells in the thymus. Next, thymic stromal cells are divided into mesenchymal cells and epithelial cells. First, in thymic organ culture with 2Guo-treated thymus and $CD4^-CD8^-$ T lymphocyte progenitors, T lymphocyte progenitors can differentiate into $CD4^+$ or $CD8^+$ mature T lymphocytes (A). Next, in thymic organ culture without mesenchymal cells, few T lymphocytes can be differentiated from T lymphocyte progenitors (B). However, in thymic organ culture with dGuo-treated epithelial cells and mesenchymal cells such as NIH3T3 fibroblasts or lung fibroblasts that are not derived from thymus, T lymphocyte progenitors can differentiate into $CD4^+$ or $CD8^+$ mature T lymphocytes (C). In contrast, thymic organ culture without mesenchymal cells can generate a small number of mature T lymphocytes that are differentiated from more mature $CD4^+CD8^+$ T lymphocytes than $CD4^-CD8^-$T lymphocyte progenitors.

Mesenchymal cells support the differentiation of thymic epithelial cells by producing cytokines such as EGF and FGF [67-70]. Mutation of the *Fgfr* gene expressed on thymic epithelial cells causes abnormal differentiation of the thymic epithelium [71]. Previously, to clarify the role of mesenchymal cells in thymic development, fetal thymic organ cultures at E12.5 were examined under conditions with or without mesenchymal cells surrounding the thymus [72] (Fig. 4). In the absence of mesenchymal cells, thymic abnormalities featuring a lack of mature T lymphocytes were observed (Fig. 4B), indicated that mesenchymal cells are critical for the differentiation of the thymic epithelial cells (Fig. 4B). Using $CD4^+CD8^+$ DP T lymphocytes instead of less mature $CD4^-CD8^-$ DN T lymphocytes, mature T lymphocytes were induced from the thymus without mesenchymal cells in culture [73] (Fig. 4D). These results suggest that the differentiation from $CD4^-CD8^-$ DN T lymphocytes, but not from $CD4^+CD8^+$ DP T lymphocytes, depends on mesenchymal cells. This may mean that mesenchymal cells do not affect the differentiation of thymic epithelial cells to support T lymphocyte differentiation from $CD4^+8^+$ DP to $CD4^+$ or $CD8^+$ SP mature T lymphocytes. However, the significant reduction of mature T lymphocytes in the thymus without mesenchymal cells may result from inadequate proliferation or differentiation of thymic epithelium that supports T lymphocyte differentiation.

Contribution of NC-Derived Cells to Thymic Organogenesis

DiGeorge syndrome is characterized by defects of aortic arch patterning and conotruncal heart, thymic and parathyroid aplasia/hypoplasia, and craniofacial anomalies [12-15]. This syndrome is thought to result from the aberrant distribution or migration of NC-derived cells during their organogenesis [14,32,74]. In patients or mice with DiGeorge syndrome, mature T lymphocytes are severely reduced [12,13]. These findings support the notion that NC-derived cells in the fetal thymus contribute to organogenesis of the thymus.

In our experiments, we detected a few NC-derived cells in E17.5 thymus (data not shown). Both cortical and medullar epithelia are organized in the E17.5 thymus, and T lymphocyte progenitors proliferate and differentiate there [17,18,75]. At late embryonic stages, T lymphocyte progenitors invade into the thymus through the blood vessels in the cortical-medullar junction [76,77]. At this stage, NC-derived cells are rarely detected, and thus NC-derived cells might be not required for the immigration of T lymphocyte progenitors into the thymus. Since NC-derived cells are rarely present in the postnatal thymus, NC-derived cells might not play roles in the postnatal thymus.

Classification of Mesenchymal Cells in the Developing Thymus

NC-derived cells are known to contribute to blood vessel formation. They differentiate into pericytes surrounding vascular endothelial cells derived from mesoderm [1,10,78]. Although blood vessels are not formed in the thymus in the middle stage (E14.5), mesenchymal cells surrounding the fetal thymus express adhesion molecules that control migration and homing of hematopoietic cells. We were able to distinguish between mesenchymal cells from NC-derived cells as $LacZ^+$ and non-NC-derived cells as $LacZ^-$ (data not shown). To compare the derivatives of mesoderm and NC in the thymus, mesenchymal cells around the E13.5 fetal thymus in *P0-cre/Rosa26R* mice were isolated by enzymatic treatment. Interestingly, both $LacZ^+$ and $LacZ^-$ cells were present in this fraction [9]. These observations account for the fact that the NC-derived mesenchyme is distinct from mesoderm-derived mesenchyme in terms of cell surface molecules and gene expression.

Roles of Thymic NC-Derived Cells

Thymic organ cultures in which thymic mesenchymal cells are replaced with 3T3 fibroblasts or lung fibroblasts give rise to mature T lymphocytes [73,79]. Under these conditions, the effect of immigration of T lymphocyte progenitors can be excluded (Fig. 4C). Regardless of the origin of mesenchymal cells, mesenchymal cells function in T lymphocyte differentiation. However, dysfunction of NC cells results in significantly reduced generation of mature T lymphocytes [12-15]. Therefore, NC-derived thymic mesenchymal cells might play a role in the immigration of T lymphocyte progenitors into the thymus.

Recently, it was reported that thymic epithelial cells derived from endoderm regenerate the thymus without mesenchymal cells when they are transplanted in kidney capsules [80]. This study was performed by using thymic epithelial cells at the middle stage (E14.5-E15.5). In this stage, the fate of these epithelial cells may already be determined, and this may explain why a functional thymus can be generated in the absence of mesenchymal cells. If so, the role of NC-derived cells is to determine the fate of the endoderm into the thymus.

The Differentiation Potential of Thymic NC-Derived Cells

NC cells are generated from dorsal tip of the neural tube on about E9.0 in mice [1-4]. It is reported that these NC cells include neural crest stem cells that are self-renewing multipotent cells, can differentiate into multi-lineages cells [81,82]. Although NC cells are consisited of heterogenous populations with the potential for differentiation, the fate of NC cells is thought to be largely determined during their migration [1,83,84]. Furthermore, it was reported that after migration, NC-derived cells lose the multi-potential to differentiate into various cell lineages in early differentiation stages and thus multipotent NC stem cells are rarely present in the peripheral tissues [83,85](Fig.5). Especially, the potential to differentiate into melanocytes was easily lost in the early stage. These reports suggest that cells retaining the potential to differentiate into both melanocytes and other cells lineages might be multipotent NC cells [83-85]. Interestingly, melanocytes were induced in the E13.5 and E15.5, but not E18.5, thymi of wild-type mice (Fig. 6A).

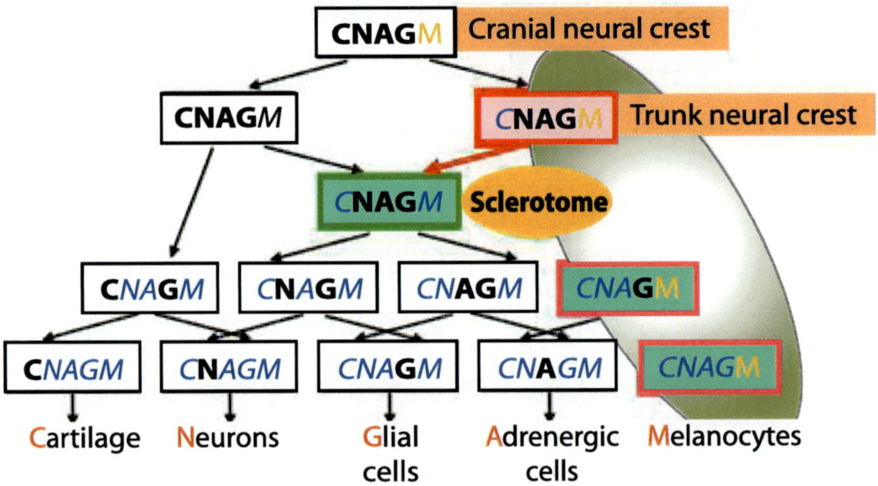

Figure 5. Fate determination of NC cells. NC cells (Trunk NC and Cranial NC) with multipotency are generated from the dorsal tip of the neural tube at about embryonic day 9.0 (E9.0) in mice. Although NC cells are consisted of heterogenous poplations with the potential for differentiation, the fate of NC cells is thought to be mostly determined during migration, and NC cells lose the potential to differentiate into several cell lineages during migration. Especially, NC cells lose the potential to differentiate into melanocytes in early differentiation stages when they migrate inside the sclerotome, and multipotent NC stem cells are rarely present in peripheral tissues. Thus, cells with the potential to differentiate into melanocytes are present in the early differentiation stages or in the skin. In this figure, bold letters indicate that NC cells have the potential to differentiate into these lineages. On the other hand, italic letters indicate that NC cells have lost the potential to differentiate into these lineages. C: cartilage, N: neurons, G: glial cells, A: adrenergic cells, M: melanocytes.

Furthermore, we performed a colony formation assay to clarify the potential of thymic NC-derived cells. Colonies containing both neurons and melanocytes or glial cells and melanocytes were induced from thymic cells prepared at E14.5 but not E17.5 in the presence of endothelin 3 (data not shown) [9]. These results suggest that multipotent NC-derived cells are present in the developing thymus. Recently, Anderson et al. reported that NC stem cells with a potential to differentiate into glial cells, neurons, and myofibroblasts are present in the gut and sciatic nerve of fetal rats [81,82]. Therefore, a small number of multipotent NC stem cells might be retained in the peripheral tissues.

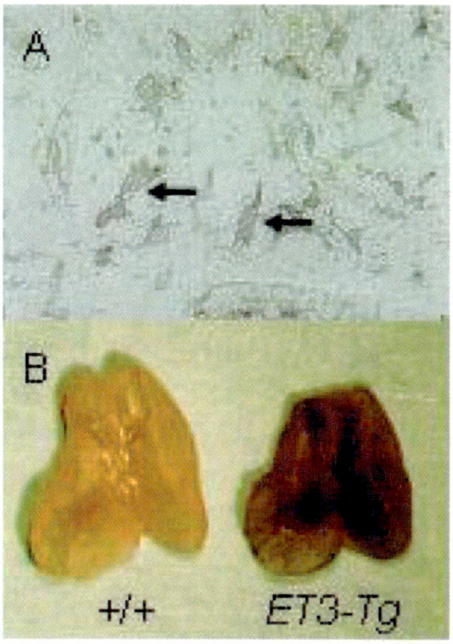

Figure 6. Presence of NC-derived cells with the potential to differentiate into melanocytes in the developing thymus. (A) Melanocytes (arrows) are induced from the E14.5 thymus of wild-type mice in the presence of ET3. (B) Transgenic mice carrying the *ET3* gene under the control of hk14 so that the expression of ET3 could be induced in thymic epithelial cells showed melanocytosis in the thymus.

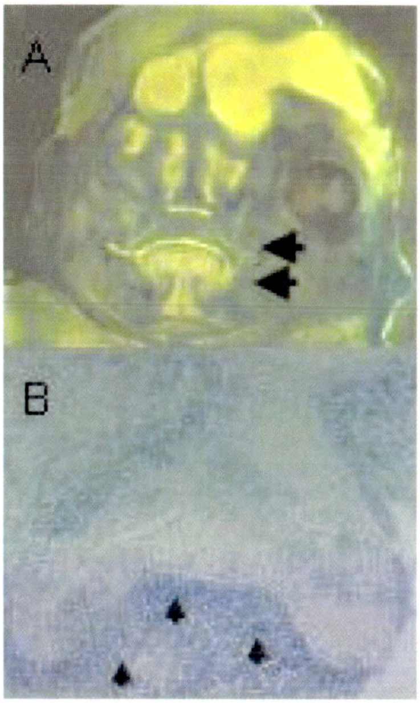

Figure 7. Contribution of NC-derived cells to the cranium/face and dental mesenchyme. Cells that expressed the *P0* gene were detected by staining of E14.5-day-old *P0-Cre/Rosa26R*-double *Tg* mice with LacZ (A). LacZ-positive cells were detected in the craniofacial region (arrowhead), including the mandible and maxilla of *P0-Cre/Rosa26R* mice (A). No LacZ-positive cells were observed in *Rosa26R* mice (data not shown). A large number of LacZ-positive NC-derived cells were present in the dental mesenchyme around the enamel organ of E14.5 *P0-Cre/Rosa26R* mice (B: arrowheads). A section with LacZ staining (A) and both LacZ and Nuclear Fast Red staining (B).

The Effect of ET3 on Thymic NC Cells

To confirm that NC-derived cells can differentiate into melanocytes in vivo, we made transgenic mice carrying the gene for endothelin 3 which is essential for melanocyte development, and demonstrated that the transgene is expressed in the thymic epithelium (*ET3-Tg*). In these mice, melanocytosis is observed in the skin and thymus (Fig. 6B). Furthermore, multipotent NC-derived cells are still maintained in the E17.5 thymus of *ET3-Tg* but not wild-type mice. NC-derived cells are decreased in the normal thymus at a late embryonic stage, but enforced expression of *ET3* leads to the maintenance of multipotent NC-derived cells in the developing thymus (Yamazaki unpublished data). Thus, ET3 might play roles in the maintenance of multipotency in NC-derived cells and/or their differentiation into melanoblasts and melanocytes in the thymus. Our observation is in agreement with reports using an avian system that ET3 plays a role in the maintenance of multipotent NC cells, or induces the proliferation of bipotent NC-derived cells into melanocytes and glial cells [86,87].

Roles of NC-Derived Cells in Heart and Tooth Development

NC-derived cells are rarely detected in the mature thymus. However, NC-derived cells participate in the organogenesis of the heart and jaw, and form their components [8,9,88]. NC-derived cells in the heart differentiate into the aortic bulb and aortic arch, and the dentine of teeth and bone and cartilage of the mandible and maxilla are derived from NC cells in the jaw [8,88,89]. Aberrant distribution of the cardiac NC–derived cells in the fetal heart also leads to deformity of regions of the heart other than the aortic bulb and aortic arch. Therefore, NC cells play important roles in not only the formation of components of organs, but also in the regulation of morphogenesis of organs [74].

To investigate the distribution of NC-derived cells in the heart and jaw, we traced NC-derived cells as LacZ-positive cells using *P0-cre/Rosa26R* mice. NC-derived cells in the cranio-face, including the teeth, were present at E14.5 (Fig. 7A), and they were still present in the heart and teeth at late embryonic stages (data not shown) [9,89]. Most of the dental mesenchyme in the tooth bud at E14.5 consisted of LacZ-positive NC-derived cells (Fig. 7B).

NC-Derived Cells in the Developing Tooth

NC-derived cells are well known to participate in tooth development [1,57,88-91]. Therefore, in experiments using antibodies directed against several surface receptors, the characteristics of NC-derived cells in the tooth bud were assessed. The NC-derived cells in the E13 tooth bud were shown to be enriched in the PDGFRα(+), β1-integrin (+), αV-integrin (+), CD45 (-) and α4-integrin (-) cell populations (Yamazaki unpublished data). Similar phenotypes were also observed for thymic NC-derived cells.

In the dental mesenchymal cells at E13.5, there are NC-derived cells that differentiate into the odontoblasts in teeth. Interestingly, they also give rise to melanocytes (data not shown). Multipotent NC-derived cells might be present in the tooth bud at E13.5.

Possible Presence of NC Stem Cells in Peripheral Tissues

Hematopoietic stem cells are known to remain in adult tissues, whereas NC stem cells and neuronal stem cells have been thought to be absent in adult tissues. However, Anderson et al. reported that NC stem cells with the potential to differentiate into glial cells, neurons and myofibroblasts are present in the sciatic nerve and gut of fetal rats, though they did not address the presence of the melanocyte lineage [81,82]. The presence of neuronal stem cells in the adult brain has also been reported [92,93].

We showed that multipotent NC-derived cells are present in the dental mesenchyme and fetal thymus, and these results suggest that multipotent NC-derived cells may also be present in various adult tissues. In our *ET3-Tg* mice, melanocytosis is observed in the meninges of the spinal cord and brain, which are the initial sites of NC cell generation. However, these NC-derived cells are not observed in normal mice, perhaps because some molecule(s) necessary to support NC-derived cells are not expressed in normal mice rather than because of the absence of NC-derived cells. A report showing that melanocytes were sometimes observed in the ovary, spleen and thymus of C57BL/6 mice by Weissman et al. may support our data [94].

Recently, cells with the potential to differentiate into osteoblasts, chondrocytes and odontoblasts were reported to be present in the dental pulp [95]. Although the possibility of the contribution of non-NC-derived stem cells, so-called mesenchymal stem cells [96], must be ruled out, NC stem cells might remain in the peripheral tissues, including the dental pulp, as hematopoietic stem cells and neuronal stem cells.

Future Perspectives

NC-derived cells are present in the fetal thymus, but rarely detected in adult thymus. Similarly, NC-derived cells may be present in various organs only during fetal life. Whether NC-derived cells contribute to aspects of the development of the immune system other than the thymic organization has not been studied yet. One-third of C57BL/6 mice show melanocytosis in their spleens, implying the presence of NC-derived cells in the developing spleen [94]. NC-derived cells give rise to the cranial bone, while they are not thought to participate in long bone formation [1,10]. In the adult bone marrow, lymphoid progenitors for T, B, and natural killer T cells are generated daily. Since NC-derived cells have the potential to differentiate into osteoblasts (the precursors of bone cells) and chondrocytes (which generate cartilage), it is not surprising that NC-derived cells participate in bone marrow formation. Moreover, studies of the relationship between NC-derived cells and the organogenesis of immune competent organs such as lymph nodes and Peyer's patches should be done in the future.

The nervous system is divided into the central nervous system (CNS) and peripheral nervous system (PNS), and NC cells are thought to contribute peripheral nerve cells in the PNS but not in the CNS [97]. NC cells can also differentiate into various cells surrounding the brain and spinal cord such as cells in the meninges and choroid plexus, and pericytes in the CNS [1,10,97]. Recently, Alexanian et al. reported that adult neuronal stem cells in the hippocampus can differentiate into NC-derived melanocytes and smooth muscle cells [98]. This supports the possibility that NC cells may contribute to the nerve cells in the CNS, and *vice versa*, that the CNS may contribute cells to the peripheral nerve. Mujtaba et al. reported that neuronal stem cells might be common neuronal progenitors that have the potential to differentiate into neurons and glial cells in both the CNS and PNS [97].

In this review, we discussed the presence and roles of NC-derived cells in the thymus. Usually, NC-derived cells are present and contribute to organogenesis continuously from the embryonic stage to adult life, as in the heart and cranio-face (Fig. 8A). However, in the thymus, NC-derived cells are only present in the early embryo (Fig. 8B). These findings may indicate the existence of different modes of regulating organogenesis. The model system we have developed will enable us to elucidate novel modes of regulation and roles of NC cells in organogenesis.

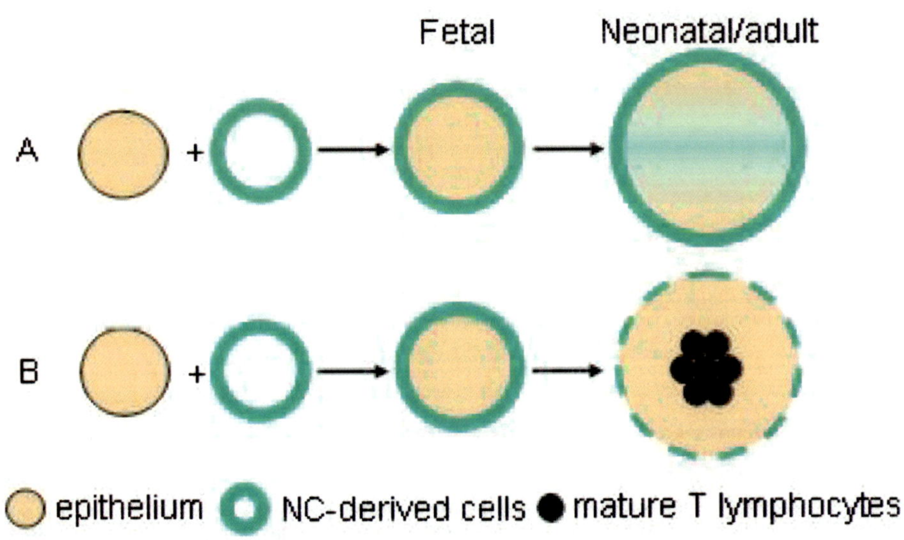

Figure 8. Hypothesis of distinct systems of NC-derived cells in organogenesis. NC-derived cells might participate in organogenesis via two modes. In the first mode (A) (for example, in the heart and cranio-face), NC-derived cells usually contribute to the organs from the embryonic stage to the adult stage and differentiate into various cells there. In the second mode (B) (for example, in the thymus), NC-derived cells contribute to the organs only in the early embryonic stage, but they are rarely present there in the late embryonic stage or adult stage.

References

[1] Le Douarin, NM., and Kalcheim, C. Cell lineage segregation during neural crest ontogeny. In: Le Douarin, NM., Kalcheim, C. (Eds.), *The Neural Crest*, Cambridge University Press, Cambridge, pp. 304-335. 1999.

[2] Anderson, DJ. Cellular and molecular biology of neural crest cell lineage determination. *Trends Genet,* 13: 276-280. 1997.

[3] Weston, JA. Phenotypic diversification in neural crest-derived cells: the time and stability of commitment during early development. *Curr Top Dev Biol,* 20:195-210. 1986.

[4] Gammill, LS., and Bronner-Fraser, M. Neural crest specification: migrating into genomics. *Nat Rev Neurosci,* 4:795-805. 2003

[5] Bockman, DE., and Kirby, ML. Dependence of thymus development on derivatives of the neural crest. *Science,* 223: 498-500. 1984.

[6] Bockman, DE., and Kirby, ML. Neural crest interactions in the development of the immune system. *J Immunol,* 135(Suppl): 766-768. 1985.

[7] Le Douarin, NM., and Jotereau, FV. Tracing of cells of the avian thymus through embryonic life in interspecific chimeras. *J Exp Med,* 142: 17-40. 1975.

[8] Jiang, X., Rowitch, DH., Soriano, P., McMahon, AP., and Sucov, HM. Fate of the mammalian cardiac neural crest. *Development,* 127: 1607-1616. 2000.

[9] Yamazaki, H., Sakata, E., Yamane, T., Yanagisawa, A., Abe, K., Yamamura, K., Hayashi, S-I., and Kunisada, T. Presence and distribution of neural crest-derived cells in the murine developing thymus and their potential for differentiation. *Int Immunol,* in press. 2005.

[10] Hall, BK. *The neural crest in development and evolution.* (Ed) Hall, BK., Springer-Velag New York, Inc. 1999.

[11] Bronner-Fraser, M. Origin and developmental potential of the neural crest. *Exp Cell Res,* 218: 405-417. 1995.

[12] Merscher, S., Funke, B., Epstein, JA., Heyer, J., Puech, A., Lu, MM., Xavier, RJ., Demay, MB., Russell, RG., Factor, S., Tokooya, K., Jore, BS., Lopez, M., Pandita, RK., Lia, M., Carrion, D., Xu, H., Schorle, H., Kobler, JB., Scambler, P., Wynshaw-Boris, A., Skoultchi, AI., Morrow, BE., and Kucherlapati, R. TBX1 is responsible for cardiovascular defects in velo-cardio-facial/DiGeorge syndrome. *Cell,* 104: 619-629. 2001.

[13] Lindsay, EA., Vitelli, F., Su, H., Morishima, M., Huynh, T., Pramparo, T., Jurecic, V., Ogunrinu, G., Sutherland, HF., Scambler, PJ., Bradley, A., and Baldini, A. Tbx1 haploinsufficieny in the DiGeorge syndrome region causes aortic arch defects in mice. *Nature,* 410: 97-101. 2001.

[14] Kochilas, L., Merscher-Gomez, S., Lu, MM., Potluri, V., Liao, J., Kucherlapati, R., Morrow, B., and Epstein, JA. The role of neural crest during cardiac development in a mouse model of DiGeorge syndrome. *Dev Biol,* 251: 157-166. 2002.

[15] Vitelli, F., Morishima, M., Taddei, I., Lindsay, EA., and Baldini, A. Tbx1 mutation causes multiple cardiovascular defects and disrupts neural crest and cranial nerve migratory pathways. *Hum Mol Genet,* 11: 915-922. 2002.

[16] Akira, S., Takeda, K., and Kaisho, T. Toll-like receptors: critical proteins linking innate and acquired immunity. *Nat Immunol,* 2: 675-680. 2001.

[17] Anderson, G., Moore, NC., Owen, JJ., and Jenkinson, EJ. Cellular interactions in thymocyte development. *Annu Rev Immunol,* 14: 73-99. 1996.

[18] Mayer, WE., Uinuk-Ool, T., Tichy, H., Gartland, LA., Klein, J., and Cooper, MD. Isolation and characterization of lymphocyte-like cells from a lamprey. *Proc Natl Acad Sci USA,* 99:14350-14355. 2002.

[19] Blackburn, CC., and Manley, NR. Developing a new paradigm for thymus organogenesis. *Nat Rev Immunol,* 4: 278-289. 2004.

[20] Anderson, G., and Jenkinson, EJ. Lymphostromal interactions in thymic development and function. *Nat Rev Immunol,* 1: 31-40. 2001.

[21] Cordier, AC., and Heremans, JF. Nude mouse embryo: ectodermal nature of the primordial thymic defect. *Scand J Immunol,* 4:193-196. 1975.

[22] Manley, NR. Thymus organogenesis and molecular mechanisms of thymic epithelial cell differentiation. *Semin Immunol,* 12: 421-428. 2000.

[23] Klug, DB., Carter, C., Gimenez-Conti, IB., and Richie, ER. Cutting edge: thymocyte-independent and thymocyte-dependent phases of epithelial patterning in the fetal thymus. *J Immunol,* 169: 2842-2845. 2002.

[24] Itoi, M., Kawamoto, H., Katsura, Y., and Amagai, T. Two distinct steps of immigration of hematopoietic progenitors into the early thymus anlage. *Int Immunol,* 13: 1203-1211. 2001.

[25] Manley, NR., and Capecchi, MR. The role of Hoxa-3 in mouse thymus and thyroid development. *Development,* 121:1989-2003. 1995.

[26] Wallin, J., Eibel, H., Neubuser, A., Wilting, J., Koseki, H., and Balling, R. Pax1 is expressed during development of the thymus epithelium and is required for normal T-cell maturation. *Development,* 122: 23-30. 1996.

[27] Hetzer-Egger, C., Schorpp, M., Haas-Assenbaum, A., Balling, R., Peters, H., and Boehm, T. Thymopoiesis requires Pax9 function in thymic epithelial cells. *Eur J Immunol,* 32:1175-1181. 2002.

[28] Su, DM., and Manley, NR. Hoxa3 and pax1 transcription factors regulate the ability of fetal thymic epithelial cells to promote thymocyte development. *J Immunol,* 164: 5753-5760. 2000.

[29] Vitelli, F., Taddei, I., Morishima, M., Meyers, EN., Lindsay, EA., and Baldini, A. A genetic link between Tbx1 and fibroblast growth factor signaling. *Development,* 129: 4605-4611. 2002.

[30] Macatee, TL., Hammond, BP., Arenkiel, BR., Francis, L., Frank, DU., and Moon, AM. Ablation of specific expression domains reveals discrete functions of ectoderm- and endoderm-derived FGF8 during cardiovascular and pharyngeal development. *Development,* 130: 6361-6374. 2003.

[31] Wendling, O., Dennefeld, C., Chambon, P., and Mark, M. Retinoid signaling is essential for patterning the endoderm of the third and fourth pharyngeal arches. *Development,* 127: 1553-1562. 2000.

[32] Conway, SJ, Henderson, DJ., and Copp, AJ. Pax3 is required for cardiac neural crest migration in the mouse: evidence from the splotch (Sp2H) mutant. *Development,* 124: 505-514. 1997.

[33] Soriano, P. The PDGF alpha receptor is required for neural crest cell development and for normal patterning of the somites. *Development,* 124: 2691-2700. 1997.

[34] Kurihara, Y., Kurihara, H., Maemura, K., Kuwaki, T., Kumada, M., and Yazaki, Y. Impaired development of the thyroid and thymus in endothelin-1 knockout mice. *J Cardiovasc Pharmacol,* 26 (Suppl3): 13-16. 1995.

[35] Xu, PX., Zheng, W., Laclef, C., Maire, P., Maas, RL., Peters, H., and Xu, X. Eya1 is required for the morphogenesis of mammalian thymus, parathyroid and thyroid. *Development,* 129: 3033-3044. 2002.

[36] Balciunaite, G., Keller, MP., Balciunaite, E., Piali, L., Zuklys, S., Mathieu, YD., Gill, J., Boyd, R., Sussman, DJ., Hollander, GA. Wnt glycoproteins regulate the expression of FoxN1, the gene defective in nude mice. *Nat Immunol,* 3: 1102-1108. 2002.

[37] Gordon, J., Bennett, AR., Blackburn, CC., and Manley, NR. Gcm2 and Foxn1 mark early parathyroid- and thymus-specific domains in the developing third pharyngeal pouch. *Mech Dev,* 103: 141-143. 2001.

[38] Blackburn, CC., Augustine, CL., Li, R., Harvey, RP., Malin, MA., Boyd, RL., Miller, JF., and Morahan, G. The nu gene acts cell-autonomously and is required for differentiation of thymic epithelial progenitors. *Proc Natl Acad Sci USA,* 93: 5742-5746. 1996.

[39] Nehls, M., Kyewski, B., Messerle, M., Waldschutz, R., Schuddekopf, K., Smith, AJ., and Boehm, T. Two genetically separable steps in the differentiation of thymic epithelium. *Science,* 272: 886-889. 1996.

[40] Suniara, RK., Jenkinson, EJ., and Owen, JJ. Studies on the phenotype of migrant thymic stem cells. *Eur J Immunol,* 29: 75-80. 1999.

[41] Wilkinson, B., Owen, JJ., and Jenkinson, EJ. Factors regulating stem cell recruitment to the fetal thymus. *J Immunol,* 162: 3873-3881. 1999.

[42] Auerbach, R. Morphogenetic interactions in the development of the mouse thymus gland. *Dev Biol,* 2: 271-284. 1960.

[43] Moore, MA., and Owen, JJ. Experimental studies on the development of the thymus. *J Exp Med,* 126: 715-726. 1967.

[44] Owen, JJ., McLoughlin, DE., Suniara, RK., and Jenkinson, EJ. The role of mesenchyme in thymus development. *Curr Top Microbiol Immunol,* 251: 133-137. 2000.

[45] Van Vliet, E., Jenkinson, EJ., Kingston, R., Owen, JJ., and Van Ewijk, W. Stromal cell types in the developing thymus of the normal and nude mouse embryo. *Eur J Immunol,* 15: 675-681. 1985.

[46] Bennett, AR., Farley, A., Blair, NF., Gordon, J., Sharp, L., and Blackburn, CC. Identification and characterization of thymic epithelial progenitor cells. *Immunity,* 16: 803-814. 2002.

[47] Blackburn, CC., Manley, NR., Palmer, DB., Boyd, RL., Anderson, G., and Ritter, MA. One for all and all for one: thymic epithelial stem cells and regeneration. *Trends Immunol,* 23: 391-395. 2002.

[48] Wilkinson, RW., Anderson, G., Owen, JJ., and Jenkinson, EJ. Positive selection of thymocytes involves sustained interactions with the thymic microenvironment. *J Immunol,* 155: 5234-5240. 1995.

[49] Ueno, T., Hara, K., Willis, MS., Malin, MA., Hopken, UE., Gray, DH., Matsushima, K., Lipp, M., Springer, TA., Boyd, RL., Yoshie, O., and Takahama, Y. Role for CCR7 ligands in the emigration of newly generated T lymphocytes from the neonatal thymus. *Immunity,* 16: 205-218. 2002.

[50] Le Lievre, CS., and Le Douarin, NM. Mesenchymal derivatives of the neural crest: analysis of chimaeric quail and chick embryos. *J Embryol Exp Morphol,* 34:125-154. 1975.

[51] Kuratani, S., and Bockman, DE. Impaired development of the thymic primordium after neural crest ablation. *Anat Rec,* 228: 185-190. 1990.

[52] Kuratani, S., and Bockman, DE. The participation of neural crest derived mesenchymal cells in development of the epithelial primordium of the thymus. *Arch Histol Cytol,* 53: 267-273. 1990.

[53] Bronner-Fraser, M. Analysis of the early stages of trunk neural crest migration in avian embryos using monoclonal antibody HNK-1. *Dev Biol,* 115: 44-55. 1986.

[54] Li, J., Chen, F., and Epstein, JA. Neural crest expression of Cre recombinase directed by the proximal Pax3 promoter in transgenic mice. *Genesis,* 26: 162-164. 2000.

[55] Yamauchi, Y., Abe, K., Mantani, A., Hitoshi, Y., Suzuki, M., Osuzu, F., Kuratani, S., Yamamura, K. A novel transgenic technique that allows specific marking of the neural crest cell lineage in mice. *Dev Biol,* 212: 191-203. 1999.

[56] Pietri, T., Eder, O., Breau, MA., Topilko, P., Blanche, M., Brakebusch, C., Fassler, R., Thiery, JP., and Dufour, S. Conditional beta1-integrin gene deletion in neural crest cells causes severe developmental alterations of the peripheral nervous system. *Development,* 131:3871-3883. 2004.

[57] Yamazaki, H., Kurino, T., Sakata, E., Yoshino, M., and Hayashi, S-I. Tooth development and tooth regeneration using tooth germ, dental pulp cells, neural crest cells and embryonic stem cells. *Recent Res Devel Biophys Biochem,* 3: 907-925. 2003.

[58] Kanazawa, N., Nakamura, T., Tashiro, K., Muramatsu, M., Morita, K., Yoneda, K., Inaba, K., Imamura, S., and Honjo, T. Fractalkine and macrophage-derived chemokine: T cell-attracting chemokines expressed in T cell area dendritic cells. *Eur J Immunol,* 29: 1925-1932. 1999.

[59] Salomon, DR., Mojcik, CF., Chang, AC., Wadsworth, S., Adams, DH., Coligan, JE., and Shevach, EM. Constitutive activation of integrin alpha 4 beta 1 defines a unique stage of human thymocyte development. *J Exp Med.,* 179: 1573-1584. 1994.

[60] Godfrey, DI., Kennedy, J., Suda, T., and Zlotnik, A. A developmental pathway involving four phenotypically and functionally distinct subsets of CD3-CD4-CD8-triple-negative adult mouse thymocytes defined by CD44 and CD25 expression. *J Immunol,* 150: 4244-4252. 1993.

[61] Savagner, P., Imhof, BA., Yamada, KM., and Thiery, JP. Homing of hemopoietic precursor cells to the embryonic thymus: characterization of an invasive mechanism induced by chemotactic peptides. *J Cell Biol,* 103: 2715-2727. 1986.

[62] Wadsworth, S., Halvorson, MJ., Chang, AC., and Coligan, JE. Multiple changes in VLA protein glycosylation, expression, and function occur during mouse T cell ontogeny. *J Immunol,* 150: 847-857. 1993.

[63] Banwell, CM., Partington, KM., Jenkinson, EJ., and Anderson, G. Studies on the role of IL-7 presentation by mesenchymal fibroblasts during early thymocyte development. *Eur J Immunol,* 30: 2125-2129. 2000.

[64] Harman, BC., Jenkinson, EJ, and Anderson, G. Entry into the thymic microenvironment triggers Notch activation in the earliest migrant T cell progenitors. *J Immunol,* 170: 1299-1303. 2003.

[65] Schmitt, TM., and Zuniga-Pflucker, JC. Induction of T cell development from hematopoietic progenitor cells by delta-like-1 in vitro. *Immunity,* 17: 749-756. 2002.

[66] Staal, FJ., and Clevers, HC. Wnt signaling in the thymus. *Curr Opin Immunol,* 15: 204-208. 2003.

[67] Shinohara, T., and Honjo, T. Epidermal growth factor can replace thymic mesenchyme in induction of embryonic thymus morphogenesis in vitro. *Eur J Immunol,* 26: 747-752. 1996.

[68] Anderson, G., Anderson, KL., Tchilian, EZ., Owen, JJ., and Jenkinson, EJ. Fibroblast dependency during early thymocyte development maps to the CD25+ CD44+ stage and involves interactions with fibroblast matrix molecules. *Eur J Immunol,* 27: 1200-1206. 1997.

[69] Tsai, PT., Lee, RA., and Wu, H. BMP4 acts upstream of FGF in modulating thymic stroma and regulating thymopoiesis. *Blood,* 102: 3947-3953. 2003.

[70] Erickson, M., Morkowski, S., Lehar, S., Gillard, G., Beers, C., Dooley, J., Rubin, JS., Rudensky, A., and Farr, AG. Regulation of thymic epithelium by keratinocyte growth factor. *Blood,* 100: 3269-3278. 2002.

[71] Revest, JM., Suniara, RK., Kerr, K., Owen, JJ., and Dickson, C. Development of the thymus requires signaling through the fibroblast growth factor receptor R2-IIIb. *J Immunol,* 167: 1954-1961. 2001.

[72] Suniara, RK., Jenkinson, EJ., and Owen, JJ. An essential role for thymic mesenchyme in early T cell development. *J Exp Med.* 191: 1051-1056. 2000.

[73] Anderson, G., Jenkinson, EJ., Moore, NC., and Owen, JJ. MHC class II-positive epithelium and mesenchyme cells are both required for T-cell development in the thymus. *Nature,* 362: 70-73. 1993.

[74] Epstein, JA., Li, J., Lang, D., Chen, F., Brown, CB., Jin, F., Lu, MM., Thomas, M., Liu, E., Wessels, A., and Lo, CW. Migration of cardiac neural crest cells in Splotch embryos. *Development,* 127:1869-1878. 2000.

[75] Jenkinson, WE., Jenkinson, EJ., and Anderson, G. Differential requirement for mesenchyme in the proliferation and maturation of thymic epithelial progenitors. *J Exp Med,* 198: 325-332. 2003.

[76] Petrie, HT. Cell migration and the control of post-natal T-cell lymphopoiesis in the thymus. *Nat Rev Immunol,* 3: 859-866. 2003.

[77] Petrie, HT. Role of thymic organ structure and stromal composition in steady-state postnatal T-cell production. *Immunol Rev,* 189: 8-19. 2002.

[78] Etchevers, HC., Vincent, C., Le Douarin, NM., and Couly, GF. The cephalic neural crest provides pericytes and smooth muscle cells to all blood vessels of the face and forebrain. *Development,* 128: 1059-1068. 2001.

[79] Itoi, M., and Amagai, T. Inductive role of fibroblastic cell lines in development of the mouse thym us anlage in organ culture. *Cell Immunol,* 183: 32-41. 1998.

[80] Gill, J., Malin, M., Hollander, GA., and Boyd, R. Generation of a complete thymic microenvironment by MTS24(+) thymic epithelial cells. *Nat Immunol,* 3: 635-642. 2002.

[81] Morrison, SJ., White, PM., Zock, C., and Anderson, DJ. Prospective identification, isolation by flow cytometry, and in vivo self-renewal of multipotent mammalian neural crest stem cells. *Cell,* 96: 737-749. 1999.

[82] Kruger, GM., Mosher, JT., Bixby, S., Joseph, N., Iwashita, T., and Morrison, SJ. Neural crest stem cells persist in the adult gut but undergo changes in self-renewal, neuronal subtype potential, and factor responsiveness. *Neuron,* 35: 657-669. 2002.

[83] Henion, PD., and Weston, JA. Timing and pattern of cell fate restrictions in the neural crest lineage. *Development,* 124: 4351-4359. 1997.

[84] Luo, R., Gao, J., Wehrle-Haller, B., and Henion, PD. Molecular identification of distinct neurogenic and melanogenic neural crest sublineages. *Development*, 130: 321-330. 2003.

[85] Ito, K., and Sieber-Blum, M. Pluripotent and developmentally restricted neural crest-derived cells in posterior visceral arches. *Dev Biol,* 156: 191-200. 1993.

[86] Lahav, R., Dupin, E., Lecoin, L., Glavieux, C., Champeval, D., Ziller, C., and Le Douarin, NM. Endothelin 3 selectively promotes survival and proliferation of neural crest-derived glial and melanocytic precursors in vitro. *Proc Natl Acad Sci USA,* 95: 15214-15219. 1998.

[87] Stone, JG., Spirling, LI., and Richardson, MK. The neural crest population responding to endothelin-3 in vitro includes multipotent cells. *J Cell Sci,* 110: 1673-1682. 1997.

[88] Chai, Y., Jiang, X., Ito, Y., Bringas, P Jr., Han, J., Rowitch, DH., Soriano, P., McMahon, AP., and Sucov, HM. Fate of the mammalian cranial neural crest during tooth and mandibular morphogenesis. *Development,* 127: 1671-1679. 2000.

[89] Yamazaki, H., and Shin-Ichi, Hayashi. Contribution of neural crest cells in tooth development and the possibility of tooth regeneration. *J Oral Sci,* 46: 509-518. 2004.

[90] Imai, H., Osumi-Yamashita, N., Ninomiya, Y., and Eto, K. Contribution of early-emigrating midbrain crest cells to the dental mesenchyme of mandibular molar teeth in rat embryos. *Dev Biol,* 176: 151-165. 1996.

[91] Lumsden, AG. Spatial organization of the epithelium and the role of neural crest cells in the initiation of the mammalian tooth germ. *Development,* 103 (Suppl):155-169. 1988.

[92] Nakatomi, H., Kuriu, T., Okabe, S., Yamamoto, S., Hatano, O., Kawahara, N., Tamura, A., Kirino, T., and Nakafuku, M. Regeneration of Hippocampal Pyramidal Neurons after Ischemic Brain Injury by Recruitment of Endogenous Neural Progenitors. *Cell,* 110: 429-441, 2002.

[93] Rietze, RL., Valcanis, H., Brooker, GF., Thomas, T., Voss, AK., and Bartlett, PF. Purification of a pluripotent neural stem cell from the adult mouse brain. *Nature,* 412: 736-739. 2001.

[94] Weissman, IL. Genetic and histochemical studies on mouse spleen black spots. *Nature,* 215: 315. 1967.

[95] Miura, M., Gronthos, S., Zhao, M., Lu, B., Fisher, LW., Robey, PG., and Shi, S. SHED: Stem cells from human exfoliated deciduous teeth. *Proc Natl Acad Sci USA,* 100: 5807-5812. 2003.

[96] Jiang, Y., Jahagirdar, BN., Reinhardt, RL., Schwartz, RE., Keene, CD., Ortiz-Gonzalez, XR., Reyes, M., Lenvik, T., Lund, T., Blackstad, M., Du, J., Aldrich, S., Lisberg, A., Low, WC., Largaespada, DA., and Verfaillie, CM. Pluripotency of mesenchymal stem cells derived from adult marrow. *Nature,* 418: 41-49. 2002.

[97] Mujtaba, T., Mayer-Proschel, M., and Rao, MS. A common neural progenitor for the CNS and PNS. *Dev Biol,* 200: 1-15. 1998.

[98] Alexanian, AR., Sieber-Blum, M. Differentiating adult hippocampal stem cells into neural crest derivatives. *Neuroscience,* 118: 1-5. 2003.

In: Trends in Stem Cell Research
Editor: Erik V. Greer, pp. 169-190

ISBN 1-59454-315-1

Chapter VII

Stem Cells of the Seminiferous Epithelium

Dirk G. de Rooij *

Departments of Endocrinology, Faculty of Biology, Utrecht University and of Cell Biology, UMC U, Utrecht, The Netherlands

Abstract

The spermatogenic lineage starts with the formation of primordial germ cells (PGCs) that migrate to the genital ridges. There they become enclosed by Sertoli cell precursors in seminiferous cords and differentiate into gonocytes. After birth the gonocytes give rise to adult type spermatogonial stem cells and various types of spermatogonia. Spermatogonial stem cells are at the beginning of the adult spermatogenic process. In non-primate mammals this are single cells (A_s spermatogonia). Their daughter cells either migrate away from each other and become two new stem cells or stay together, connected by an intercellular bridge. The latter designates these cells to differentiate and ultimately become spermatozoa.

Spermatogonial stem cells can be counted by morphologically distinguishing these cells in whole mounts of seminiferous tubules or by way of a functional assay, the spermatogonial stem cell transplantation technique. To work with spermatogonial stem cells it would be nice if one could purify and/or culture these cells. Fortunately, the search for specific markers that would facilitate the purification of these cells is progressing rapidly. Furthermore, pure spermatogonial stem cells proved difficult to culture, better results have been obtained with co-cultures of Sertoli cells with mouse and bovine spermatogonia. Nevertheless no culture protocol through which one can study the regulation of proliferation and differentiation of these cells is yet available.

---

* Dirk G. de Rooij, Ph.D. Department of Endocrinology, Faculty of Biology, Utrecht University; H.R. Kruytgebouw, Padualaan 8 3584 CH Utrecht The Netherlands; Tel. +31 (0)30 2533084; Fax +31 (0)30 2532837; E-mail d.g.derooij@bio.uu.nl; http://www.bio.uu.nl/endocrinology/

Infertile mutant and genetically modified mice have indicated the presence of regulatory mechanisms governing the proliferation, differentiation, migration and survival of PGCs, gonocytes and adult spermatogonial stem cells. For example, a major factor that plays a role in regulating stem cell fate is glial cell line derived neurotrophic factor (GDNF) produced by Sertoli cells which inhibits stem cell differentiation. Furthermore, the *Plzf* gene encoded protein, expressed in spermatogonial stem cells, also inhibits their differentiation. Finally, spermatogonial stem cells are mainly localized to those tubule areas that border on the interstitial tissue. Apparently, also factor(s) from the interstitium, possibly testosterone or FSH, inhibit stem cell differentiation. Recent progress in understanding the regulation of spermatogonial stem cell behavior is discussed.

Finally, spermatogonial multiplication and stem cell renewal in primates is discussed which differs from that in other mammals.

Keywords: spermatogenesis, spermatogonial stem cells, testis, GDNF, transplantation, cell culture, cell lines

The Seminiferous Epithelium

Spermatogenesis in mammals takes place in seminiferous tubules (reviewed in [1]). These tubules are lined by peritubular myoid cells and contain both germ cells at all steps of their development into spermatozoa, as well as a somatic cell type, the Sertoli cells (Figure 1). The Sertoli cells are situated on the basal membrane of the tubules but have their cytoplasmic ramifications throughout the epithelium up to the tubule lumen. These cells support the developing germ cells by the secretion of growth factors [2] and other factors like lactate and pyruvate that some types of germ cells need for their metabolism [3]. In between the tubules there is interstitial tissue containing Leydig cells that produce testosterone (Figure 1).

The process of spermatogenesis starts with a series of mitotic divisions carried out by a cell type called spermatogonia, that are situated on the basal membrane of the seminiferous tubules. The last mitotic division renders the next cell type, the spermatocytes, that go through S phase, then pass through the lengthy prophase of the first meiotic division and subsequently carry out the two meiotic divisions to give rise to haploid spermatids. Initially, spermatids have a round shape but then elongate to become spermatozoa that leave the seminiferous tubules through the tubule lumen [1].

Stem cells are at the basis of spermatogenesis and are both able to selfrenew and to give rise to differentiating daughter spermatogonia. This dual capacity of spermatogonial stem cells ensures the long-lasting ability of the testes to produce spermatozoa. In recent years knowledge about these important cells is growing rapidly, especially in our understanding of the regulation of the behavior of these cells. The progress in this field will be reviewed.

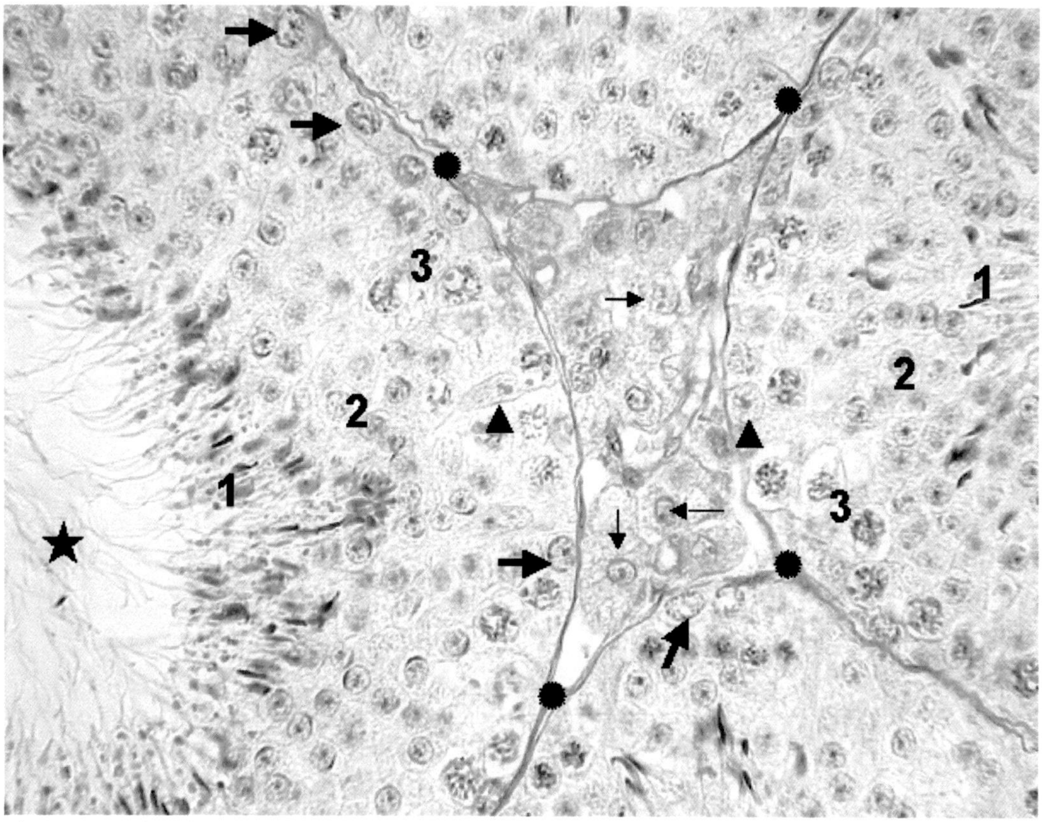

Figure 1. Section through various seminiferous tubules of an adult mouse. Asterisk - Tubule lumen through which the spermatozoa are transported out of the testis; 1 - Layer with spermatids that already have acquired most of the spermatozoal shape and that will be released into the tubule lumen within a couple of days; 2 - Round spermatids that yet have to start their elongation into spermatozoa; 3 - Spermatocytes in the prophase of the first meiotic division; Fat arrows - Spermatogonia on the basal membrane of the seminiferous tubules; Thin arrows - Leydig cells in the interstitial tissue in between the seminiferous tubules; Multipointed stars - The tubule basal membrane in between these stars, bordering on the interstitial tissue, represents the area in which most of the spermatogonial stem cells reside. This area can be seen as the stem cell niche in the seminiferous epithelium.

Development of the Seminiferous Epithelium

Spermatogonial stem cells originate from primordial germ cells (PGCs) which in turn derive from epiblast cells (embryonal ectoderm) [4]. By 7 days post coitum (p.c.) in the mouse embryo, about 100 alkaline phosphatase positive PGCs can be detected in the extraembryonal mesoderm posterior to the definitive primitive streak [5]. Later in development the PGCs migrate from the base of the allantois, along the hindgut to finally reach the genital ridges. The PGCs proliferate during migration and by day 13 of fetal life in the mouse, when these cells have reached the genital ridges, their numbers have increased to about 10,000 in each gonad [6].

Primordial germ cells are single cells that under certain culture conditions can form colonies of cells which morphologically resemble undifferentiated embryonic stem cells (ES cells; [7]). These cells can be maintained on feeder layers for extended periods of time and can give rise to embryoid bodies and to multiple differentiated cell phenotypes in monolayer culture and in tumors in nude mice. Primordial-germ-cell-derived ES cells can also contribute to chimeras when injected into host blastocysts [7]. Clearly, PGCs are stem cells, still having the capacity to renew themselves and to differentiate in various directions.

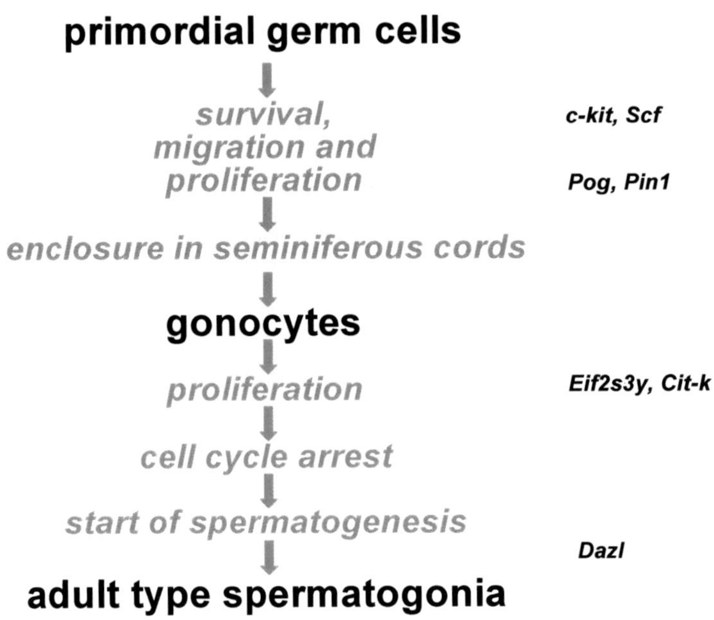

Figure 2. Scheme of the order of events during the development of primordial germ cells and gonocytes up till the start of spermatogenesis. To the right some genes have been indicated that play a role at specific developmental steps.

Once arrived in the genital ridges, the PGCs become enclosed by the differentiating Sertoli cells, and seminiferous cords are formed. The germ cells present within the seminiferous cords differ morphologically from PGCs and are called gonocytes [8-10] or various subsequent types of prospermatogonia [11]. In rats and mice, after formation the gonocytes proliferate for a few days and then become arrested in the G_0/G_1 phase of the cell cycle [8, 10-13] (Figure 2). Shortly after birth the gonocytes resume proliferation to give rise to A spermatogonia [9, 10, 13-15]. This event marks the start of spermatogenesis (Figure 3).

At the mitoses of the gonocytes, cytokinesis is often not complete and many gonocytes remain interconnected by intercellular bridges [16, 17]. As discussed below, in the adult testis the first visible sign of differentiation of stem cells is the formation of a pair of cells interconnected by an intercellular bridge. These paired spermatogonia are destined to ultimately become spermatozoa. Likely, the gonocytes are a heterogenous population of cells only some of which still have stem cell properties, while the rest are destined to become differentiating cells directly at the start of spermatogenesis after birth.

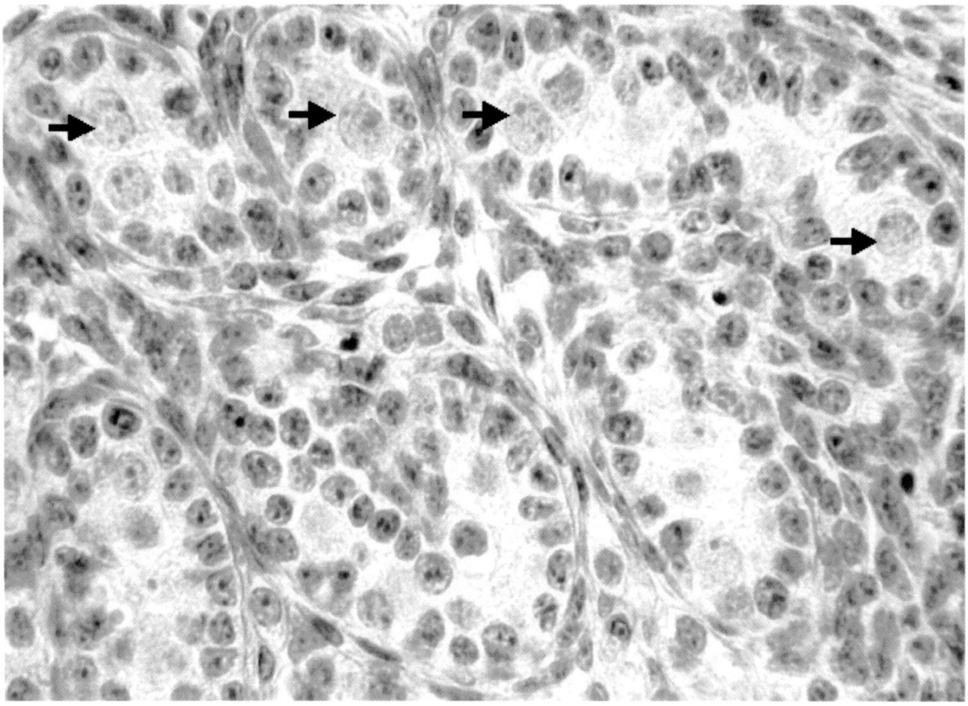

Figure 3. Photograph of a section of a neonatal mouse testis showing seminiferous cords, still lacking a lumen, and quiescent gonocytes (arrows) that will start spermatogenesis a couple of days later. The cords are lined by Sertoli cell precursors that are still dividing.

Regulatory Mechanisms in the Development PGCs and Gonocytes

Regulatory mechanisms govern the behavior of PGCs and gonocytes. Some of the genes and encoded proteins involved have become apparent from the effects of gene deficiency in mutant and genetically modified mice [18].

Primordial Germ Cells

A number of natural mutations affect the behavior of PGCs. The most well known involve the c-Kit receptor and its ligand Stem Cell Factor (SCF), the *W/W* and *Sl/Sl* mutants, respectively. There are many mutant alleles with a varying severity of the phenotype but in many there is an impairment of fertility because of an (almost) total absence of germ cells, in addition to defects in other stem cell systems [19-21]. In *W/W* and *Sl/Sl* mice, normal numbers of PGCs are present at day 8.5 p.c. but few are able to migrate to the gonads [22, 23]. In vitro experiments have shown that the action of the c-kit/SCF pathway is necessary for PGC survival but not the proliferation of these cells [24, 25]. (Figure 3).

Mice with the insertional mutation *gcd* show only few PGCs in the fetal male and female gonads [26]. The gene mutated in *gcd* mice is *Pog* (*proliferation of germ cells*) which encodes a protein with a plant homeodomain (PHD) at its C terminus [27, 28]. PHD motifs are common in proteins involved in chromatin remodeling. The Pog protein is essential for PGC proliferation but probably has no role in spermatogonial proliferation in the adult testis.

Pin1, a peptidyl–prolyl isomerase, is also involved in the regulation of PGC proliferation. In *Pin1 -/-* mice, PGCs go through fewer cell cycles and strikingly fewer PGCs are found in the gonads [29]. Adult *Pin1 -/-* males are subfertile and many seminiferous tubules are markedly depleted from germ cells.

Finally, there are two mutations in which the specific effect on PGCs is not yet clear. In mice homozygous for a mutation of unknown etiology called *atrichosis (at/at)*, very few if any germ cells are present in male and female gonads [30] suggesting problems at the PGC level. Another mutation concerns connexin 43. Connexins are a family of proteins that form intercellular channels of gap junctions. Mice homozygous for a null mutation in *Gja 1,* the gene encoding connexin 43, have a deficiency of germ cells at birth that can be traced back to day 11.5 p.c., implying that not enough PGCs reach the gonadal ridges [31].

Gonocytes

Gonocytes proliferate during a couple days after their formation and then become quiescent till after birth, when the gonocytes give rise to adult type A spermatogonia. From the rate of the appearance of the subsequent germ cell types after the start of spermatogenesis it can be inferred that gonocytes at their first division produce A2 spermatogonia implicating that part of the gonocytes should differentiate into A1 spermatogonia before their first division. As spermatogenesis continues after the first wave, A_s, A_{pr} and A_{al} spermatogonia should be formed too by gonocytes [32].

In *XOSxr*^{b} mice, gonocyte proliferation is decreased and the gene on the Y chromosome involved was called *Spy.* Recently, this gene was found to be similar to *Eif2s3y* [33, 34]. Eif2s3y encodes a subunit of the eukaryotic translation factor Eif2 (Figure 3).

An interesting model is a mouse deficient for citron kinase (Cit-k), a myotonin-related protein acting downstream of the GTPase Rho in cytokinesis control during proliferation. Multinucleated cells and extensive apoptosis among gonocytes was seen in *Cit-k -/-* mice [35]. Cytokinesis is a very special process in spermatogenesis as incomplete cytokinesis predestines cells for the differentiation pathway. After birth in *Cit-K -/-* mice, apoptosis was also seen among the few A spermatogonia formed by the gonocytes. Apparently, the Cit-k protein is necessary for all those cell divisions of gonocytes and spermatogonia that involve the formation of intercellular bridges.

At day 9 after birth in *Deleted in azoospermia-like (Dazl)* knock out mice no preleptotene spermatocytes are present, indicating that gonocytes are unable to differentiate into A1 spermatogonia that will follow the differentiation track. Instead, at day 9 after birth only A spermatogonia and no spermatocytes are present [36] (Figure 3). The *DAZ* gene encodes a RNA binding protein the precise action of which in germ cells is still unknown but apparently is indispensable in gonocytes for the formation of A1 spermatogonia.

Spermatogonial stem cells are single cells located on the basal membrane of the seminiferous tubules and are called A-single (A_s) spermatogonia (review [37]). These cells either divide into two new single cells or into a pair of spermatogonia (A_{pr}) that do not complete cytokinesis and stay connected by an intercellular bridge [38, 39](Figure 4). In all further divisions, starting with the pair, cytokinesis will also be incomplete, leading to the formation of increasingly large syncytia of germ cells. As A_{pr} spermatogonia are morphologically similar to A_s spermatogonia, the intercellular bridge can be taken as the first visible sign of the entrance of the cells into the differentiation pathway.

The A_{pr} spermatogonia divide further to form chains of 4, 8, up to occasionally 32 A-aligned (A_{al}) spermatogonia. The A_{al} spermatogonia can go through a differentiation step and become so-called A1 spermatogonia. This differentiation step involves slight morphological changes [40] and brings about changes in cell cycle characteristics of the spermatogonia [41-44]. In most non-primate mammals there are six divisions following the formation of A1 spermatogonia, the last of which giving rise to spermatocytes. In total, there are about 10 spermatogonial divisions between the spermatogonial stem cells and the formation of spermatocytes [37] (Figure 4).

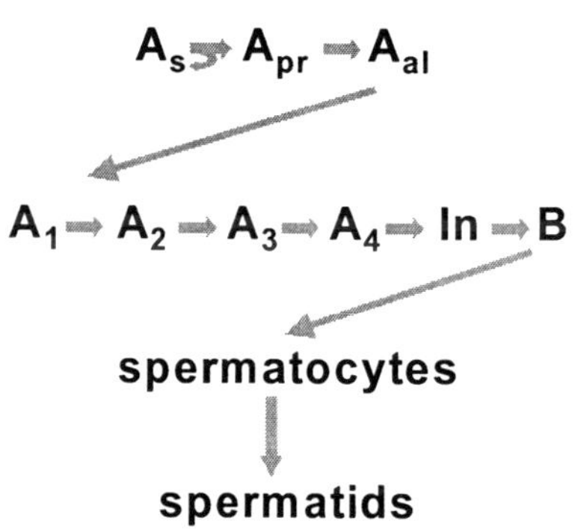

Figure 4. Scheme of spermatogonial multiplication and stem cell renewal in rats and mice.

Regulation of Stem Cell Renewal and Differentiation

Like in all other renewing tissues, the seminiferous epithelium is able to react to (stem) cell loss by enhanced stem cell renewal in order to replace lost stem cells. After a high dose of irradiation, surviving spermatogonial stem cells almost only self-renew during at least their first 6 divisions, leading to a rapid recovery of stem cell numbers [45]. This indicates that there are mechanisms that can inhibit stem cell differentiation and/or enhance self-renewal in situations of cell loss.

In several renewing tissues, stem cells occupy specific areas. For example in the intestine, stem cells reside near the bottom of the crypts [46] and stem cells in the bone marrow also occupy specific niches [47]. Until recently, in the seminiferous epithelium no such niches were found for spermatogonial stem cells. Now it has become clear that most spermatogonial stem cells are present in those areas of seminiferous tubules that border on interstitial tissue [48] (Figure 1). Apparently, the interstitial tissue affects stem cell behavior in such a way that differentiation is less likely to occur when stem cells lie close to it. Interestingly, high testosterone levels have been found to prevent spermatogonial differentiation [49-51]. Possibly, testosterone levels, which will be the highest in the areas bordering on the interstitial tissue, also have a role in regulating stem cell behavior. However, germ cells do not possess androgen receptors so that testosterone can only indirectly affect spermatogonia via peritubular myoid cells or (more likely) Sertoli cells that both express this receptor.

The ratio between self-renewal and differentiation of spermatogonial stem cells being under the control of regulatory mechanisms, the question arises which molecular pathways are involved. Recent data indicate the involvement of glial cell line derived neurotrophic factor (GDNF). Normally, GDNF is secreted by Sertoli cells [52] while a subset of spermatogonia express both receptors for this growth factor, Ret and GFR-alpha1 [53]. Ectopic expression of GDNF in spermatogonia induces the formation of large clusters of single type A spermatogonia, while normal spermatogenesis is suppressed. Moreover, in mice overexpressing GDNF in spermatogonia, germ cell tumors that resemble human seminoma, are formed at about one year of age [54]. GDNF deficient mice die during the first postnatal day [55] whereas heterozygotes survive. In heterozygotes, spermatogenesis deteriorates with age as germ cells become depleted [53]. It was concluded that GDNF has a role in the regulation of self-renewal and differentiation of spermatogonial stem cells. Too high levels of GDNF inhibit stem cell differentiation and cause an accumulation of stem cells and low levels stimulate differentiation and cause stem cell depletion (Figure 5).

In the experiments described by Meng et al the GDNF was produced by spermatogonia while this growth factor normally is produced by Sertoli cells [53]. A more natural situation was created by Yomogida et al [56] as they transfected mouse Sertoli cells. Indeed, increased expression of GDNF in Sertoli cells also caused the appearance of stem cell clusters in mouse seminiferous tubules [56]. Interestingly, the production of GDNF by Sertoli cells is stimulated by FSH [57]. This indicates that the spermatogonial stem cell niche may also be maintained by the higher levels of FSH in the interstitial tissue.

An important recent finding is that in the classical spontaneous mouse mutant *luxoid*, adult males exhibit a progressive loss of spermatogonial stem cells with age. This mouse contains a nonsense mutation in the gene encoding Plzf, a transcriptional repressor that regulates the epigenetic state of undifferentiated cells [58]. In accordance with this, spermatogonial stem cells from *lu/lu* mice are unable to colonize the testis of a recipient mouse upon transplantation. This indicates that expression of the *Plzf* gene in spermatogonial stem cells plays a key role in the regulation of self-renewal and differentiation of these cells [58] (Figure 5). The BTB/POZ domain of Plzf is able to recruit members of the mammalian Polycomb family, e.g. BMI1, which in turn recruit histone deacetylases to discrete nuclear foci [59, 60]. Interestingly, BMI1 has recently been shown to be required for the maintenance of hemopoietic and leukaemic stem cells [61, 62]. Furthermore, fusion of the BTB/POZ domain of Plzf to RARalpha in humans with acute promyelocytic leukemia is thought to maintain myeloid cells in an undifferetiated state through the recruitment of other nuclear co-repressors and HDACs [63]. Thus, as discussed by Buaas et al [58], Plzf is thought to directly regulate the epigenetic repression of chromatin domains required for cell differentiation [60].

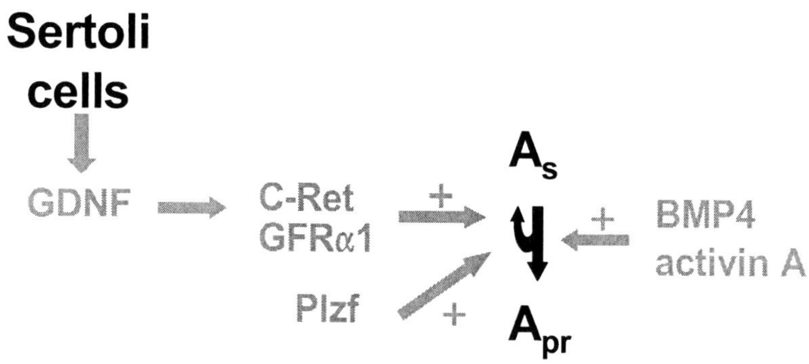

Figure 5. Summary of the present knowledge on the molecular regulation of spermatogonial stem cell self-renewal and differentiation. To the left are the factors that promote self-renewal and to the right those that promote differentiation.

Spermatogonial Stem Cell Transplantation

Spermatogonial stem cells in non-primate mammals are unique in that they can be distinguished morphologically in whole mounts of seminiferous tubules and counted [37]. However, this is a very laborious procedure and as soon as the cells are purified they can no longer be identified. Therefore, it was a real breakthrough when a functional assay for these cells was developed, the spermatogonial stem cell transplantation technique [64, 65]. With this technique a cell suspension can be checked for the presence of functional spermatogonial stem cells, for example allowing the development of purification protocols. In this procedure, germ cells of one mouse are transplanted into the testes of a recipient mouse, the endogenous spermatogenesis of which is depleted because of the *Wv/Wv* mutation or by treatment of the mice with the alkylating agent busulfan. An alternative method to kill the endogenous spermatogonial stem cells in the recipient mice uses fractionated X-irradiation (local testicular doses of 1.5 and 12 Gy, 24 hr apart). The latter protocol reliably causes more than 95% depletion [66]. After transplantation, the donor stem cells repopulate the seminiferous epithelium of the recipient mice. Interestingly, also rat spermatogonial stem cells are able to repopulate the mouse testis and produce normal rat spermatogenesis in the mouse [67, 68]! However, stem cells from other species transplanted into mouse testes either produce defective spermatogenesis (hamster [69]) or initiate repopulation by spermatogonia only, as these cells apparently fail to develop further (rabbit and dog [70]; bull [71]). Recently, the possibility of carrying out spermatogonial stem cell transplantation in larger domestic animals has been shown for the pig, goat and bull [72-74, 75].

The spermatogonial stem cell transplantation technique has been instrumental to answer the question whether particular genes, required for normal spermatogenesis, are expressed in germ cells or in Sertoli cells. For this purpose spermatogonial stem cells from mutant or genetically modified mice are transplanted into wild type testes or vice versa (e.g. [58, 66, 76-78]).

Side Population Stem Cells in the Testis

In recent years it has been shown possible to purify stem cells of various tissues on the basis of their ability to rapidly efflux the fluorescent DNA-binding dye Hoechst 33342 [79]. These cells were called side population cells (SP) and could be identified readily by using dual-wavelength flow cytometric analysis. In the bone marrow the SP cells represent 0.05-0.1 % of the nucleated cells [80]. With the help of transplantation assays it was shown that SP cells in the bone marrow have a longterm reconstituting ability and are hemopoietic stem cells [79]. Interestingly, SP cells were also found in skeletal muscle, brain, mammary gland, prostate and liver, and were suggested to be stem cells there too [81-86]. In contrast, SP cells isolated from the mouse epidermis and from the adult rat kidney do not seem to possess stem cell properties for epidermal and kidney cell types, respectively [87, 88]. The enhanced exclusion of the Hoechst dye in SP cells is caused by the expression in these cells of ABC transporter proteins like breast cancer resistance protein 1 (BCRP1) or multidrug resistance protein 1 (MDR1). As yet there is no complete picture of the role of these proteins but it may play a protective role in the sensitivity of SP cells to cytotoxic substrates [89] and to hypoxia [90].

In the testis contrasting conclusions have been drawn. Testicular SP cells were first described by Kubota et al. [91]. Upon transplantation of the isolated SP cells, these cells appeared not to be able to colonize the testes of recipient mice and it was concluded that testicular SP cells do not represent spermatogonial stem cells. Testicular SP cells were also isolated by two other groups and both did find a positive transplantation assay, indicating the presence of spermatogonial stem cells in their cell suspensions [92, 93]. Looking for differences in experimental design, Kubota et al used germ cells from cryptorchid testes as a starting material in which only early type spermatogonia are present [94], while Falciatori et al and Lassale et al used young mice and normal adult mice, respectively. The reason for the different findings is not clear but, as suggested by both of the latter groups, may lie in the use of a different starting material. One puzzling finding was that while in both publications about 1 % of the total cells were SP cells, the enrichment of spermatogonial stem cells, as determined by the transplantation assay, was only about 15 fold. This suggested the presence of SP cells incapable of colonizing the seminiferous epithelium in the SP population.

Interestingly, recent data indicate that the testicular SP also contains Leydig stem cells [95]. So, as suggested by Lo et al [95] the testicular SP population might be heterogenous, harboring SP cells of multiple origins.

Purification of Spermatogonial Stem Cells

In the adult mouse testis, there are about 35.000 spermatogonial stem cells which is 0.03% of all germ cells [96].Various techniques have been developed to purify the total population of A spermatogonia, achieving a purity varying between 85 to 98% [15, 97, 98]. Unfortunately, in the mouse only about 3% of the A spermatogonia are stem cells [96] and it will not likely be much different in other animals. Hence, although a 100-fold enrichment of stem cells can be achieved by purifying A spermatogonia, the purity is still very low. To further increase the purity, a method has been developed to isolate spermatogonia from vitamin A deficient animals [99]. In vitamin A deficient rats and mice, spermatogenesis is arrested at the differentiation step of A_{al} into A1 spermatogonia and the testes of these animals only contain A_s, A_{pr} and A_{al} spermatogonia [100]. Starting from testes of vitamin A deficient animals, theoretically a cell population containing about 10% stem cells can be obtained [96, 99]. However, the recent experimental model of Yomogida et al [56], in which mouse Sertoli cells were made to produce excessive amounts of GDNF now seems to provide the superior starting material to purify spermatogonial stem cells. In these mice the seminiferous tubules contain numerous clusters of normal stem cells and a very small number of other germ cells. A very high purity of stem cells may well be obtained with these testes.

Certain markers have been found that can be used to enrich spermatogonial stem cells [101, 102]. Using antibodies against beta1-integrin and alpha6-integrin and negatively selecting for the c-kit receptor, which is not present on spermatogonial stem cells [103], a 40-fold enrichment of spermatogonial stem cells from testicular germ cells could be accomplished [101]. A recently described additional possibility to further purify spermatogonial stem cells from testis cell suspensions is the surface marker CD9 [104]. This is a type III membrane with four transmembrane domains, expressed on many cell types. A more comprehensive approach to find spermatogonial stem cell markers has been instigated by Hamra et al [105] using microarray techniques. These authors found early growth response protein 3 (Egr3) to be a marker for spermatogonial stem cells.

Taken together, considerable progress is being made both in the development of experimental models in which the testes are enriched for spermatogonial stem cells and in finding markers for these cells.

Spermatogonial Cultures and Cell Lines

Clearly, it would be a breakthrough in spermatogonial stem cell research when one could culture spermatogonial stem cells in such a way that the regulation of their proliferation and differentiation could be studied. Alternatively, the same could be done when cell lines of spermatogonial stem cells would be available that could still be induced to differentiate. A number of attempts to reach either of these goals have been made.

First, attempts have been made to culture pure populations of stem cells in the absence of serum or a feeder layer. Unfortunately, this approach has been less successful in that few cells survive one week of culture [98, 106]. Nevertheless, in a longterm culture of a mixed germ cell suspension, in the presence of serum and on a feeder layer, some spermatogonial stem cells did survive and were able to repopulate a recipient mouse testis after transplantation [107]. Recent results indicate that the maintenance of spermatogonial stem cells in culture can be improved by the addition of glial cell line derived neurotrophic factor (GDNF) and that the presence of Sertoli cells, activin A and BMP4 is deleterious to stem cells probably through the induction of differentiation [108].

Long-term survival (25 days) and proliferation of mouse spermatogonia cultured with Sertoli cells and in the presence of serum has also been achieved [109]. Recently, a bovine spermatogonia / Sertoli cell co-culture system was developed allowing survival, proliferation and differentiation up to cells showing characteristics of spermatids, during at least 4 months [110]. However, the course of stem cell numbers with time has not yet been established in the cultures.

Second, rat spermatogonial stem cell lines have been established by immortalizing isolated type A spermatogonia by transfection with pSV3-neo (containing the SV40 large T-antigen) [111]. The cell lines express germ cell characteristics, have the appearance of A spermatogonia but lack c-kit expression. Through over a 100 passages the cell lines proved to be stable. Furthermore, the cell lines were able to colonize a host mouse testis after transplantation, indicating their stem cell capacity. However, no differentiation into further advanced spermatogenic cells was described as yet.

Another recently described cell line is the one developed by Feng et al [112]. In this case A spermatogonia were immortalized by telomerase transfection. This cell line is c-kit positive and can be induced to differentiate up to spermatids by adding SCF to the culture. It still has to be established whether or not this cell line has stem cell properties as several reports indicate that spermatogonial stem cells are c-kit negative [103, 113].

In conclusion, to date there is no method to readily propagate spermatogonial stem cells in culture and to easily study the regulation of the proliferation and differentiation of these cells. Furthermore, the cell lines established still seem deficient in their differentiation capacity or may already be more advanced than the stem cell stage. Clearly, more will have to done in this area.

Spermatogonial Stem Cells in Primates

Spermatogonial multiplication and stem cell renewal in primates has been described in much less detail than in rodents. As in rodents, in primates there are A and B spermatogonia but the composition of the type A spermatogonial population is less defined. The A spermatogonia have been subdivided into A_{pale} and A_{dark} spermatogonia according to their nuclear staining with hematoxylin [114] (Figure 6). The A_{pale} spermatogonia divide once every cycle of the seminiferous epithelium. In the human, this means that they divide only once every 16 days [115, 116]. A_{dark} spermatogonia in the normal situation are quiescent cells and are supposed to be reserve (stem) cells.

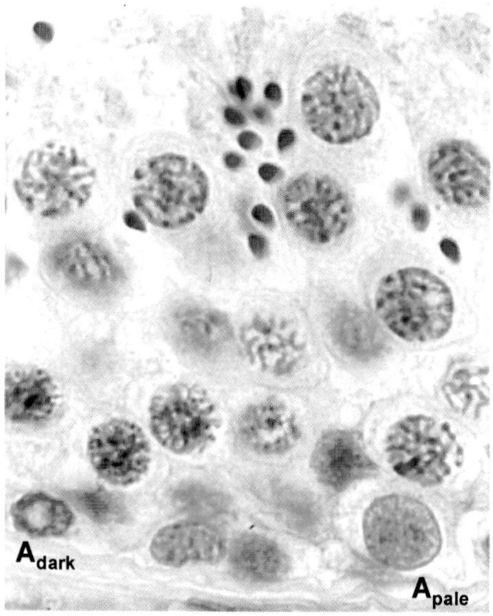

Figure 6. Photograph of part of a human seminiferous tubule, showing A_{pale} and A_{dark} spermatogonia as indicated.

An important question is whether A_{pale} and A_{dark} are spermatogonial stem cells. To answer this question, an obvious approach is to look at the topographical arrangement of these spermatogonia on the basal membrane. In the normal seminiferous epithelium the density of A_{pale} and A_{dark} spermatogonia is relatively high. Therefore the clones are too close together to decide whether or not A_{pale} and A_{dark} spermatogonia consist of clones of 1 or 2^n cells comparable to spermatogonia in non-primate mammals. However, such a clonal arrangement could be observed for both A_{pale} and A_{dark} spermatogonia during repopulation after irradiation when spermatogonial density is much lower [117]. In that situation singles, pairs and chains of A_{pale} and A_{dark} spermatogonia can be observed in tubule whole mounts. As it seems unlikely that the spermatogonial compartment would be principally different between primates and non-primate mammals it has been suggested that only the single cells among the A_{pale} and A_{dark} spermatogonia have stem cell properties [117, 118] (Figure 7).

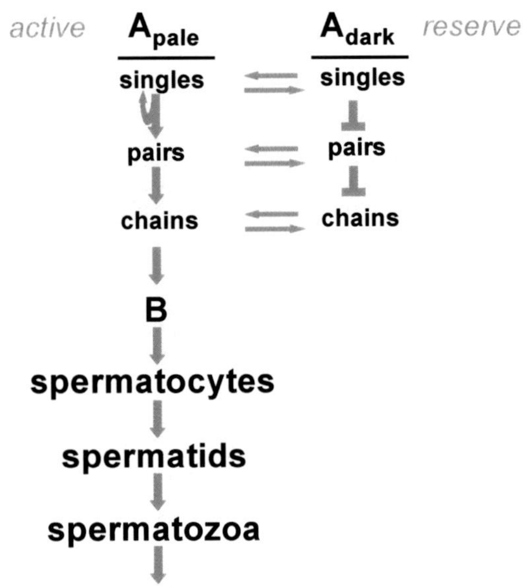

Figure 7. Proposed scheme of spermatogonial multiplication and stem cell renewal in primates. The A_{dark} spermatogonia do not proliferate but transform into A_{pale} first. Likely only the single A_{pale} and A_{dark} spermatogonia have stem cell properties.

Now what is the functional difference between A_{pale} and A_{dark} spermatogonia? It has generally been assumed that A_{dark} spermatogonia are reserve (stem) cells. The first real evidence for such a reserve function came from a study in irradiated monkeys. In rhesus monkeys, shortly after irradiation, it was found that the number of A_{pale} spermatogonia decreased to a minimum at about day 9 after irradiation, while there was no change in the number of A_{dark} spermatogonia [119]. This pattern can be explained by the different proliferative activity of these cells. Irradiation kills the cells when they divide and consequently the proliferating A_{pale} die and the quiescent A_{dark} survive. However, after longer intervals than 9 days, the A_{dark} spermatogonia also decreased in number concomitantly with a transient rise in the number of A_{pale}. Finally, both cell types dropped to very low numbers. The transient nature of the increase in A_{pale} after more than 9 days after irradiation can be explained by assuming that after the decline in A_{pale} numbers, the A_{dark} spermatogonia are activated. The activation first causes the A_{dark} to acquire the A_{pale} appearance and accompanying proliferative activity. Then having become A_{pale}, they try to divide but due to the lethal radiation damage, acquired during the time they were A_{dark}, proper division fails and the cells enter apoptosis.

In summary, the A_{pale} and A_{dark} spermatogonia in primates are topographically arranged in singles, pairs and chains of spermatogonia. Unless spermatogonial renewal and multiplication in primates is totally different from that in non-primate mammals only the singles among them are stem cells and are comparable to the A_s in non-primates (Figure 7). Only the A_{pale} spermatogonia are able to proliferate and do so once every epithelial cycle. From this it can be deduced that the formation of a chain of 8 from a differentiating stem cell will take 3 cycles, i.e. 48 days in the human. The A_{dark} spermatogonia are quiescent and only are activated after cell loss. When the latter occurs they become A_{pale} spermatogonia first and then start to proliferate. During repopulation by surviving stem cells new A_{dark} spermatogonia are set aside again [117]. Comparing primate to non-primate spermatogenesis it is clear that there is much less proliferative activity of the stem cells in the primate seminiferous epithelium, even by the active stem cells, the A_{pale}. A quiescent stem cell compartment, the A_{dark}, is missing in non-primate mammals. The low proliferative activity of spermatogonial stem cells in primates seems advantageous since this lowers the chance of errors in DNA duplication during S phase. The less stem cells divide the better [120].

Clearly, research on spermatogonial stem cells in primates is lagging behind that in other mammals. One reason of course is the difficulty of getting testis material and the other is that the presence of the A_{dark} population and the very slow proliferation rate makes things more difficult. Hopefully, the knowledge about these cells in non-primate mammals which is rapidly growing will for a large part also hold for primates.

Concluding Remarks

While until recently the emphasis of spermatogonial research was more on the regulation of proliferation and differentiation of the population of A_s, A_{pr} and A_{al} spermatogonia as a whole, now specific data on the molecular regulation of spermatogonial stem cell behavior are rapidly emerging. GDNF produced by Sertoli cells and its receptors on the stem cells, and Plzf, the gene involved in the *luxoid* mutation and expressed in stem cells, seem directly involved in the regulation of stem cell renewal and differentiation. Furthermore, testosterone clearly has an indirect role and may even be responsible for the intriguing fact that spermatogonial stem cells are preferably present in those areas of seminiferous tubules that border on interstitial tissue. These findings together with the possibility to do functional tests for stem cell potential by way of the spermatogonial stem cell transplantation technique, as well as the progress that is recently made in developing culture techniques and in finding markers for spermatogonial stem cells open this field for new approaches to establish the regulatory mechanisms that govern spermatogonial stem cell renewal and differentiation.

References

[1] Russell LD, Ettlin RA, Hikim APS, Clegg ED. *Histological and histopathological evaluation of the testis*: Cache River Press, Clearwater, Fl. USA; 1990.

[2] Jegou B. The Sertoli-germ cell communication network in mammals. *Int Rev Cytol.* 1993;147:25-96.

[3] Jutte NH, Jansen R, Grootegoed JA, Rommerts FF, Clausen OP, van der Molen HJ. Regulation of survival of rat pachytene spermatocytes by lactate supply from Sertoli cells. *J Reprod Fertil.* 1982;65:431-438.

[4] Lawson KA, Pederson RA. Clonal analysis of cell fate during gastrulation and early neurulation in the mouse. *Ciba Foundation Symposium 165. Post implantation development in the mouse.* New York: John Whiley & Sons; 1992:3-26.

[5] Ginsburg M, Snow MH, McLaren A. Primordial germ cells in the mouse embryo during gastrulation. *Development.* 1990;110:521-528.

[6] Tam PP, Snow MHL. Proliferation and migration of primordial germ cells during compensatory growth in mouse embryos. *J Embryol Exp Morphol.* 1981;64:133-147.

[7] Resnick JL, Bixler LS, Cheng L, Donovan PJ. Longterm proliferation of mouse primordial germ cells in culture. *Nature.* 1992;359:550-551.

[8] Clermont Y, Perey B. Quantitative study of the cell population of the seminiferous tubules of immature rats. *Am J Anat.* 1957;100:241-268.

[9] Sapsford CS. Changes in the cells of the sex cords and the seminiferous tubules during development of the testis of the rat and the mouse. *Austr J Zool.* 1962;10:178-192.

[10] Huckins C, Clermont Y. Evolution of gonocytes in the rat testis during late embryonic and early post-natal life. *Arch Anat Histol Embryol.* 1968;51:341-354.

[11] Hilscher B, Hilscher W, Bulthoff-Ohnolz B, et al. Kinetics of gametogenesis. I. Comparative histological and autoradiographic studies of oocytes and transitional prospermatogonia during oogenesis and prespermatogenesis. *Cell Tissue Res.* 1974;154:443-470.

[12] Franchi LL, Mandl AM. The ultrastructure of germ cells in foetal and neonatal male rats. *J Embryol Exp Morphol.* 1964;12:289-308.

[13] Vergouwen RPFA, Jacobs SG, Huiskamp R, Davids JAG, de Rooij DG. Proliferative activity of gonocytes, Sertoli cells and interstitial cells during testicular development in mice. *J Reprod Fertil.* 1991;93:233-243.

[14] Novi AM, Saba P. An electron microscopic study of the development of rat testis in the first 10 postnatal days. *Z Zellforsch Mikrosk Anat.* 1968;86:313-326.

[15] Bellve AR, Cavicchia JC, Millette CF, O'Brien DA, Bhatnagar YM, Dym M. Spermatogenic cells of the prepuberal mouse. Isolation and morphological characterization. *J Cell Biol.* 1977;74:68-85.

[16] Zamboni L, Merchant H. The fine morphology of mouse primordial germ cells in extragonadal locations. *Am J Anat.* 1973;137:299-335.

[17] Moore TJ, de Boer-Brouwer M, van Dissel-Emiliani FM. Purified gonocytes from the neonatal rat form foci of proliferating germ cells in vitro. *Endocrinology.* 2002;143:3171-3174.

[18] de Rooij DG. Specific arrests of spermatogenesis in genetically modified and mutant mice. *Cytogenet Genome Res.* 2003;103:267-276.

[19] Coulombre JL, Russell ES. Analysis of the pleiotropism at the W locus in the mouse. The effects of W and W^v substitutions upon post natal development of germ cells. *J Exp Zool.* 1954;126:277-296.

[20] Russell ES. Hereditary anemias of the mouse: a review for geneticists. *Adv Genet.* 1979;20:357-459.

[21] Geissler EN, McFarland EC, Russell ES. Analysis of pleiotropism at the dominant white-spotting (W) locus of the house mouse: a description of ten new W alleles. *Genetics.* 1981;97:337-361.

[22] Mintz B, Russell ES. Gene-induced embryological modifications of primordial germ cells in the mouse. *J Exp Zool.* 1957;134:207-237.

[23] McCoshen JA, McCallion DJ. A study of the primordial germ cells during their migratory phase in Steel mutant mice. *Experientia.* 1975;31:589-590.

[24] Dolci S, Williams DE, Ernst MK, et al. Requirement for mast cell growth factor for primordial germ cell survival in culture. *Nature.* 1991;352:809-811.

[25] Godin I, Deed R, Cooke J, Zsebo K, Dexter M, Wylie CC. Effects of the steel gene product on mouse primordial germ cells in culture. *Nature.* 1991;352:807-809.

[26] Pellas TC, Ramachandran B, Duncan M, Pan SS, Marone M, Chada K. Germ-cell deficient (gcd), an insertional mutation manifested as infertility in transgenic mice. *Proc Natl Acad Sci U S A.* 1991;88:8787-8791.

[27] Agoulnik AI, Lu B, Zhu Q, et al. A novel gene, Pog, is necessary for primordial germ cell proliferation in the mouse and underlies the germ cell deficient mutation, gcd. *Hum Mol Genet.* 2002;11:3047-3053.

[28] Lu B, Bishop CE. Late onset of spermatogenesis and gain of fertility in POG-deficient mice indicate that POG is not necessary for the proliferation of spermatogonia. *Biol Reprod.* 2003;69:161-168.

[29] Atchison FW, Capel B, Means AR. Pin1 regulates the timing of mammalian primordial germ cell proliferation. *Development.* 2003;130:3579-3586.

[30] Hummel KP. Atrichosis. *Mouse News Lett.* 1966;34:31.

[31] Juneja SC, Barr KJ, Enders GC, Kidder GM. Defects in the germ line and gonads of mice lacking connexin43. *Biol Reprod.* 1999;60:1263-1270.

[32] de Rooij DG. Stem cells in the testis. *Int J Exp Pathol.* 1998;79:67-80.

[33] Sutcliffe MJ, Burgoyne PS. Analysis of the testes of H-Y negative XOSxrb mice suggests that the spermatogenesis gene (Spy) acts during the differentiation of the A spermatogonia. *Development.* 1989;107:373-380.

[34] Mazeyrat S, Saut N, Grigoriev V, et al. A Y-encoded subunit of the translation initiation factor Eif2 is essential for mouse spermatogenesis. *Nat Genet.* 2001;29:49-53.

[35] Di Cunto F, Imarisio S, Camera P, Boitani C, Altruda F, Silengo L. Essential role of citron kinase in cytokinesis of spermatogenic precursors. *J Cell Sci.* 2002;115:4819-4826.

[36] Schrans-Stassen BH, Saunders PT, Cooke HJ, de Rooij DG. Nature of the spermatogenic arrest in Dazl -/- mice. *Biol Reprod.* 2001;65:771-776.

[37] de Rooij DG, Russell LD. All you wanted to know about spermatogonia but were afraid to ask. *J Androl.* 2000;21:776-798.

[38] Fawcett DW, Ito S, Slautterback DL. The occurrence of intercellular bridges in groups of cells exhibiting synchronous differentiation. *J Biophys Biochem Cytol.* 1959;5:453-460.

[39] Weber JE, Russell LD. A study of intercellular bridges during spermatogenesis in the rat. *Am J Anat.* 1987;180:1-24.

[40] Chiarini-Garcia H, Russell LD. High-resolution light microscopic characterization of mouse spermatogonia. *Biol Reprod.* 2001;65:1170-1178.

[41] Huckins C. The spermatogonial stem cell population in adult rats. II. A radioautographic analysis of their cell cycle properties. *Cell Tissue Kinet.* 1971;4:313-334.

[42] Huckins C. Cell cycle properties of differentiating spermatogonia in adult Sprague-Dawley rats. *Cell Tissue Kinet.* 1971;4:139-154.

[43] Lok D, de Rooij DG. Spermatogonial multiplication in the Chinese hamster. I. Cell cycle properties and synchronization of differentiating spermatogonia. *Cell Tissue Kinet.* 1983;16:7-18.

[44] Lok D, Jansen MT, de Rooij DG. Spermatogonial multiplication in the Chinese hamster. II. Cell cycle properties of undifferentiated spermatogonia. *Cell Tissue Kinet.* 1983;16:19-29.

[45] van Beek MEAB, Meistrich ML, de Rooij DG. Probability of self-renewing divisions of spermatogonial stem cells in colonies, formed after fission neutron irradiation. *Cell Tissue Kinet.* 1990;23:1-16.

[46] Potten CS. Stem cells in gastrointestinal epithelium: numbers, characteristics and death. *Philos Trans R Soc Lond B Biol Sci.* 1998;353:821-830.

[47] Schofield R. The stem cell system. *Biomed Pharmacother.* 1983;37:375-380.

[48] Chiarini-Garcia H, Hornick JR, Griswold MD, Russell LD. Distribution of type A spermatogonia in the mouse is not random. *Biol Reprod.* 2001;65:1179-1185.

[49] Shuttlesworth GA, de Rooij DG, Huhtaniemi I, et al. Enhancement of A spermatogonial proliferation and differentiation in irradiated rats by GnRH antagonist administration. *Endocrinology.* 2000;141:37-49.

[50] Tohda A, Matsumiya K, Tadokoro Y, et al. Testosterone suppresses spermatogenesis in juvenile spermatogonial depletion (jsd) mice. *Biol Reprod.* 2001;65:532-537.

[51] Shetty G, Wilson G, Huhtaniemi I, Boettger-Tong H, Meistrich ML. Testosterone inhibits spermatogonial differentiation in juvenile spermatogonial depletion mice. *Endocrinology.* 2001;142:2789-2795.

[52] Trupp M, Ryden M, Jornvall H, et al. Peripheral expression and biological activities of GDNF, a new neurotrophic factor for avian and mammalian peripheral neurons. *J Cell Biol.* 1995;130:137-148.

[53] Meng X, Lindahl M, Hyvonen ME, et al. Regulation of cell fate decision of undifferentiated spermatogonia by GDNF. *Science.* 2000;287:1489-1493.

[54] Meng XJ, de Rooij DG, Westerdahl K, Saarma M, Sariola H. Promotion of seminomatous tumors by targeted overexpression of glial cell line-derived neurotrophic factor in mouse testis. *Cancer Res.* 2001;61:3267-3271.

[55] Pichel JG, Shen L, Sheng HZ, et al. Defects in enteric innervation and kidney development in mice lacking GDNF. *Nature.* 1996;382:73-76.

[56] Yomogida K, Yagura Y, Tadokoro Y, Nishimune Y. Dramatic Expansion of Germinal Stem Cells by Ectopically- Expressed Human Glial Cell Line-Derived Neurotrophic Factor in Mouse Sertoli Cells. *Biol Reprod.* 2003;69:1303-1307.

[57] Tadokoro Y, Yomogida K, Ohta H, Tohda A, Nishimune Y. Homeostatic regulation of germinal stem cell proliferation by the GDNF/FSH pathway. *Mech Dev.* 2002;113:29-39.

[58] Buaas FW, Kirsh AL, Sharma M, et al. Plzf is required in adult male germ cells for stem cell self-renewal. *Nat Genet.* 2004;36:647-652.

[59] Reid A, Gould A, Brand N, et al. Leukemia translocation gene, PLZF, is expressed with a speckled nuclear pattern in early hematopoietic progenitors. *Blood.* 1995;86:4544-4552.

[60] Barna M, Merhoub T, Costoya JA, et al. PLZF mediates transcriptional repression of HoxD gene expression through chromatin modeling. *Dev Cell.* 2002;3:499-510.

[61] Lessard J, Sauvageau G. Bmi-1 determines the proliferative capacity of normal and leukaemic stem cells. *Nature.* 2003;423:255-260.

[62] Park I, Qian D, Kiel M, et al. Bmi-1 is required for maintenance of adult self-renewing haematopoietic stem cells. *Nature.* 2003;423:302-305.

[63] Lin RJ, Nagy P, Inoue S, Shao W, Miller WH, Evans RM. Role of the histone deacetylase complex in acute promyelocytic leukaemia. *Nature.* 1998;391:811-814.

[64] Brinster RL, Avarbock MR. Germline transmission of donor haplotype following spermatogonial transplantation. *Proc Natl Acad Sci USA.* 1994;91:11303-11307.

[65] Brinster RL, Zimmermann JW. Spermatogenesis following male germ-cell transplantation. *Proc Natl Acad Sci USA.* 1994;91:11298-11302.

[66] Creemers LB, Meng X, Den Ouden K, et al. Transplantation of germ cells from glial cell line-derived neurotrophic factor-overexpressing mice to host testes depleted of endogenous spermatogenesis by fractionated irradiation. *Biol Reprod.* 2002;66:1579-1584.

[67] Clouthier DE, Avarbock MR, Maika SD, Hammer RE, Brinster RL. Rat spermatogenesis in mouse testis. *Nature.* 1996;381:418-421.

[68] Russell LD, Brinster RL. Ultrastructural observations of spermatogenesis following transplantation of rat testis cells into mouse seminiferous tubules. *J Androl.* 1996;17:615-627.

[69] Ogawa T, Dobrinski I, Avarbock MR, Brinster RL. Xenogeneic spermatogenesis following transplantation of hamster germ cells to mouse testes. *Biol Reprod.* 1999;60:515-521.

[70] Dobrinski I, Avarbock MR, Brinster RL. Transplantation of germ cells from rabbits and dogs into mouse testes. *Biol Reprod.* 1999;61:1331-1339.

[71] Izadyar F, Creemers LB, den Ouden K, de Rooij DG. Culture and transplantation of bovine spermatogonial stem cells. In: Robaire B, Chemes HE, Morales CR, eds. *Andrology in the 21st century.* Englewood: Medimond; 2001:149-155.

[72] Izadyar F, den Ouden K, Stout T, et al. Successful allogeneic transplantation of bovine spermatogonial stem cells. *Biology of Reproduction.* 2002;66, Suppl 1:291.

[73] Honaramooz A, Behboodi E, Blash S, Megee SO, Dobrinski I. Germ cell transplantation in goats. *Mol Reprod Dev.* 2003;64:422-428.

[74] Honaramooz A, Megee SO, Dobrinski I. Germ cell transplantation in pigs. *Biol Reprod.* 2002;66:21-28.

[75] Izadyar F, DenOuden K, Stout T, et al. Autologous and homologous transplantation of bovine spermatogonial stem cells. *Reproduction.* 2003;126:765-774.

[76] Johnston DS, Russell LD, Friel PJ, Griswold MD. Murine germ cells do not require functional androgen receptors to complete spermatogenesis following spermatogonial stem cell transplantation. *Endocrinology.* 2001;142:2405-2408.

[77] Ohta H, Yomogida K, Tadokoro Y, Tohda A, Dohmae K, Nishimune Y. Defect in germ cells, not in supporting cells, is the cause of male infertility in the jsd mutant mouse: proliferation of spermatogonial stem cells without differentiation. *Intl J Androl.* 2001;24:15-23.

[78] Boettger-Tong HL, Johnston DS, Russell LD, Griswold MD, Bishop CE. Juvenile spermatogonial depletion (Jsd) mutant seminiferous tubules are capable of supporting transplanted spermatogenesis. *Biol Reprod.* 2000;63:1185-1191.

[79] Goodell MA, Brose K, Paradis G, Conner AS, Mulligan RC. Isolation and functional properties of murine hematopoietic stem cells that are replicating in vivo. *J Exp Med.* 1996;183:1797-1806.

[80] Goodell MA, Rosenzweig M, Kim H, et al. Dye efflux studies suggest that hematopoietic stem cells expressing low or undetectable levels of CD34 antigen exist in multiple species. *Nat Med.* 1997;3:1337-1345.

[81] Gussoni E, Soneoka Y, Strickland CD, et al. Dystrophin expression in the mdx mouse restored by stem cell transplantation. *Nature.* 1999;401:390-394.

[82] Asakura A, Seale P, Girgis-Gabardo A, Rudnicki MA. Myogenic specification of side population cells in skeletal muscle. *J Cell Biol.* 2002;159:123-134.

[83] Welm BE, Tepera SB, Venezia T, Graubert TA, Rosen JM, Goodell MA. Sca-1pos cells in the mouse mammary gland represent an enriched progenitor cell population. *Dev Biol.* 2002;245:42-56.

[84] Murayama A, Matsuzaki Y, Kawaguchi A, Shimazaki T, Okano H. Flow cytometric analysis of neural stem cells in the developing and adult mouse brain. *J Neurosci Res.* 2002;69:837-847.

[85] Shimano K, Satake M, Okaya A, et al. Hepatic oval cells have the side population phenotype defined by expression of ATP-binding cassette transporter ABCG2/BCRP1. *Am J Pathol.* 2003;163:3-9.

[86] Bhatt RI, Brown MD, Hart CA, et al. Novel method for the isolation and characterisation of the putative prostatic stem cell. *Cytometry.* 2003;54A:89-99.

[87] Triel C, Vestergaard ME, Bolund L, Jensen TG, Jensen UB. Side population cells in human and mouse epidermis lack stem cell characteristics. *Exp Cell Res.* 2004/4/15 2004;295:79-90.

[88] Iwatani H, Ito T, Imai E, et al. Hematopoietic and nonhematopoietic potentials of Hoechst^{low}/side population cells isolated from adult rat kidney. *Kidney Int.* 2004;65:1604-1614.

[89] Zhou S, Morris JJ, Barnes Y, Lan L, Schuetz JD, Sorrentino BP. Bcrp1 gene expression is required for normal numbers of side population stem cells in mice, and confers relative protection to mitoxantrone in hematopoietic cells in vivo. *Proc Natl Acad Sci USA.* 2002;99:12339-12344.

[90] Krishnamurthy P, Ross DD, Nakanishi T, et al. The stem cell marker Bcrp/ABCG2 enhances hypoxic cell survival through interactions with heme. *J Biol Chem.* 2004;279:24218-24225.

[91] Kubota H, Avarbock MR, Brinster RL. Spermatogonial stem cells share some, but not all, phenotypic and functional characteristics with other stem cells. *Proc Natl Acad Sci USA.* May 27, 2003 2003;100:6487-6492.

[92] Falciatori I, Borsellino G, Haliassos N, et al. Identification and enrichment of spermatogonial stem cells displaying side-population phenotype in immature mouse testis. *FASEB J.* 2004;18:376-378.

[93] Lassalle B, Bastos H, Louis JP, et al. `Side Population' cells in adult mouse testis express Bcrp1 gene and are enriched in spermatogonia and germinal stem cells. *Development.* 2004;131:479-487.

[94] de Rooij DG, Okabe M, Nishimune Y. Arrest of spermatogonial differentiation in jsd/jsd, Sl17H/ Sl17H, and cryptorchid mice. *Biol Reprod.* 1999;61:842-847.

[95] Lo KC, Lei Z, Rao CV, Beck J, Lamb DJ. De novo testosterone production in luteinizing hormone receptor knockout mice after transplantation of Leydig stem cells. *Endocrinology.* 2004;145:4011-4015.

[96] Tegelenbosch RA, de Rooij DG. A quantitative study of spermatogonial multiplication and stem cell renewal in the C3H/101 F1 hybrid mouse. *Mutat Res.* 1993;290:193-200.

[97] Morena AR, Boitani C, Pesce M, De Felici M, Stefanini M. Isolation of highly purified type A spermatogonia from prepubertal rat testis. *J. Androl.* 1996;17:708-717.

[98] Dirami G, Ravindranath N, Pursel V, Dym M. Effects of stem cell factor and granulocyte macrophage-colony stimulating factor on survival of porcine type A spermatogonia cultured in KSOM. *Biol Reprod.* 1999;61:225-230.

[99] van Pelt AM, Morena AR, van Dissel-Emiliani FM, et al. Isolation of the synchronized A spermatogonia from adult vitamin A- deficient rat testes. *Biol Reprod.* 1996;55:439-444.

[100] Van Pelt AM, De Rooij DG. The origin of the synchronization of the seminiferous epithelium in vitamin A-deficient rats after vitamin A replacement. *Biol Reprod.* 1990;42:677-682.

[101] Shinohara T, Avarbock MR, Brinster RL. beta(1)- and alpha(6)-integrin are surface markers on mouse spermatogonial stem cells. *Proc Natl Acad Sci USA.* 1999;96:5504-5509.

[102] Shinohara T, Brinster RL. Enrichment and transplantation of spermatogonial stem cells. *International Journal of Andrology.* 2000;23 Suppl. 2:89-91.

[103] Schrans-Stassen BH, van de Kant HJ, de Rooij DG, van Pelt AM. Differential expression of c-kit in mouse undifferentiated and differentiating type A spermatogonia. *Endocrinology.* 1999;140:5894-5900.

[104] Kanatsu-Shinohara M, Toyokuni S, Shinohara T. CD9 Is a Surface Marker on Mouse and Rat Male Germline Stem Cells. *Biol Reprod.* 2004;70:70-75.

[105] Hamra FK, Schultz N, Chapman KM, et al. Defining the spermatogonial stem cell. *Dev Biol.* 2004;269:393-410.

[106] Creemers LB, Izadyar F, van Pelt AMM, de Rooij DG. Culture of highly purified mouse A spermatogonia. *Reproduction.* 2002;124:791-799.

[107] Nagano M, Avarbock MR, Leonida EB, Brinster CJ, Brinster RL. Culture of mouse spermatogonial stem cells. *Tissue Cell.* 1998;30:389-397.

[108] Nagano M, Ryu BY, Brinster CJ, Avarbock MR, Brinster RL. Maintenance of Mouse Male Germ Line Stem Cells In Vitro. *Biol Reprod.* 2003;68:2207-2214.

[109] van der Wee KS, Johnson EW, Dirami G, Dym TM, Hofmann MC. Immunomagnetic isolation and long-term culture of mouse type A spermatogonia. *J Androl.* 2001;22:696-704.

[110] Izadyar F, Den Ouden K, Creemers LB, Posthuma G, Parvinen M, De Rooij DG. Proliferation and differentiation of bovine type a spermatogonia during long-term culture. *Biol Reprod.* 2003;68:272-281.

[111] van Pelt AM, Roepers-Gajadien HL, Gademan IS, Creemers LB, de Rooij DG, van Dissel-Emiliani FM. Establishment of cell lines with rat spermatogonial stem cell characteristics. *Endocrinology.* 2002;143:1845-1850.

[112] Feng LX, Chen Y, Dettin L, et al. Generation and in vitro differentiation of a spermatogonial cell line. *Science.* 2002;297:392-395.

[113] Shinohara T, Orwig KE, Avarbock MR, Brinster RL. Spermatogonial stem cell enrichment by multiparameter selection of mouse testis cells. *Proc Natl Acad Sci USA.* 2000;97:8346-8351.

[114] Clermont Y. Spermatogenesis in man. A study of the spermatogonial population. *Fertil Steril.* 1966;17:705-721.

[115] Heller CG, Clermont Y. Spermatogenesis in man: an estimate of its duration. *Science.* 1963;140:184-185.

[116] Clermont Y. Renewal of spermatogonia in man. *Am J Anat.* 1966;118:509-524.

[117] van Alphen MMA, van de Kant HJG, de Rooij DG. Repopulation of the seminiferous epithelium of the rhesus monkey after X irradiation. *Radiat Res.* 1988;113:487-500.

[118] de Rooij DG. Proliferation and differentiation of undifferentiated spermatogonia in the mammalian testis. In: Potten CS, ed. *Stem cells. Their identification and characterization.* Edinburgh: Churchill Livingstone; 1983:89-117.

[119] van Alphen MMA, van de Kant HJG, de Rooij DG. Depletion of the spermatogonia from the seminiferous epithelium of the rhesus monkey after X irradiation. *Radiat Res.* 1988;113:473-486.

[120] Lajtha LG. Stem cell concepts. *Differentiation.* 1979;14:23-34.

In: Trends in Stem Cell Research
Editor: Erik V. Greer, pp. 191-218

ISBN 1-59454-315-1

Chapter VIII

Bone Marrow Failure in Systemic Autoimmune Diseases

Helen A. Papadaki[*]
Department of Hematology, University of Crete School of Medicine, Crete, Greece

Abstract

The author summarizes the available evidence suggesting that humoral or cellular immune mechanisms may damage the bone marrow hematopoietic progenitor cell compartment and may also affect the bone marrow stromal cell hematopoiesis supporting capacity in patients with systemic lupus erythematosus and rheumatoid arthritis. Since there is currently much interest in exploring the role of hematopoietic stem cell transplantation as an alternative treatment approach in patients with systemic autoimmune diseases refractory to conventional medication including systemic lupus erythematosus and rheumatoid arthritis, data on patients' bone marrow function are of particular importance. An injured bone marrow may potentially impact on harvesting of stem cells and their engraftment potential.

Introduction

The growth and development of hematopoietic cells in the bone marrow (BM) occurs in intimate association with a heterogeneous population of cells and their biosynthetic products including cytokines and extracellular matrix components that, collectively, constitute the marrow microenvironment [1,2]. In addition to the native cells of mesenchymal and hematopoietic origin, the BM microenvironment comprises periphery-derived immune cells,

[*] Helen A Papadaki, MD, PhD, Assistant Professor of Hematology; Department of Hematology, University Hospital of Heraklion, P.O.Box 1352, Heraklion, Crete, Greece. Tel +30 2810 394629, Fax +30 2810 394632; E-mail: epapadak@med.uoc.gr

such as macrophages, monocytes and lymphocytes, that either settle in the BM or continuously traffic between the BM and peripheral blood [3]. It is, therefore, evident that normal hematopoiesis and maintenance of homeostasis in the BM is normally dependent on the immune system and that an abnormal immune milieu may potentially affect the growth and survival characteristics of BM hematopoietic stem cells and their progeny by inducing intricate cell-to-cell interactions and/or pro-inflammatory cytokine production [4]. A number of reports have demonstrated that a variety of cytokines and effector molecules can influence the function of both immune and hematopoietic cells, therefore advancing our knowledge on the interactions of these cellular systems. However, important questions remain to be answered before we can fully appreciate the intricacies of the relationship between immune and hematopoietic cells and their impact on BM function and pathophysiology of varying disease states.

The most extreme paradigm of BM dysfuntion due to an immune-mediated damage at the level of pluripotent hematopoietic stem cell leading to BM failure is acquired aplastic anemia [5,6]. Although the precise nature of the immune process remains elusive in terms of the presence of either a primary stem cell defect triggering a secondary immune response or a primary autoimmune organ-specific T-cell attack against BM stem cells, the underlying pathogenetic mechanisms that operate in aplastic anemia clearly implicate activated cytotoxic T-lymphocytes and inhibitory lymphokines that induce cell surface and intracellular pathways of apoptosis [7,8]. The beneficial effect of immunosuppressive therapy with anti-lymphocyte globulin (ALG) and cyclosporin A (CsA) in a significant proportion of aplastic anemia patients regarding the BM function and blood cytopenias supports further the concept of the immune-mediated nature of BM failure in this disease [5;9-12].

The immune model of aplastic anemia has provided a powerful prism through which to focus unanswered questions on the pathophysiology of blood cytopenias accompanying systemic autoimmune diseases not attributable to antibody-mediated damage of peripheral blood elements. There is, therefore, substantial evidence suggesting that patients with active systemic autoimmune disorders such as rheumatoid arthritis (RA) and systemic lupus erythematosus (SLE) may display impaired BM hematopoiesis as a consequence of the aberrant local immune cell activation [13-15]. Experimental data based mainly on BM culture assays have shown that the presence of autoreactive inflammatory cells in patients' BM may affect the hematopoiesis supporting capacity of BM stroma and may also damage hematopoiesis at several stages of differentiation from the early progenitors to the mature progeny, not only by mediating a direct immune destruction but also by inducing an indirect effect due to the release of pro-inflammatory cytokines that render hematopoietic cells sensitive to apoptotic stimuli [3,16].

In the present review, we summarize the available evidence suggesting that humoral or cellular immune mechanisms may damage the BM hematopoietic progenitor cell compartment and may also affect the BM stromal cell hematopoiesis supporting capacity in patients with SLE and RA. Since there is currently much interest in exploring the role of hematopoietic stem cell transplantation as an alternative treatment approach in patients with systemic autoimmune diseases refractory to conventional medication including SLE and RA, data on patients' BM function are of particular importance [17-29]. An injured BM may potentially impact on harvesting of stem cells and their engraftment potential.

The Long-Term BM Culture System

The complex interactions between the hematopoietic progenitor cells and their marrow microenvironment in systemic autoimmune diseases have been extensively studied using long-term BM cultures (LTBMCs). This liquid culture system that was initially established by Dexter in 1976 using murine BM cells and subsequently adapted in humans by Gartner and Kaplan in 1980, allows the formation of an adherent layer consisting mainly of BM microenvironmental cells that induce the survival, proliferation and differentiation of hematopoietic cells in the supernatant providing, therefore, an appropriate experimental tool to investigate interrelationships between marrow microenvironment and hematopoiesis [30,31].

In brief, BM mononuclear cells (BMMCs) from marrow aspirates are seeded in medium supplemented with fetal calf serum, horse serum and hydrocortisone and incubated for several weeks at 33^0C with half medium change once a week. It has been shown that hydrocortisone modifies the nature of adherent cells for a better hematopoetic supporting capacity while the lower than usual culture temperature regulates cell division allowing stem cell renewal to be balanced with differentiation [32,33]. During the first week of culture an adherent layer progressively appears and reaches confluence after 3-4 weeks, consisting mainly of macrophages, resembling resident tissue macrophages, and cells of mesenhymal origin such as fibroblast-like cells, adipocytes and endothelial cells that allow the growth and maintenance of hematopoietic progenitor and precursor cells in the supernatant for up to 12 weeks without addition of exogenous growth factors (Figure 1). After this time period the adherent layer detaches, piecemeal or whole, indicating termination of the culture. During development of LTBMCs, foci of hematopoiesis occur in the adherent layer consisting of close-packed granulomonocytic progenitors called "cobblestone areas" overlaid by stromal cells called "blanket cells" and considerable cell movement occurs with mature granulocytes and late erythroid progenitors migrating in the supernatants after interacting with fixed macrophage in the adherent layer [34]. Notably, the LTBMCs is not the appropriate system to maintain megakaryocytic progenitor and precursor cells. Furthermore, it is of interest that mature T-lymphocytes, B-lymphocytes and plasma cells are also present and persist in the adherent layer of the culture [35]. In this review we adopt the term "stromal cells" to define all the kind of cells found in the adherent layer apart from the granulomonocytic and erythroid progenitor and precursor cell populations.

It has been shown that the long-term survival of hematopoietic cells in the LTBMC system depends on the production of cytokines by the stromal cells and also by the intimate contact of the hematopoietic cells to the adherent cell population. Accordingly, LTBMCs faithfully replicate marrow hematopoiesis. Each week one can evaluate the number and the clonogenic potential of the nonadherent cells using standard short-term culture assays in semi-sold culture media, reflecting essentially both the initial hematopoietic progenitor cell input and the stromal cell support. To discern between a primary hematopoietic progenitor or stromal cell abnormality in cases of defective nonadherent cell growth, the stromal cell hematopoiesis supporting capacity can be evaluated by studying the time course for generation of confluence in the adherent layer, the morphology and composition of the adherent cell population as well as the cytokine production and release in cell-free LTBMC supernatants or the mRNA cytokine expression in the stromal cell extracts [36-38]. For an independent evaluation of the stromal cell and hematopoietic progenitor cell function, crossover LTBMC experiments can be performed by seeding purified normal BM $CD34^+$ hematopoietic cells onto allogeneic irradiated LTBMC adherent cell layers from patients and control subjects (to evaluate the stromal cell function) or, alternatively, by seeding purified $CD34^+$ hematopoietic cells from patients and control subjects onto allogeneic irradiated LTBMC adherent cell layers from normal controls (to evaluate the hematopoietic progenitor cell function). The cultures are monitored for the survival and growth of the non-adherent cells over a period of 5 additional weeks [39]. Notably, CD34 is the surrogate cell surface marker for the identification of the hematopoietic stem/progenitor cells [34].

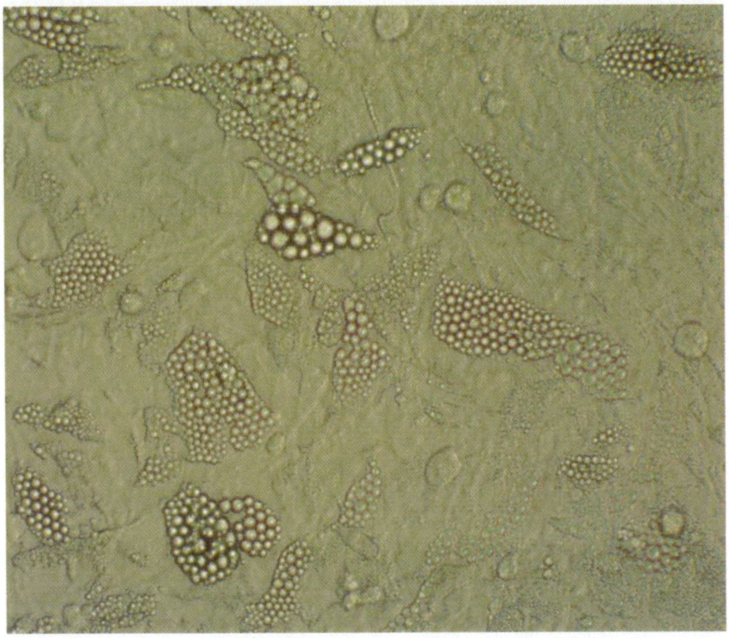

Figure 1. Long-term bone marrow culture stromal layer. Adherent fibroblasts, macrophages, and adipocytes (vacuolated fat cells) of a confluent culture at week 4, are evident.

Consequently, studies using standard or irradiated LTBMCs in patients with systemic autoimmune diseases are considered instrumental to identify abnormal interactions between hematopoietic progenitor cells and their microenvironment and define potential intrinsic hematopoietic progenitor cell abnormalities or defective stromal support of hematopoiesis in these patients.

BM Failure in Systemic Lupus Erythematosus

SLE has been considered as the prototype of human autoimmune disease. Abnormal activation of both B and T cells and the aberrant production of pro-inflammatory cytokines have been critically involved in the initiation and progress of tissue pathology and organ damage in the disease [40]. Hematological complications in the disease usually involve peripheral destruction of blood cells due to humoral or cellular cytotoxic mechanisms [41,42]. BM findings, however, from SLE patients with cytopenias are not always compatible with peripheral cell destruction since a compensatory marrow hyperplasia, as anticipated in such cases, is only occasionally identified. In contrast, several lines of evidence have shown hypoplastic BM in SLE. Specifically, Feng *et al* [43] by assessing the BM morphology in a series of 23 SLE patients with pancytopenia described striking marrow hypoplasia with dysplastic features in 39% of the cases studied. Notably, neither feature was related to cytotoxic therapy since the majority of the patients had been previously treated with steroids only. Similarly, Lorand-Metze *et al* [44] and Pereira *et al* [45] have described BM hypoplasia with foci of necrosis and presence of dysplastic features affecting one or more cell lineages, in approximately 50% of BM specimens from 21 SLE patients with peripheral blood cytopenias. Data from the above studies indicate that the BM might be a target organ of activated T-cells in SLE.

T-Cell Function in SLE

T-cells have been considered primarily responsible for the pathogenesis of SLE by expanding or inducing B cell hyperactivity and subsequent antibody production against self-antigens. Alterations in T-cell function in SLE have been extensively reviewed [40,46,47]. We shall briefly recapitulate the principal observations on T-lymphocyte function that may have impact on patients' hematopoiesis.

Solid evidence suggests that T-cells have an important role in the pathophysiology of SLE by inducing intricate cell-to-cell interactions and pro-inflammatory cytokine production. It has been demonstrated that in the inflamed tissues such as the skin and kidneys, most infiltrating lymphocytes belong to the T-cell lineage while B-cells are rarely found [48]. Among the pathogenetic mechanisms implicated in immune dysregulation and local inflammation in SLE, abnormalities concerning the $CD8^+$ T-suppressor lymphocyte activity have long been postulated [49]. It has been suggested that impaired $CD8^+$ lymphocyte function at sites of inflammation is possibly related to an imbalanced cytokine secretion pattern of these cells favoring interferon-γ (IFNγ) and interleukin-6 (IL-6) production [50]. Data are conflicting, however, regarding the type of $CD4^+$ T-helper (Th) lymphocyte predominance in SLE. It has been widely recognized that functionally polarized responses can be detected within the $CD4^+$ Th lymphocyte subpopulations on the basis of cytokine production upon activation. Th_1 cells mainly produce IFNγ, Th_2 cells produce IL-4, IL-5, and IL-10, while Th_0 cells display mixed cytokine production [51]. The balance between these cross-talking Th_1- and Th_2-derived cytokines seems to be important for normal immune reactions. Based on animal models of SLE, several lines of evidence have suggested a role of Th_2-type immune reactions in the pathogenesis of the manifestations of SLE [52,53] whereas a biphasic Th-cell activation pattern, in terms of an initial production of Th_1-type cytokines followed by Th_2-type cytokine induction during the course of the disease, has also been proposed [54]. The weigh of evidence, however, indicates a pivotal role of Th_1-type responses in SLE with predominance of IFNγ-producing T-cells in both the inflamed tissues and peripheral blood of the affected subjects [48,55-57] Significantly increased levels of IFNγ in the sera of patients with active disease have also been reported [58]. Parenthetically, IFNγ is a well-recognised inhibitor of both early and late stages of hematopoiesis mainly by inducing Fas up-modulation and apoptotic cell death but also by blocking the cell-cycle progression through the interferon regulatory factor-1 (IRF-1) transduction pathway [59-62].

The presence of activated T-cells in the peripheral blood of SLE patients has been well documented by a number of studies demonstrating enhanced expression of cell-surface activation markers such as Fas and Fas ligand (FasL), CD40 ligand (CD40L), IL-2 receptor and adhesion molecules, increased shedding of soluble forms of these molecules and elevated levels of pro-inflammatory cytokines in patient sera [40,63-66]. Among the markers of T-cell activation in SLE, Fas receptor and its ligand have been reported to play a prominent role in triggering lymphocyte apoptosis resulting in lymphopenia in the setting of a process known as activation induced cell death [67,68]. The expression of FasL on activated T-lymphocytes, however, may have a more expanded role in regulation of BM hematopoiesis in SLE since hematopoietic stem and progenitor cells have been reported sensitive to Fas-induced apoptosis [69,70]. Thus, the T-cell trafficking and homing in the marrow microenvironment might injure hematopoiesis not only by means of a direct cell-mediated immune suppression but also in terms of an indirect effect due to the release of pro-inflammatory cytokines such as IFNγ and expression of pro-apoptotic molecules including FasL.

T-Cell Implication in SLE BM Progenitor Cell Abnormalities

Antibody-mediated immune destruction of BM hematopoietic progenitor cells at several stages of differentiation has long been recognized in SLE patients presenting pure red cell aplasia or other immune cytopenias [4,71-72], while humoral suppression of hematopoiesis at the level of primitive hematopoietic stem cell has also been suggested in rare cases of SLE associated with aplastic anemia [73-77]. Although the target antigen on the hematopoietic stem and progenitor cells has never been identified, studies demonstrating that patient sera may suppress the hematopoietic clonogenic progenitor cell growth in a complement-dependent or independent mechanism, undoubtedly demonstrate the presence of humoral factors that probably affect hematopoiesis in SLE [4,71-77]. The strongest evidence, however, for an antibody-mediated immune suppression of hematopoietic progenitor cell growth in SLE is the observed clinical response following plasmapheresis [75,76] or intravenous immunoglobulin infusion [78].

T-cell inhibition of hematopoiesis, however, has been recognized as a more common mechanism of BM failure in SLE than the antibody formation. Laboratory support for this hypothesis arises from co-culture experiments in which T-cells from patients' peripheral blood or BM has been shown to suppress the colony formation of hematopoietic progenitor cells *in vitro*. Removal of T-cells from patient BM samples has been occasionally reported to improve their *in vitro* colony formation. Yamasaki *et al* have described a series of 25 SLE patients with anemia or neutropenia attributed to T-cell hematopoiesis suppressor activity since patient T-lymphocytes inhibited autologous or allogeneic BM erythroid or myeloid colony formation *in vitro* [79,80]. Koyanagawa [81] has also described $CD8^+$-cell-mediated erythroid colony forming cell inhibition in 3 patients with SLE and anemia, while Nagasawa *et al* [82] and Sakurai *et al* [83] have similarly demonstrated a T-cell mediated suppression of megakaryocytic progenitor cell growth in 2 patients with SLE and severe thrombocytopenia. In addition to the above studies indicating a T-cell induced inhibition of the committed hematopoietic progenitor cell growth in SLE, there are numerous case reports providing laboratory evidence for a more primitive progenitor cell inhibition by autoreactive cells in the disease, resulting in aplastic anemia [84,85] or severe cytopenia [86,87]. Interestingly, Otsuka *et al* have shown a significant relationship between the degree of peripheral blood cytopenias and colony numbers in the BM of patients with SLE [86]. Moreover, they demonstrated that the significant suppression of multilineage colony formation by allogeneic normal primitive hematopoietic progenitor cells (mixed colony forming units, CFU-Mix) in the presence of T-cells from SLE patients with active disease was reversed when the disease became inactive as a result of corticosteroid treatment. In favor of a T-cell-mediated progenitor cell damage in SLE is the effective treatment of cytopenic SLE patients with immunosuppressive agents.

All these reports strongly suggest that autoreactive lymphocytes may damage hematopoiesis and may result in a varying degree of BM failure in SLE. The mechanisms, however, of the T-cell mediated effects on hematopoietic progenitor cell growth remain largely speculative. Based on data from aquired aplastic anemia that has long been considered as the prototype of human BM failure syndromes associated with apoptotic death of hematopoietic stem cells as a result of T-cell attack [5,6,88], we have studied the reserve and survival characteristics of BM hematopoietic progenitor cells in SLE patients with active disease using flow cytometry and *in vitro* culture assays [3,13,16]. We have shown that SLE patients display significantly lower number of $CD34^+$ hematopoietic progenitor cells compared to healthy controls, due to a decrease of both the committed $CD34^+/CD38^+$ and the more primitive $CD34^+/CD38^-$ cells (Figure 2). In keeping with the low $CD34^+$ cell number in the patients was the defective clonogenic potential of the BMMCs in short-term and long-term culture conditions concerning both the erythroid (Burst forming units-Erythroid, BFU-E) and the myeloid (Granylocyte-macrophage colony forming units, CFU-GM) progenitors (Figure 2 and Figure 3). The degree of reduction in the $CD34^+/CD38^+$, $CD34^+/CD38^-$, CFU-GM and BFU-E cell compartments was of the same magnitude indicating a non-stage specific defect of SLE hematopoiesis. To investigate whether the decrease of $CD34^+$ cells in the patients was due to accelerated apoptotic cell death, we used flow cytometry and the fluorescent DNA-binding dye 7-amino actinomycin D (7AAD)[89] and, indeed, we found increased proportion of apoptotic cells within the $CD34^+$ cell fraction of patient BMMCs compared to controls. Since Fas antigen expression on hematopoietic progenitor cells has been implicated in the apoptotic depletion of these cells in certain BM failure syndromes such as aplastic anemia [90], myelodysplastic syndromes [91] and chronic idiopathic neutropenia [38], we evaluated the expression of Fas on patient $CD34^+$ cells. As anticipated, there were more Fas^+ cells within the $CD34^+$ compartment of SLE patients compared to healthy controls and the percentage of apoptotic cells were significantly higher among the $CD34^+/Fas^+$ than the $CD34^+/Fas^-$ cells, suggesting that Fas expression is actively involved in the apoptotic depletion of patient progenitor cells. The highly significant correlation between the proportion of apoptotic cells and the proportion of Fas$^+$ cells within the CD34$^+$ cell compartment corroborates this assumption.

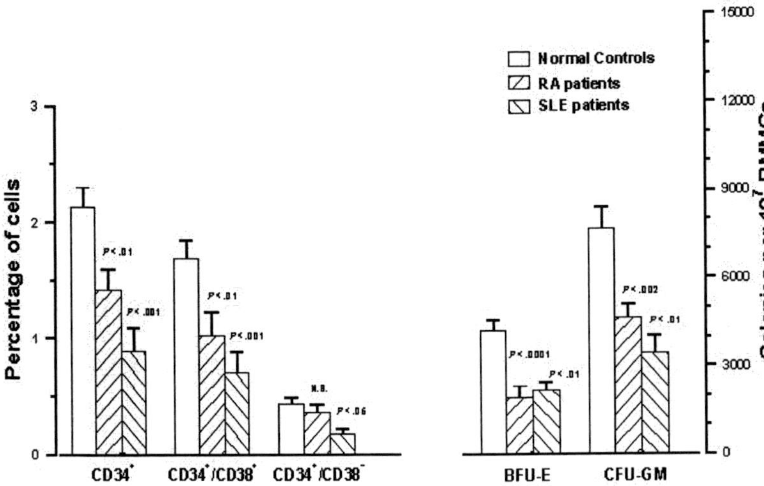

Figure 2. Bone marrow CD34$^+$ cells and clonogenic progenitor cells in RA and SLE patients and healthy control subjects. The left bars represent the mean percentages (± SEM) of CD34$^+$ cells in 25 RA and 7 SLE patients and 37 healthy controls obtained in 2-color flow cytometric analysis of BMMCs. The right bars represent the mean colony values (± SEM) obtained from the BMMCs in the clonogenic progenitor cell assay. We cultured 10^5 BMMCs from the patients and controls in 1mL methylellulose 0.9% in Iscove modified Dulbecco medium supplemented with 30% fetal calf serum, 1% bovine serum albumin, 10^{-4} mol/L mercaptoethanol, 0.075% sodium bicarbonate, and 2mmol/L L-glutamine in the presence of 5ng GM-CSF (Granulocyte-macrophage colony stimulating factor), 50ng interleukin(IL)-3 and 2IU erythropoietin for CFU-GM and BFU-E colony formation. Colonies enumerated on day 14 according to established criteria [32]. *P* values represent comparison between patients and controls by means of the 2-tailed Student t-test. N.S.; non-statistically significant difference.

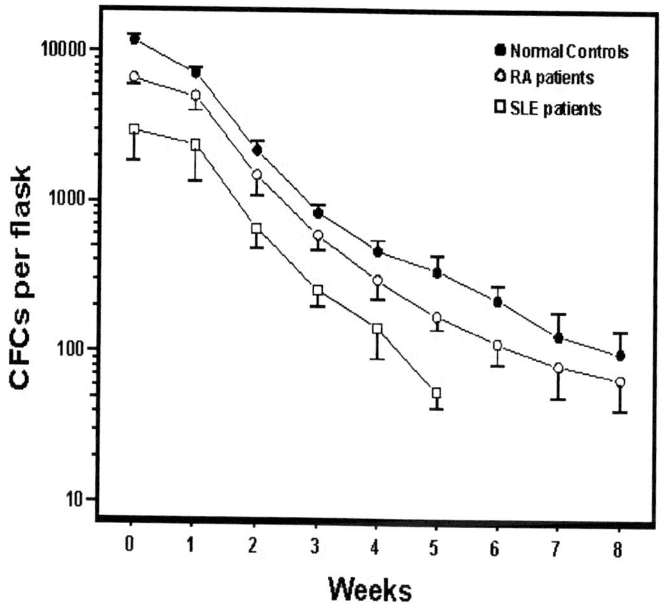

Figure 3. Long-term bone marrow cultures. Dots represent the mean frequency (± SEM) of colony forming cells (CFC) (CFU-GM plus BFU-E) obtained weekly in the nonadherent cell fraction of RA (n=25) and SLE (n=7) patients and healthy controls (n=25), throughout 8 weeks of culture. Notably, cultures from SLE patients failed to produce colonies beyond week 5. Comparison between the distributions of patients and healthy controls revealed a statistically significant difference (*P*<.001; 2-way analysis of variance).

It is well established that hematopoietic progenitor cells do not express Fas antigen under steady-state conditions [70,90,92]. However, it has been shown that pro-inflammatory cytokines such as TNFα and IFNγ abnormally produced by immune cells within the BM microenvironment may induce Fas up-regulation on marrow CD34$^+$ cells rendering them susceptible to FasL-mediated apoptosis [70,90,92]. To evaluate the production of the above cytokines in patients' marrow microenvironment we have used the standard LTBMC system and evaluated cytokine levels in the supernatants on confluence (weeks 3-4) [3,13]. TNF□ levels did not differ significantly between SLE patients and healthy controls as evaluated by using an enzyme-linked immunosorbent assay (ELISA). Alvarado-de la Barrera *et al* however, have reported increased TNFα mRNA expression in total BM cell extracts from SLE patients, indicating the presence of activated immune cells in SLE BM [93]. The fact, however, that the study was performed using total BM cells and not on purified cell populations, does not allow drawing firm conclusions regarding the exact cell type responsible for TNFα production in SLE BM, since not only T-lymphocytes but also monocytes may produce the cytokine upon activation [94]. In contrast to TNFα,IFNγ as well as FasL were undetectable in LTBMC supernatants of our SLE patients or control subjects by means of ELISA. However, we have been able to show IFNγ and FasL mRNA expression in immunomagnetically sorted CD3$^+$ cells from the BMMC fraction of 3 SLE patients with neutropenia. Notably, neither IFNγ nor FasL mRNA expression was detected in the immunomagnetically sorted BM CD14$^+$ cells of these patients [16]. Nor was detected IFNγ or FasL mRNA expression in the isolated CD3$^+$ and CD14$^+$ BM cell populations of 10 healthy controls studied. These data strongly suggest that activated T-lymphocytes specifically localized in patients' BM probably injure hematopoietic stem and progenitor cells by inducing apoptosis through cell surface molecules such as FasL and overproduction of pro-inflammatory cytokines such as IFNγ The recovery obtained in the number and survival characteristics of BM CD34$^+$ cells in one of our patients following intensive immunosuppression with autologous stem cell rescue, provides additional inferential evidence for a T-cell mediated damage of hematopoiesis in SLE. In particular, this patient was studied at day +90 following autologous stem cell transplantation (ASCT) and results were compared to pre-transplantation data. A significant increase was obtained in the number of CD34$^+$ cells following transplantation that was mainly due to the increase in the CD34$^+$/CD38$^+$ committed progenitor cell numbers. Furthermore, the proportion of CD34$^+$ cells expressing the Fas death receptor was normalized after transplantation and the reduction was associated with a significant decrease in the proportion of apoptotic cells within the CD34$^+$ cell compartment [13].

The possibility of a drug-related effect on the hematopoietic progenitor cell reserve and function of SLE patients previously treated with azathioprine and cyclophosphamide cannot be ruled out. The fact, however, that BM abnormalities may also occur in patients previously treated with corticosteroids or cyclosporin-A or even in previously untreated patients suggests that the therapy-related effect may be a contributing rather than the pathogenetic factor of BM failure in SLE. On the other hand, corticosteroids and cyclosporin-A have been reported to induce apoptosis in mitogen-stimulated peripheral blood mononuclear cells [95] but they do not damage the hematopoietic stem cells. In contrast, there is strong evidence for a positive effect of corticosteroids on the proliferation and differentiation of BM progenitor cells [96-98].

T-Cell Involvement in SLE BM Stromal Cell Abnormalities

Atta *et al* have reported that SLE patients may occasionally develop autoantibodies against fibronectin, a main extracellular matrix element of the BM microenvironment [99]. Their potential role, however, in the development of BM failure in SLE by interrupting hematopoietic cell-to-matrix interactions has not been investigated. We have evaluated the hematopoiesis supporting capacity of BM microenvironment in patients with SLE using LTBMCs [13]. We found that the number of clonogenic progenitor cells detected weekly in the non-adherent cell fraction was significantly lower in patient LTBMCs compared to the normal controls. Since this abnormality may simply reflect the lower CD34$^+$ progenitor cell number in patients' BMMC fraction rather than the inhibitory effect of BM stromal cells, we used a two-stage culture system in which confluent LTBMCs from SLE patients and healthy controls were irradiated and recharged with normal allogeneic $CD34^+$ cells from the same control subject. The number of non-adherent cells and their clonogenic potential were significantly lower in SLE patients, compared to the controls, all over a period of 5 weeks after the $CD34^+$ cell inoculation. Results from this study definitely demonstrate the impaired hematopoiesis supporting capacity of BM stroma in SLE while provide inferential evidence for the potential role of autoreactive lymphocytes in this abnormality. The findings of Berneman *et al* who extensively studied the nature of the adherent hematopoietic cells in LTBMCs over a period of 8 weeks using immunofluorescence, morphological and cytochemical examination and flow cytometry, substantiate the latter hypothesis [35]. Specifically, the investigators have demonstrated that on week 4, T-lymphocytes constitute the majority of the adherent non-macrophage hematopoietic cells in LTBMCs. It is of interest, that in sharp contrast to the liquid cultures where BM T-lymphocytes are unable to survive for long without stimulation, in LTBMC conditions T-cells persist and survive more than 8 weeks without the addition of any exogenous stimulatory factor. Consequently, we may extrapolate our results from the BMMC study and assume that IFNγ and FasL expressing T-cells may also persist in LTBMC conditions and probably affect the hematopoiesis supporting capacity of patient stromal cells. The study of Otsuka *et al* demonstrating that BM fibroblast cultures (BM fibroblast colony forming units, CFU-F) from SLE patients develop and produce growth factors similarly to normal cultures, corroborate further the assumption that the impaired hematopoiesis supporting capacity of BM microenvironment in SLE is probably due to autoreactive lymphocytes rather than to mesenhymal cell abnormalities [100]. In favor of this hypothesis is the restoration of the impaired hematopoiesis supporting capacity of BM stromal cells in a patient with SLE studied following intensive immunosupression with autologous stem cell support [13]. This particular patient displayed markedly impaired support of normal hematopoiesis in the two-stage LTBMC system described above, indicated by the low number of non-adherent cells and their defective clonogenic potential that completely normalized after ASCT. We have speculated that once the autoreactive lymphocytes were eliminated, the stromal cell function recovered.

A Synthesis on the Current Knowledge on BM Function in SLE

Data from clinical and experimental studies provide evidence that patients with active SLE display quantitative and functional abnormalities of BM hematopoietic progenitor cells and impaired hematopoiesis supporting capacity of BM stroma probably due to the presence of autoreactive lymphocytes that abnormally produce pro-inflammatory cytokines and pro-apoptotic mediators. Consequently, in addition to immune-mediated peripheral blood cell destruction, defective BM function due to immune dysregulation may also contribute to cytopenias frequently seen in patients with active disease.

BM Failure in RA

RA is a systemic autoimmune disease mainly characterized by chronic destructive polyarthritis [101]. Although the main target tissue of the disease is the synovium, evidence has been accumulated that BM may be also actively involved in the pathophysiology of RA by providing immune and inflammatory cells in the affected synovial membranes [102-104]. It has been suggested that during the inflammatory stages of the disease, BM progenitor cells or even BM derived immature mesenchymal cells enter the joint microenvironment through mechanisms similar to those of mature inflammatory cells, and gradually replace normal synovial cells.[105,106] Interestingly, an extreme hypothesis has been made suggesting that RA might even begin within the BM and might be considered as a BM disorder [107,108].

Nevertheless, since it is well established that lymphocytes, macrophages, and fibroblast-like cells found in rheumatoid synovium originate from or have their counterpart in the BM and these cells are capable of producing a variety of cytokines that may act not only as mediators of inflammation but also as regulators of hematopoiesis, it is anticipated that BM even if not primarily affected may display impaired function in RA. This might be the consequence of abnormal interactions between immune and hematopoietic cells or the result of an aberrant pro-inflammatory cytokine production potentially affecting the survival, proliferation and optimal growth of hematopoietic cells. On the other hand, TNFα which is considered as the pivotal factor for the induction and maintenance of the synovial damage in RA is detected at elevated levels not only in the inflamed synovial fluid but also in patient sera and serum cytokine concentration has been shown to correlate with disease activity [109,110]. Although conflicting data have been reported relative to the effects of TNFα on the *in vitro* hematopoiesis, ranging from potential inhibition to stimulation depending on the cytokine concentration and the synergistic effect of other cytokines, it has been clearly shown that high concentrations of TNFα inhibit progenitor cell growth regardless of the cytokines used to stimulate progenitor cell proliferation *in vitro* [111]. Clearly, TNFα is a main component of BM failure syndromes including aplastic anemia, myelodysplastic syndromes and hypoplastic type chronic idiopathic neutropenia [5,112,113,38]. It is, therefore, possible that increased systemic TNFα production in RA may affect the reserve and functional characteristics of hematopoietic progenitor and precursor cells in the BM.

Evidence for a TNFα-Mediated Suppression of Erythropoiesis in RA

The most striking evidence for impaired BM hematopoiesis in patients with active RA is anemia associated with the disease, the so-called anemia of chronic disease (ACD). Its pathogenesis is multifactorial however, impaired BM erythropoiesis due to increased local or systemic TNFα production appears to play a prominent role by modulating macrophage iron metabolism, affecting mature BM erythroid cell survival, and signaling a direct and/or indirect negative effect on BM erythroid progenitor cell growth [114-123].

To further investigate the pathophysiology of impaired erythropoiesis in RA and probe the role of TNFα in ACD we have evaluated the number and survival characteristics of BM cells at sequential stages of erythroid development using flow cytometry and *in vitro* culture assays in RA patients with active disease [15]. We have based our analysis on the knowledge that the BM $CD34^+/CD71^+$ cell compartment includes the early erythroid progenitor cells and that the CD36 cell surface marker is expressed on erythroid progenitor and early precursor erythroid cells but is lost during the subsequent erythroid differentiation [124]. Furthermore, glycophorin A (GlycoA) is expressed on mature erythroid cells but is not present on the early progenitors [125]. Accordingly, the $CD36^+$/GlycoA+ and $CD36^-/GlycoA^+$ cells represent the early and mature precursor cells of the erythroid development, respectively.

We have found that RA patients display low number of erythroid progenitor cells defined by the $CD34^+/CD71^+$ phenotype, low number of mature precursor cells defined by the $CD36^-/GlycoA^+$ phenotype and normal number of early precursor cells defined by the $CD36^+/GlycoA^+$ phenotype. Since apoptotic mechanisms have been reported to control erythropoiesis in normal and pathologic conditions,[126-128] we explored whether the decrease of BM erythroid cells in RA patients was due to apoptotic cell death by evaluating the percentage of apoptotic cells within the BM erythroid cell compartments using 7-AAD staining and flow cytometry. We found that patient $CD34^+/CD71^+$ and $CD36^+/GlycoA^+$ cells contained significantly higher proportion of apoptotic cells compared to the normal controls while no statistically significant difference was found between patients and normal controls in the percentage of apoptotic cells detected in the $CD36^-/GlycoA^+$ cell compartment. In favor of the increased rate of apoptotic cell death within patient BM erythroid compartments was the low frequency of BFU-E in the BMMC fraction and the low clonogenic potential of the purified $CD34^+$ BM cells concerning the erythroid colony formation. Notably, a subset analysis showed that the above abnormalities concerning the proportion of the $CD34^+/CD71^+$ erythroid progenitor and $CD36^-/GlycoA^+$ precursor cells, the percentage of apoptotic cells within the $CD34^+/CD71^+$ and $CD36^+/GlycoA^+$ cell compartments and the BFU-E clonogenic progenitor cell numbers were more prominent among patients with ACD.

The local TNFα production in patients' BM was evaluated using the LTBMC system. The levels of TNFα in LTBMC supernatants were significantly higher in RA patients, compared to the normal controls, and TNFα concentration was markedly elevated among patients with ACD compared to the non-anemic patients. Interestingly, individual TNFα values inversely correlated with the values of hemoglobin and the number of BFU-E and positively with the percentage of apoptotic $CD34^+/CD71^+$ and $CD36^+/GlycoA^+$ cells suggesting that the increased local TNFα production by BM stromal cells probably accounts for the apoptotic depletion of patient erythroid progenitor and early precursor cells and is possibly involved in the pathogenesis of ACD in RA. To further investigate this hypothesis we incubated normal BM cells in the absence or presence of recombinant human TNFα (rhTNFα) at varying concentrations. Following a 48 hours incubation, a significant, concentration-dependent increase in the percentage of apoptotic cells was obtained in the $CD34^+/CD71^+$ and $CD36^+/GlycoA^+$ cell compartments in the cultures treated with rhTNFα compared to the untreated cultures. The fact that no statistically significant difference was found between treated and untreated with rhTNFα cultures in the percentage of apoptotic cells detected in the $CD36^-/GlycoA^+$ cell compartment coupled with the lack of correlation between TNFα and apoptotic CD36$^-$/GlycoA$^+$ cells, suggest that TNFα induces apoptosis in the erythroid progenitor and early precursor cell compartments but does not affect the mature erythroid precursor cells. We hypothesize that the protective effect of endogenous erythropoietin (EPO) in the more mature stages of erythroid development where the hormone merely exerts its action,[34] counteracts the negative effect of TNFα on erythroid cell survival. In keeping with previous reports suggesting normal EPO response to anemia in RA,[129-131] we have also found that RA patients display normal endogenous EPO production indicating that mechanism(s) other that EPO suppression affects the BM erythroid cell survival.

The pivotal role of TNFα in the pathogenesis of anemia in RA was further substantiated by studying a number of RA patients pre- and post-treatment with the anti-TNFα monoclonal antibody cA2 (Infliximab) [132-137]. This monoclonal antibody is a chimeric mouse/human IgG_{1kappa} antibody that binds with high affinity to soluble and membrane forms of human TNFα. Binding to soluble TNFα results in neutralization of its activity whereas binding to the membrane form of TNFα causes lysis of the cell by activation of complement and induction of antibody-mediated cellular toxicity [138]. The levels of TNFα in LTBMC supernatants decreased dramatically following treatment and this reduction was associated with normalization of the proportion of the CD34$^+$/CD71$^+$ and CD36$^-$/GlycoA$^+$ cells, the frequency of BFU-E, and the percentage of apoptotic CD34$^+$/CD71$^+$ and CD36$^+$/GlycoA$^+$ cells. Interestingly, following 6 doses of cA2 a significant increase of hemoglobin levels was observed in the total group of anemic patients compared to their baseline values that was more prominent among patients with ACD. These data confirm and extend the observations of Davis *et al* suggesting an improvement of anemia in RA patients following treatment with cA2 while highlight further the inhibitory effect of TNFα on patients' erythropoiesis [131].

Defective BM Hematopoietic Progenitor Cells in RA

Immune-mediated suppression of myelopoiesis concerning merely later stages of the granulocytic development has been recognized in a subset of patients with Felty's syndrome. In particular, it has been reported that expansion of large granular CD8$^+$/CD57$^+$ lymphocytes in patients BM or neutralizing antibodies against granulocyte-colony stimulating factor (G-CSF) may suppress the granulocytic precursor cells resulting in neutropenia typically characterized the above syndrome [139,140]. The earlier stages, however, of hematopoiesis have not been extensively studied in RA patients other than those with Felty's syndrome.

We have evaluated the number and functional characteristics of BM hematopoietic stem cells in RA patients with active disease using flow cytometry, clonogenic assays and LTBMCs [14]. The patients displayed significantly lower proportion of CD34$^+$ cells within the BMMCs, compared to the controls mainly due to the reduction of the committed CD34$^+$/CD38$^+$ cells rather than the more primitive CD34$^+$/CD38$^-$ cells (Figure 2). To explore whether the decrease of CD34$^+$ cells in RA patients was due to increased apoptotic cell death, we evaluated the proportion of apoptotic cells within the CD34$^+$ cell fraction using 7-AAD staining. Indeed, we found that patient $CD34^+$ cells contained significantly higher number of apoptotic cells compared to the controls while no statistically significant difference was found between patients and control subjects in the percentage of apoptotic cells detected in the non-$CD34^+$ BMMC fraction. In keeping with the low $CD34^+$ cell numbers in the patients was the low frequency of clonogenic progenitor cells obtained by BMMCs in short-term and long-term culture conditions (Figure 2 and Figure 3).

Since, as mentioned above, Fas antigen expression has been associated with apoptosis of BM hematopoietic progenitor cells, we evaluated the expression of this molecule on patient $CD34^+$ cells. The proportion of Fas^+ cells in the gate of $CD34^+$ cells was significantly higher in the group of patients than in the group of controls and a highly significant positive correlation was noted between the proportion of Fas^+ cells and the percentage of apoptotic cells within the $CD34^+$ cell population. Notably, the increased apoptosis found in patients' $CD34^+$ cell compartment was mainly due to the increased proportion of apoptotic cells within the $CD34^+/Fas^+$ cell fraction since no statistically significant difference could be demonstrated between patients and normal controls in the proportion of apoptotic cells within the $CD34^+/Fas^-$ cell fraction. These data suggest that Fas antigen expression probably accounts for the increased apoptosis in patients' $CD34^+$ cell compartment. The highly significant inverse correlation noted between the percentage of $CD34^+$ cells and the proportion of both, Fas^+ cells and apoptotic cells indicates that Fas antigen expression and increased apoptosis are probably involved in the reduction of $CD34^+$ cells in RA patients. Furthermore, TNFα values in LTBMC supernatants correlated inversely with the proportion of $CD34^+$ cells and the number of clonogenic progenitor cells and positively with the proportion of $CD34^+/Fas^+$ cells and the proportion of $CD34^+/7AAD^+$ (apoptotic) cells suggesting that the increased TNFα production by patient BM stromal cells probably accounts for the quantitative and functional abnormalities found in patients' progenitor cell compartment. The significant increase in the percentage of $CD34^+$ cells and their clonogenic potential in association with the significant reduction in the proportion of apoptotic cells and the percentage of Fas^+ cells within the $CD34^+$ cell compartment and following anti-TNFα treatment (cA2) highlights further the prominent role of TNFα in patients' impaired hematopoiesis.

The long-term exposure to anti-rheumatic agents such a methotrexate might be a contributing factor affecting BM progenitor cell reserve and function in RA patients. However, it has been shown that the myelosuppression occasionally seen during treatment with methotrexate affects exclusively late stages of hematopoietic development but not the early myeloid and progenitor cells [141,142]. On the other hand, 23.1% of the patients included in our study had never received cytotoxic agents or immune suppressants indicating that a drug-induced damage is not the major factor affecting hematopoiesis in RA patients. In keeping with this assumption is the low number of $CD34^+$ cells, the increased apoptosis and expression of Fas antigen in the $CD34^+$ compartment and the low progenitor cell recovery in short-term and LTBMCs in the subgroup of patients who had never been previously treated with immune suppressants compared to normal subjects, suggesting that a drug-induced damage of patient progenitor cells is unlikely.

Abnormal BM Stromal Cell Function in RA

To evaluate the capacity of BM stromal layers from RA patients to support hematopoiesis we have used the standard and the two-way LTBMC system [14]. In standard LTBMCs, typical stromal layers consisting of fibroblast-like cells, adipocytes, macrophages and cobblestone areas were formed over the first 3 weeks in patients as well as in healthy controls. The clonogenic progenitor cells, however, detected weekly in the non-adherent cell fraction were significantly lower in RA patients, compared to the controls, suggesting a possible defect in the capacity of patient stromal cells to support the autologous hematopoiesis. In addition, patient LTBMC stromal layers failed to support the normal hematopoiesis as indicated by the number of non-adherent cells and the frequency of clonogenic progenitor cells detected weekly in the non-adherent cell fraction of irradiated LTMBCs from RA patient recharged with normal $CD34^+$ cells.

To investigate the possible involvement of TNFα in the defective stromal cell function in RA, we evaluated the hematopoiesis supporting capacity of patient stromal cells following treatment with cA2. TNFα levels in LTBMC supernatants on confluence were found significantly reduced following treatment with the antibody, compared to pre-treatment values. Importantly, this reduction was associated with a restoration of the impaired stromal cell function as indicated by the number of clonogenic progenitor cells in the non-adherent cell fraction of both standard and recharged LTBMCs. These data suggest that increased TNFα production by BM stroma probably accounts for the impaired ability of patient stromal cells to sustain hematopoiesis. Moreover, the exogenous weekly addition of an anti-TNFα neutralizing antibody in LTBMCs increased significantly, in comparison to the untreated cultures, the colony forming cell numbers in RA patients studied prior the *in vivo* anti-TNFα treatment but not in patients post-treatment, highlighting further the role of TNFα in abnormal stromal cell function in RA.

It is of interest the stromal layers from the patients reached confluence earlier than those of normal controls indicating increased number and/or enhanced proliferating capacity of patient adherent cells because of an unknown intrinsic abnormality or a response to abnormal local cytokine production. This finding is in agreement with previous reports suggesting increased proportion and proliferative capacity of cells of monocyte-macrophage lineage in the BMMC fraction of RA patients [143,144]. These cells primarily participate in the adherent layer formation in LTBMC assays and their synovial counterparts are considered to be the central inflammatory cells producing TNFα in RA. Interestingly, recent studies suggest that fibroblast-like cells in the synovial tissue and the BM microenvironment share a common origin in RA and may contribute to the disease pathogenesis by producing inflammatory cytokines including TNFα [106,145]. Previous studies suggesting that human BM stromal cell lines derived from RA patients display increased capacity to support the murine pre-B DW34 cell growth, indicate further an abnormal activity of cells in the marrow microenvironment in RA irrespectively of the underlying pathogenetic mechanism [146].

Synopsis of the Data on BM Function in RA

There is now strong evidence suggesting that patients with active RA produce abnormally high amounts of TNFα in the BM microenvironment. Data from the *in vitro* and *ex vivo* study of hematopoiesis in RA patients and inferential evidence from clinical trials using anti-TNFα suggest a TNFα-induced marrow failure in RA indicated by the low frequency and accelerated apoptosis of BM $CD34^+$ cells, defective clonogenic potential of marrow progenitor cells, and impaired hematopoiesis supporting capacity of BM stroma. Furthermore, the TNFα-induced accelerated apoptosis of BM erythroid progenitor and precursor cells largely contributes to the pathogenesis of ACD in RA.

Conclusions and Practical Issues

Immune dysregulation in systemic autoimmune disorders such as RA and SLE may not only injure peripheral blood cells but also influence BM hematopoiesis by affecting the survival characteristics of hematopoietic progenitor and precursor cells and the capacity of BM stromal cell to support hematopoiesis. Immune damage may involve intricate cell-to-cell interactions and/or inflammatory cytokine production triggering cellular apoptotic mechanisms. Whether inflammatory cells and their products primarily target the BM or simply reside in the marrow microenvironment rendering hematopoietic cells the "innocent bystanders" of the immune process remains unknown. On the other hand, an interesting hypothesis has been proposed by Ikehara [108,147] suggesting a primary BM stem cell abnormality in autoimmune disorders. In particular, Ikehara using animal models of autoimmune diseases was able to treat autoimmune-prone mice with normal hematopoietic stem cells but also to transfer organ specific and systemic autoimmune diseases from autoimmune-prone mice to normal recipients by transplanting hematopoietic stem cells [108,147]. Therefore, on the basis of experimental data it has been proposed that autoimmune diseases might be considered as "stem cell disorders", a hypothesis that has not adequately addressed in humans.

From the practical point of view and irrespectively of a primary stem cell defect or a secondary BM damage, the quantitative and functional abnormalities of the hematopoietic stem and progenitor cells may give an explanation for the recently described difficulties in stem cell collection procedure in SLE and RA patients undergoing autologous stem cell transplantation. Specifically, Burt *et al* by studying retrospectively data from the outcome of stem cell collection in 187 patients with autoimmune diseases including 37 RA and 19 SLE patients, described that RA patients display low progenitor cell yield while SLE patients display the lowest $CD34^+$ cell recovery or even complete failure in $CD34^+$ cell mobilization [148]. The investigators suggest that stem cell yield is probably dependent on the disease-specific cytokine and chemokine milieu as well as on BM microenvironment alterations while the effect of previous cytotoxic treatment, although not entirely excluded, seems unlikely. Interestingly, the knowledge that SLE patients display low peripheral blood progenitor cell number under steady-state conditions goes back more than a decade although the underlying pathogenetic mechanisms were at that time undefined [149]. In keeping with the results of Burt *et al* are the data of Snowden *at al* on the composition of peripheral blood stem and progenitor cell harvests from patients with severe active RA demonstrating significantly lower $CD34^+$ cell numbers compared to healthy donors [150]. Moreover, the recent report by Shaughnessy *et al* describing graft failure in a patient with SLE treated with high-dose immunosuppression followed by autologous and allogeneic stem cell rescue might imply a defect in the hematopoiesis supporting capacity of patient BM stromal cells [151].

Another practical issue arising from the study of BM in SLE and RA patients is the elucidation of the pathophysisology of the accompanying cytopenias not attributable to conventional immune mechanism. In particular, the BM failure resulting from the complex interaction between autoreactive T-cells and the hematopoietic progenitor cells in patients with SLE and the suppressive effect of the abnormal local production of TNFα on hematopoiesis of patients with RA might be major contributors to peripheral cytopenias involving several cell lines. Prolonged therapy with immunesuppressants might also be an additional factor contributing to marrow failure. Furthermore, the identification of the effector mechanisms operating in the hematopoietic progenitor cell damage in SLE and RA implicating the immune-mediated up-regulation of death receptors and induction of apoptotic pathways, may support further the use of new therapeutic treatment approaches for these disease states, the so-called "biological" therapies targeting to diminish lymphocyte activation and blockade pro-inflammatory cytokine production and action [152,153].

References

[1] Charbord P. Microenvironmental cell populations essential for the support of hematopoietic stem cells. In: Zon LI, editor. *Hematopoiesis. A developmental approach.* New York: Oxford University Press, Inc, 2001: 691-701.

[2] Charbord P. Mediators involved in the control in hematopoiesis by the microenvironment. In: Zon LI, editor. *Hematopoiesis. A developmental approach.* New York: Oxford University Press, Inc, 2001: 702-717.

[3] Papadaki HA, Marsh JC, Eliopoulos GD. Bone marrow stem cells and stromal cells in autoimmune cytopenias. *Leuk Lymphoma* 2002; 43:753-760.

[4] Cline MJ, Golde DW. Immune suppression of hematopoiesis. *Am J Med* 1978; 64:301-310.

[5] Young NS. *Bone marrow failure syndromes.* 1st ed. Philadelphia: W.B. Saunders Company, 2000.

[6] Young NS. Hematopoietic cell destruction by immune mechanisms in acquired aplastic anemia. *Semin Hematol* 2000; 37:3-14.

[7] Zoumbos NC, Gascon P, Djeu JY, Young NS. Interferon is a mediator of hematopoietic suppression in aplastic anemia in vitro and possibly in vivo. *Proc Natl Acad Sci U S A* 1985; 82:188-192.

[8] Gordon-Smith EC, Marsh JC, Gibson FM. Views on the pathophysiology of aplastic anaemia. *Int J Hematol* 2002; 76 Suppl 2:163-166.

[9] Young NS. Immunosuppressive treatment of acquired aplastic anemia and immune-mediated bone marrow failure syndromes. *Int J Hematol* 2002; 75:129-140.

[10] Marsh JC. Results of immunosuppression in aplastic anaemia. *Acta Haematol* 2000; 103:26-32.

[11] Ball SE. The modern management of severe aplastic anaemia. *Br J Haematol* 2000; 110:41-53.

[12] Brodsky RA. High-dose cyclophosphamide for aplastic anemia and autoimmunity. *Curr Opin Oncol* 2002; 14:143-146.

[13] Papadaki HA, Boumpas DT, Gibson FM, Jayne DR, Axford JS, Gordon-Smith EC, Marsh JC, Eliopoulos GD. Increased apoptosis of bone marrow CD34(+) cells and impaired function of bone marrow stromal cells in patients with systemic lupus erythematosus. *Br J Haematol* 2001; 115:167-174.

[14] Papadaki HA, Kritikos HD, Gemetzi C, Koutala H, Marsh JC, Boumpas DT, Eliopoulos GD. Bone marrow progenitor cell reserve and function and stromal cell function are defective in rheumatoid arthritis: evidence for a tumor necrosis factor alpha-mediated effect. *Blood* 2002; 99:1610-1619.

[15] Papadaki HA, Kritikos HD, Valatas V, Boumpas DT, Eliopoulos GD. Anemia of chronic disease in rheumatoid arthritis is associated with increased apoptosis of bone marrow erythroid cells: improvement following anti-tumor necrosis factor-alpha antibody therapy. *Blood* 2002; 100:474-482.

[16] Papadaki HA. Autoreactive T-lymphocytes are implicated in the pathogenesis of bone marrow failure in patients with systemic lupus erythematosus. *Leuk Lymphoma* 2003; 44:1301-1307.

[17] Marmont A. Stem cell therapy for severe autoimmune disease: The present situation. *Haema* 1998; 1:21-26.

[18] Burt RK, Traynor A, Burns W. Hematopoietic stem cell transplantation of multiple sclerosis, rheumatoid arthritis, and systemic lupus erythematosus. *Cancer Treat Res* 1999; 101:157-184.

[19] Burt RK, Traynor AE. Hematopoietic stem cell transplantation: a new therapy for autoimmune disease. *Oncologist* 1999; 4:77-83.

[20] Van Laar JM, Tyndall A. Intense immunosuppression and stem-cell transplantation for patients with severe rheumatic autoimmune disease: a review. *Cancer Control* 2003; 10:57-65.

[21] Cohen Y, Polliack A, Nagler A. Treatment of refractory autoimmune diseases with ablative immunotherapy using monoclonal antibodies and/or high dose chemotherapy with hematopoietic stem cell support. *Curr Pharm Des* 2003; 9:279-288.

[22] Openshaw H, Nash RA, McSweeney PA. High-dose immunosuppression and hematopoietic stem cell transplantation in autoimmune disease: clinical review. *Biol Blood Marrow Transplant* 2002; 8:233-248.

[23] Bingham SJ, Snowden JA, Emery P. Autologous blood stem cell transplantation as therapy for autoimmune diseases. *Ann Med* 2000; 32:615-621.

[24] Snowden JA, Brooks PM. Hematopoietic stem cell transplantation in rheumatic diseases. *Curr Opin Rheumatol* 1999; 11:167-172.

[25] Zandman-Goddard G, Shoenfeld Y. Novel Approaches to Therapy for SLE. *Clin Rev Allergy Immunol* 2003; 25:105-112.

[26] Verburg RJ, Toes RE, Fibbe WE, Breedveld FC, Van Laar JM. High dose chemotherapy and autologous hematopoietic stem cell transplantation for rheumatoid arthritis: a review. *Hum Immunol* 2002; 63:627-637.

[27] Traynor AE, Barr WG, Rosa RM, Rodriguez J, Oyama Y, Baker S, Brush M, Burt RK. Hematopoietic stem cell transplantation for severe and refractory lupus. Analysis after five years and fifteen patients. *Arthritis Rheum* 2002; 46:2917-2923.

[28] Traynor AE, Barr WG, Rosa RM, Rodriguez J, Oyama Y, Baker S, Brush M, Burt RK. Hematopoietic stem cell transplantation for severe and refractory lupus. *Arthritis Rheum* 2002; 46:2917-2923.

[29] Marmont AM. Lupus: tinkering with haematopoietic stem cells. *Lupus* 2001; 10:769-774.

[30] Dexter TM, Testa NG. Differentiation and proliferation of hemopoietic cells in culture. *Methods in Cell Biology* 1976; 14:387-405.

[31] Gartner S, Kaplan HS. Long-term culture of human bone marrow cells. *Proc Natl Acad Sci U S A* 1980; 77:4756-4759.

[32] Coutinho LH, Gilleece MH, De Wynter EA, Will A, Testa NG. Clonal and long-term cultures using human bone marrow. In: Testa NG, Molineux G, editors. *Haemopoiesis. A practical approach.* Oxford: Oxford University Press, 1993: 75-106.

[33] Charbord P. The maintenance of hemopoietic stem cells in human long-term marrow culture. In: Levitte D, Mertelsmann R, editors. *Hematopoietic stem cells. Biology and therapeutic applications.* New York: Marcel Dekker, Inc., 1995: 151-169.

[34] Garland JM, Rudin CE. Introduction to the hematopoietic system. In: Garland JM, Quesenberry PJ, Hilton DJ, editors. *Colony-stimulating factors. Molecular and cellular biology.* New York: Marcel Dekker, Inc., 1997: 1-34.

[35] Berneman ZN, Chen Z, Van Bockstaele D, Ramael M, Korthout M, Peetermans ME. The nature of the adherent hemopoietic cells in human long-term bone marrow cultures (HLTBMCs): Presence of lymphocytes and plasma cells next to the myelomonocytic population. *Leukemia* 1989; 3:648-661.

[36] Papadaki HA, Gibson FM, Rizzo S, Gordon-Smith EC, Marsh JC. Assessment of bone marrow stem cell reserve and function and stromal cell function in patients with autoimmune cytopenias. *Blood* 2000; 96:3272-3275.

[37] Papadaki HA, Gibson FM, Psyllaki M, Gordon-Smith EC, Marsh JC, Eliopoulos GD. Assessment of bone marrow stem cell reserve and function and stromal cell function in patients with severe congenital neutropenia. *Eur J Haematol* 2001; 67:245-251.

[38] Papadaki HA, Eliopoulos AG, Kosteas T, Gemetzi C, Damianaki A, Koutala H, Bux J, Eliopoulos GD. Impaired granulocytopoiesis in patients with chronic idiopathic neutropenia is associated with increased apoptosis of bone marrow myeloid progenitor cells. *Blood* 2003; 101:2591-2600.

[39] Marsh JC, Chang J, Testa NG, Hows JM, Dexter TM. In vitro assessment of marrow 'stem cell' and stromal cell function in aplastic anaemia. *British Journal of Haematology* 1991; 78(2):258-267.

[40] Tsokos GC, Kovacs B, Liossis SN. Lymphocytes, cytokines, inflammation, and immune trafficking. *Curr Opin Rheumatol* 1997; 9:380-386.

[41] Courtney PA, Crockard AD, Williamson K, McConnell J, Kennedy RJ, Bell AL. Lymphocyte apoptosis in systemic lupus erythematosus: relationships with Fas expression, serum soluble Fas and disease activity. *Lupus* 1999; 8:508-513.

[42] Courtney PA, Crockard AD, Williamson K, Irvine AE, Kennedy RJ, Bell AL. Increased apoptotic peripheral blood neutrophils in systemic lupus erythematosus: relations with disease activity, antibodies to double stranded DNA, and neutropenia. *Ann Rheum Dis* 1999; 58:309-314.

[43] Feng CS, Ng MH, Szeto RS, Li EK. Bone marrow findings in lupus patients with pancytopenia. *Pathology* 1991; 23:5-7.

[44] Lorand-Metze I, Carvalho MA, Costallat LT. [Morphology of bone marrow in systemic lupus erythematosus]. *Pathologe* 1994; 15(5):292-296.

[45] Pereira RM, Velloso ER, Menezes Y, Gualandro S, Vassalo J, Yoshinari NH. Bone marrow findings in systemic lupus erythematosus patients with peripheral cytopenias. *Clin Rheumatol* 1998; 17:219-222.

[46] Via CS, Handwerger BS. B-cell and T-cell function in systemic lupus erythematosus. *Curr Opin Rheumatol* 1993; 5:570-574.

[47] Theofilopoulos AN, Dummer W, Kono DH. T cell homeostasis and systemic autoimmunity. *J Clin Invest* 2001; 108:335-340.

[48] Converso M, Bertero MT, Vallario A, Caligaris-Cappio F. Analysis of T-cell clones in systemic lupus erythematosus. *Haematologica* 2000; 85:118-123.

[49] Theofilopoulos AN. The basis of autoimmunity: Part I. Mechanisms of aberrant self-recognition. *Immunol Today* 1995; 16:90-98.

[50] Filaci G, Bacilieri S, Fravega M, Monetti M, Contini P, Ghio M, Setti M, Puppo F, Indiveri F. Impairment of CD8+ T suppressor cell function in patients with active systemic lupus erythematosus. *J Immunol* 2001; 166:6452-6457.

[51] Mosmann TR, Sad S. The expanding universe of T-cell subsets: Th1, Th2 and more. *Immunol Today* 1996; 17:138-146.

[52] Naiki M, Chiang BL, Cawley D, Ansari A, Rozzo SJ, Kotzin BL, Zlotnik A, Gershwin ME. Generation and characterization of cloned T helper cell lines for anti-DNA responses in NZB.H-2bm12 mice. *J Immunol* 1992; 149:4109-4115.

[53] Goldman M, Druet P, Gleichmann E. TH2 cells in systemic autoimmunity: insights from allogeneic diseases and chemically-induced autoimmunity. *Immunol Today* 1991; 12:223-227.

[54] Segal R, Bermas BL, Dayan M, Kalush F, Shearer GM, Mozes E. Kinetics of cytokine production in experimental systemic lupus erythematosus: involvement of T helper cell 1/T helper cell 2-type cytokines in disease. *J Immunol* 1997; 158:3009-3016.

[55] Takahashi S, Fossati L, Iwamoto M, Merino R, Motta R, Kobayakawa T, Izui S. Imbalance towards Th1 predominance is associated with acceleration of lupus-like autoimmune syndrome in MRL mice. *J Clin Invest* 1996; 97:1597-1604.

[56] Masutani K, Akahoshi M, Tsuruya K, Tokumoto M, Ninomiya T, Kohsaka T, Fukuda K, Kanai H, Nakashima H, Otsuka T, Hirakata H. Predominance of Th1 immune response in diffuse proliferative lupus nephritis. *Arthritis Rheum* 2001; 44:2097-2106.

[57] Akahoshi M, Nakashima H, Tanaka Y, Kohsaka T, Nagano S, Ohgami E, Arinobu Y, Yamaoka K, Niiro H, Shinozaki M, Hirakata H, Horiuchi T, Otsuka T, Niho Y.

Thl/Th2 balance of peripheral T helper cells in systemic lupus erythematosus. *Arthritis Rheum* 1999; 42:1644-1648.

[58] al Janadi M, al Balla S, al Dalaan A, Raziuddin S. Cytokine profile in systemic lupus erythematosus, rheumatoid arthritis, and other rheumatic diseases. *J Clin Immunol* 1993; 13:58-67.

[59] Zoumbos NC, Baranski B, Young NS. Different haematopoietic growth factors have different capacity in overcoming the in vitro interferon gamma-induced suppression of bone marrow progenitor cells. *Eur J Haematol* 1990; 44:282-290.

[60] Zoumbos NC, Gascon P, Djeu JY, Trost SR, Young NS. Circulating activated suppressor T lymphocytes in aplastic anemia. *N Engl J Med* 1985; 312:257-265.

[61] Selleri C, Sato T, Anderson S, Young NS, Maciejewski JP. Interferon-gamma and tumor necrosis factor-alpha suppress both early and late stages of hematopoiesis and induce programmed cell death. *J Cell Physiol* 1995; 165:538-546.

[62] Selleri C, Maciejewski JP, Sato T, Young NS. Interferon-gamma constitutively expressed in the stromal microenvironment of human marrow cultures mediates potent hematopoietic inhibition. *Blood* 1996; 87:4149-4157.

[63] Kovacs B, Liossis SN, Dennis GJ, Tsokos GC. Increased expression of functional Fas-ligand in activated T cells from patients with systemic lupus erythematosus. *Autoimmunity* 1997; 25:213-221.

[64] Kossy MD, Berger D, Crow MK. Increased expression of CD40 ligand on systemic lupus erythematosus lymphocytes. *J Clin Invest* 1996; 98:826-837.

[65] Georgescu L, Vakkalanka RK, Elkon KB, Crow MK. Interleukin-10 promotes activation-induced cell death of SLE lymphocytes mediated by Fas ligand. *J Clin Invest* 1997; 100:2622-2633.

[66] Nozawa K, Kayagaki N, Tokano Y, Yagita H, Okumura K, Hasimoto H. Soluble Fas (APO-1, CD95) and soluble Fas ligand in rheumatic diseases. *Arthritis Rheum* 1997; 40:1126-1129.

[67] Wesselborg S, Janssen O, Kabelitz D. Induction of activation-driven death (apoptosis) in activated but not resting peripheral blood T cells. *J Immunol* 1993; 150:4338-4345.

[68] Emlen W, Niebur J, Kadera R. Accelerated in vitro apoptosis of lymphocytes from patients with systemic lupus erythematosus. *J Immunol* 1994; 152:3685-3692.

[69] Nagafuji K, Takenaka K, Shibuya T, Harada M, Niho Y. Fas antigen (CD95) and hematopoietic progenitor cells. *Leuk Lymphoma* 1996; 24:43-56.

[70] Nagafuji K, Shibuya T, Harada M, Mizuno S, Takenaka K, Miyamoto T, Okamura T, Gondo H, Niho Y. Functional expression of Fas antigen (CD95) on hematopoietic progenitor cells. *Blood* 1995; 86:883-889.

[71] Liu H, Ozaki K, Matsuzaki Y, Abe M, Kosaka M, Saito S. Suppression of haematopoiesis by IgG autoantibodies from patients with systemic lupus erythematosus (SLE). *Clin Exp Immunol* 1995; 100:480-485.

[72] Nossent JC, Swaak AJ. Pure red-cell aplasia in systemic lupus erythematosus. *Clin Rheumatol* 1988; 7:527-529.

[73] Bailey FA, Lilly M, Bertoli LF, Ball GV. An antibody that inhibits in vitro bone marrow proliferation in a patient with systemic lupus erythematosus and aplastic anemia. *Arthritis Rheum* 1989; 32:901-905.

[74] Roffe C, Cahill MR, Samanta A, Bricknell S, Durrant ST. Aplastic anaemia in systemic lupus erythematosus: a cellular immune mechanism? *Br J Rheumatol* 1991; 30:301-304.

[75] Marques JA, Rhim H, Distenfeld A. Inhibition of hematopoiesis by a plasma factor in a case of aplastic anemia associated with systemic lupus erythematosus. *P R Health Sci J* 1995; 14:293-296.

[76] Brooks BJ, Jr., Broxmeyer HE, Bryan CF, Leech SH. Serum inhibitor in systemic lupus erythematosus associated with aplastic anemia. *Arch Intern Med* 1984; 144:1474-1477.

[77] Pavithran K, Raji NL, Thomas M. Aplastic anemia complicating systemic lupus erythematosus--report of a case and review of the literature. *Rheumatol Int* 2002; 22:253-255.

[78] Ilan Y, Naparstek Y. Pure red cell aplasia associated with systemic lupus erythematosus: remission after a single course of intravenous immunoglobulin. *Acta Haematol* 1993; 89:152-154.

[79] Yamasaki K, Niho Y, Yanase T. Erythroid colony forming cells in systemic lupus erythematosus. *J Rheumatol* 1984; 11:167-171.

[80] Yamasaki K, Niho Y, Yanase T. Granulopoiesis in systemic lupus erythematosus. *Arthritis Rheum* 1983; 26:516-521.

[81] Koyanagawa Y. A possible role of T cell subsets causing anemia in systemic lupus erythematosus (SLE)]. *Hokkaido Igaku Zasshi* 1987; 62:370-380.

[82] Nagasawa T, Sakurai T, Kashiwagi H, Abe T. Cell-mediated amegakaryocytic thrombocytopenia associated with systemic lupus erythematosus. *Blood* 1986; 67:479-483.

[83] Sakurai T, Kono I, Kabashima T, Yamane K, Nagasawa T, Kashiwagi H. Amegakaryocytic thrombocytopenia associated with systemic lupus erythematosus successfully treated by a high-dose prednisolone therapy. *Jpn J Med* 1984; 23:135-138.

[84] Sumimoto S, Kawai M, Kasajima Y, Hamamoto T. Aplastic anemia associated with systemic lupus erythematosus. *Am J Hematol* 1991; 38:329-331.

[85] Kiely PD, McGuckin CP, Collins DA, Bevan DH, Marsh JC. Erythrocyte aplasia and systemic lupus erythematosus. *Lupus* 1995; 4:407-411.

[86] Otsuka T, Okamura S, Harada M, Ohhara N, Hayashi S, Yamaga S, Nagasawa K, Niho Y. Multipotent hemopoietic progenitor cells in patients with systemic lupus erythematosus. *J Rheumatol* 1988; 15:1085-1090.

[87] Sugimoto M, Wakabayashi Y, Shiokawa Y, Takaku F. Effect of peripheral blood lymphocytes from systemic lupus erythematosus patients on human bone marrow granulocyte precursor cells (colony-forming units in culture). *Stem Cells* 1982; 2:164-176.

[88] Young NS. Acquired aplastic anemia. *Ann Intern Med* 2002; 136(7):534-546.

[89] Philpott NJ, Turner AJ, Scopes J, Westby M, Marsh JC, Gordon-Smith AC, Dalgleish AG, Gibson FM. The use of amino actinomycin D in identifying apoptosis: simplicity of use and broad spectrum of application compared with other techniques. *Blood* 1996;87:2244-2251

[90] Maciejewski J, Selleri C, Anderson S, Young NS. Fas antigen expression on CD34+ human marrow cells is induced by interferon gamma and tumor necrosis factor alpha

and potentiates cytokine-mediated hematopoietic suppression in vitro. *Blood* 1995; 85:3183-3190.

[91] Gersuk GM, Beckham C, Loken MR, Kiener P, Anderson JE, Farrand A, Troutt AB, Ledbetter JA, Deeg HJ. A role for tumour necrosis factor-alpha, Fas and Fas-Ligand in marrow failure associated with myelodysplastic syndrome. *Br J Haematol* 1998; 103:176-188.

[92] Maciejewski JP, Selleri C, Sato T, Anderson S, Young NS. Increased expression of Fas antigen on bone marrow CD34+ cells of patients with aplastic anaemia. *Br J Haematol* 1995; 91:245-252.

[93] Alvarado-de la Barrera C, Alcocer-Varela J, Richaud-Patin Y, Alarcon-Segovia D, Llorente L. Differential oncogene and TNF-alpha mRNA expression in bone marrow cells from systemic lupus erythematosus patients. *Scand J Immunol* 1998; 48:551-556.

[94] Rook G, Balkwill F. Cell-mediated immune reactions. In: Roitt I, Brostoff J, Male D, editors. *Immunology*. London: Mosby, 1998: 121-138.

[95] Horigome A, Hirano T, Oka K, Takeuchi H, Sakurai E, Kozaki K, Matsuno N, Nagao T, Kozaki M. Glucocorticoids and cyclosporine induce apoptosis in mitogen-activated human peripheral mononuclear cells. *Immunopharmacology* 1997; 37:87-94.

[96] Rinehart J, Keville L, Clayton S, Figueroa JA. Corticosteroids alter hematopoiesis in vitro by enhancing human monocyte secretion of granulocyte-colony stimulating factor. *Exp Hematol* 1997; 25:405-412.

[97] Von Lindern M, Zauner W, Mellitzer G, Steinlein P, Fritsch G, Huber K, Lowenberg B, Beug H. The glucocorticoid receptor cooperates with the erythropoietin receptor and c-Kit to enhance and sustain proliferation of erythtroid progenitors in vitro. *Blood* 1999; 94:550-559.

[98] Clausen J, Stockschlader M, Fehse N, Hassan HT, Gabl C, Zander AR. Blood-derived macrophage layers in the presence of hydrocortisone support myeloid progenitors in long-term cultures of CD34+ cord blood and bone marrow cells. *Ann Hematol* 2000; 79:59-65.

[99] Atta MS, Powell RJ, Todd I. The influence of anti-fibronectin antibodies on interactions involving extracellular matrix components and cells: a possible pathogenic mechanism. *Clin Exp Immunol* 1994; 96:26-30.

[100] Otsuka T, Nagasawa K, Harada M, Niho Y. Bone marrow microenvironment of patients with systemic lupus erythematosus. *J Rheumatol* 1993; 20(6):967-971.

[101] Lee DM, Weinblatt ME. Rheumatoid arthritis. *Lancet* 2001; 358: 903-911

[102] Nakagawa S, Toritsuka Y, Wakitani S, Denno K, Tomita T, Owaki E, Kimura T, Shino K, Ochi T. Bone marrow stromal cells contribute to synovial cell proliferation in rats with collagen induced arthritis. *J Rheumatol.* 1996;23:2098-2103.

[103] Tomita T, Shimaoka Y, Kashiwagi N, Hashimoto H, Kawamura S, Lee SB, Nakagawa S, Shiho O, Hayashida K, Ochi T. Enhanced expression of CD14 antigen on myeloid lineage cells derived from the bone marrow of patients with severe rheumatoid arthritis. *J Rheumatol.* 1997;24:465-469.

[104] Sen M, Lauterbach K, El-Gabalawy H, Firestein GS, Corr M, Carson DA. Expression and function of wingless and frizzled homologs in rheumatoid arthritis. *Proc Natl Acad Sci USA.* 2000;97:2791-2796.

[105] Santiago-Schwarz F, Sullivan C, Rappa D, Carsons SE. Distinct alterations in lineage committed progenitor cells exist in the peripheral blood of patients with rheumatoid arthritis and primary Sjogren's syndrome. *J Rheumatol.* 1996;23:439-446.

[106] Tomita T, Takeuchi E, Toyosaki-Maeda T, Oku H, Kaneko M, Takano H, Sugamoto K, Ohzono K, Suzuki R, Ochi T. Establishment of nurse-like stromal cells from bone marrow of patients with rheumatoid arthritis: indication of characteristic bone marrow microenvironment in patients with rheumatoid arthritis. *Rheumatology* (Oxford) 1999;38:854-863.

[107] Berthelot JM, Bataille R, Maugars Y, Prost A. Rheumatoid arthritis as a bone marrow disorder. *Semin Arthritis Rheum.* 1996;26:505-514.

[108] Ikehara S. Autoimmune diseases as stem cell disorders: normal stem cell transplant for their treatment. *Int J Mol Med.* 1998;1:5-16.

[109] Choy EH, Panayi GS. Cytokine pathways and joint inflammation in rheumatoid arthritis. *N Engl J Med.* 2001;344:907-916.

[110] Manicourt DH, Triki R, Fukuda K, Devogelaer JP, De Deuxchaisnes C, Thonar EJA. Levels of circulating tumor necrosis factor alpha and interleukin-6 in patients with rheumatoid arthritis. Relationship to serum levels of hyaluronan and antigenic keratan sulfate. *Arthritis Rheum.* 1993;36:490-499.

[111] Keller JR, Ruscetti FW, Grzegorzewski KJ, Wiltrout RH, Bartelmez SH, Sitnicka E, Jacobsen SEW. Transforming growth factor-β, tumor necrosis factor-α, and macrophage inflammatory protein-1α are bidirectional regulators of hematopoietic cell growth. In Garland JM, Quesenberry PJ, Hilton DJ, eds. *Colony-stimulating factors. Molecular and cellular biology.* New York, NY: Marcell Dekker,Inc; 1997: 445-466.

[112] Gersuk GM, Beckham C, Loken MR, Kiener P, Anderson J, Farrand A, Troutt AB, Ledbetter JA, Deeg HJ. A role for tumour necrosis factor-alpha, Fas and Fas-Ligand in marrow failure associated with myelodysplastic syndrome. *Br J Haematol.* 1998;103:176-188.

[113] Zoumbos NC. The pathogenesis of myelodysplastic syndromes. *Haema* 2001;4:151-157

[114] Means RT, Krantz SB. Progress in understanding the pathogenesis of the anemia of chronic disease. *Blood.* 1992;80:1639-1647.

[115] Means RT. Advances in the anemia of chronic disease. *Int J Hematol.* 1999;70:7-12.

[116] Alvarez-Hernandez X, Liceaga J, McKay IC, Brock JH. Induction of hypoferremia and modulation of macrophage iron metabolism by tumor necrosis factor. *Lab Invest.* 1989;61:319-322.

[117] Johnson RA, Waddelow TA, Caro J, Oliff A, Roodman GD. Chronic exposure to tumor necrosis factor in vivo preferentially inhibits erythropoiesis in nude mice. *Blood.* 1989;74:130-138.

[118] Rusten LS, Jacobsen SE. Tumor necrosis factor (TNF)-alpha directly inhibits human erythropoiesis in vitro: role of p55 and p75 TNF receptors. *Blood.* 1995;85:989-996.

[119] Means RT, Krantz SB. Inhibition of human erythroid colony-forming units by tumor necrosis factor requires beta interferon. *J Clin Invest.* 1993;91:416-419.

[120] Katevas P, Andonopoulos AP, Kourakli-Symeonidis A, Manopoulou E, Lafi T, Makri M, Zoumbos NC. Peripheral blood mononuclear cells from patients with rheumatoid

arthritis suppress erythropoiesis in vitro via the production of tumor necrosis factor alpha. *Eur J Haematol.* 1994;53:26-30.

[121] Voulgari PV, Kolios G, Papadopoulos GK, Katsaraki A, Seferiadis K, Drosos AA. Role of cytokines in the pathogenesis of anemia of chronic disease in rheumatoid arthritis. *Clin Immunol.* 1999;92:153-160.

[122] Vreugdenhil G, Lowenberg B, Van Eijk HG, Swaak AJ. Tumor necrosis factor alpha is associated with disease activity and the degree of anemia in patients with rheumatoid arthritis. *Eur J Clin Invest.* 1992;22:488-493.

[123] Baer AN, Dessypris EN, Krantz SB. The pathogenesis of anemia in rheumatoid arthritis: A clinical and laboratory analysis. *Semin Arthritis Rheum.* 1990;4:209-223.

[124] Greenwalt DE, Lipsky RH, Ockenhouse CF, Ikeda H, Tandon NN, Jamieson GA. Membrane glycoprotein CD36: A review of its roles in adherence, signal transduction, and transfusion medicine. *Blood.* 1992;80:1105-1115.

[125] Loken MR, Shah VO, Dattilio KL, Civin CI. Flow cytometric analysis of human bone marrow: I. Normal erythroid development. *Blood.* 1987;69:255-263.

[126] Dai CH, Price JO, Brunner T, Krantz SB. Fas ligand is present in human erythroid colony-forming cells and interacts with Fas induced by interferon gamma to produce erythroid cell apoptosis. *Blood.* 1998;91:1235-1242.

[127] De Maria R, Testa U, Luchetti L, et al. Apoptotic role of Fas/Fas ligand system in the regulation of erythropoiesis. *Blood.* 1998;93:796-803.

[128] Zamai L, Secchiero P, Pierpaoli S, Bassini A, Papa S, Alnemri ES, Guidotti L, Vitale M, Zauli G. TNF-related apoptosis-inducing ligand (TRAIL) as a negative regulator of normal human erythropoiesis. *Blood.* 2000;95:3716-3724.

[129] Birgegard G, Hallgren R, Caro J. Serum erythropoietin in rheumatoid arthritis and other inflammatory arthritides: relationship to anaemia and the effect of other anti-inflammatory treatment. *Br J Haematol* 1987; 65: 479-483

[130] Cazzola M, Ponchio L, de Benedetti F, Ravelli A, Rosti V, Beguin Y, Invernizzi R, Barosi G, Mantini A. Defective iron supply for erythropoiesis and adequate endogenous erythropoietin production in the anemia associated with systemic-onset juvenile chronic arthritis. *Blood.* 1996;87:4824-4830.

[131] Davis D, Charles PJ, Potter A, Feldmann M, Maini RN, Elliott MJ. Anaemia of chronic disease in rheumatoid arthritis: in vivo effects of tumour necrosis factor alpha blockade. *Br J Rheumatol.* 1997;36:950-956.

[132] Knight DM, Trinh H, Le J, Siegel S, Shealy D, McDonough M, Scallon B, Moore MA, Vilcek J, Daddona P. Construction and initial characterization of a mouse-human chimeric anti-TNF antibody. *Mol Immunol.* 1993;30:1443-1453.

[133] Elliott MJ, Maini RN, Feldmann M, Kalden JR, Antoni C, Smolen JS, Leeb B, Breedveld FC, Macfarlane JD, Bijl H, Woody JN. Randomised double-blind comparison of chimeric monoclonal antibody to tumor necrosis (cA2) versus placebo in rheumatoid arthritis. *Lancet.* 1994;344:1105-1110.

[134] Feldmann M, Brennan F, Paleolog E, Taylor P, Maini RN. Anti-tumor necrosis factor therapy of rheumatoid arthritis. Mechanism of action. *Eur Cytokine Netw* 1997; 8: 297-300

[135] Maini RN, Breedveld FC, Kalden JR, Smolen JS, Davis D, Macfarlane JD, Antoni C, Leeb B, Elliott MJ, Woody JN, Schaible TF. Therapeutic efficacy of multiple intravenous infusions of anti-tumor necrosis factor alpha monoclonal antibody combined with low-dose weekly methotrexate in rheumatoid arthritis. *Arthritis Rheum.* 1998;41:1552-1563.

[136] Valle E, Gross M, Bickston SJ. Infliximab. *Expert Opin Pharmacother* 2001; 2: 1015-1025.

[137] Taylor PC, Williams RO, Maini RN. Immunotherapy for rheumatoid arthritis. *Curr Opin Immunol* 2001; 13:611-616.

[138] Hitraya EG, Schaible TF. Anti-tumor necrosis factor therapies in Crohn's Disease and Rheumatoid Arthritis. In Ciliberto G, Savino R, eds. *Cytokine inhibitors.* New York, NY: Marcell Dekker,Inc; 2001:97-131.

[139] Coakley G, Iqbal M, Brooks D, Panayi GS, Lanchbury JS. CD8+, CD57+ T cells from healthy eldery subjects suppress neutrophil development in vitro: implications for the neutropenia of Felty's syndrome and large granular lymphocyte syndrome. *Arthritis Rheum* 2000; 43:834-843.

[140] Hellmich B, Csernok E, Schatz H, Gross WL, Schnabel A. Autoantibodies against granulocyte-colony stimulating factor in Felty's syndrome and neutropenic systemic lupus erythematosus. *Arthrits Rheum* 2002; 46:2384-2391.

[141] Blau CA, Neff T, Papayannopoulou T. The hematological effects of folate analogs: implications for using the dihydrofolate reductase gene for in vivo selection. *Hum Gene Ther.* 1996;7:2069-2078.

[142] Allay JA, Spencer HT, Wilkinson SL, Belt JA, Blakley RL, Sorrentino BP. Sensitization of hematopoietic stem and progenitor cells to trimetrexate using nucleoside transport inhibitors. *Blood.* 1997;90:3546-3554.

[143] Seitz M, Zwicker M, Pichler W, Gerber N. Activation and differentiation of myelomonocytic cells in rheumatoid arthritis and healthy individuals. Evidence for antagonistic in vitro regulation by interferon-gamma and tumor necrosis factor alpha, granulocyte monocyte colony stimulating factor and interleukin 1. *J Rheumatol* 1992; 19:1038-1044.

[144] Hirohata S, Yanagida T, Itoh K, Nakamura H, Yoshino S, Tomita T, Ochi T. Accelerated generation of CD14+ monocyte-lineage cells from the bone marrow of rheumatoid arthritis patients. *Arthritis Rheum.* 1996;39:836-843.

[145] Takeuchi E, Tomita T, Toyosaki-Maeda T, Kaneko M, Takano H, Hashimoto H, Sugamoto K, Suzuki R, Ochi T. Establisment and characterization of nurse-like stromal cell lines from synovial tissues of patients with rheumatoid arthritis. *Arthritis Rheum.* 1999;42:221-228.

[146] Kaisho T, Oritani K, Ishikawa J, Tanabe M, Muraoka O, Ochi T, Hirano T. Human bone marrow stromal cell lines from myeloma and rheumatoid arthritis that can support murine pre-B cell growth. *J Immunol.* 1992;149:4088-4095.

[147] Ikehara S. New strategies for BMT, organ transplantation and regenerative therapy. *Hematology* 2003; 8:77-81.

[148] Burt RK, Fassas A, Snowden J, Van Laar JM, Kozak T, Wulffraat NM, Nash R, Dunbar CE, Arnold R, Prentice G, Bingham S, Marmont AM, McSweeney PA.

Collection of hematopoietic stem cells from patients with autoimmune diseases. *Bone Marrow Transplant* 2001; 28:1-12.

[149] Lopez-Karpovitch X, Cardiel M, Cardenas R, Piedras J, Alarcon-Segovia D. Circulating colony-forming units of granulocytes and monocytes/macrophages in systemic lupus erythematosus. *Clin Exp Immunol* 1989; 77: 43-46.

[150] Snowden JA, Nink V, Cooley M, Zaunders J, Keir M, Wright L, Milliken ST, Brooks PM, Biggs JC. Composition and function of peripheral blood stem and progenitor cell harvests from patients with severe active rheumatoid arthritis. *Br J Haematol.* 1998;103:601-609.

[151] Shaughnessy PJ, Ririe DW, Ornstein DL, Kissack B, Bickford DJ, Molina R, Lanzkron S, Bachier C, Nanez A, LeMaistre CF. Graft failure in a patient with systemic lupus erythematosus (SLE) treated with high-dose immunosuppression and autologous stem cell rescue. *Bone Marrow Transplant* 2001;27:221-224.

[152] Isenberg D, Leckie MJ Biological treatments for systemic lupus erythematosus. Scand J Rheumatol 2002; 31:187-191.

[153] Blumberg SN, Fox DA. Rheumatoid arthritis: guidelines for emerging therapies. *Am J Manag Care* 2001;7:617-626.

In: Trends in Stem Cell Research
Editor: Erik V. Greer, pp. 219-234

ISBN 1-59454-315-1

Chapter IX

The Influences of Political Ideology on Pain Care and Stem Cell Research

Anthony J. Eksterowicz*

Professor of Political Science, James Madison University, Department of Political Science, Harrisonburg VA

Nancy B. Eksterowicz**

Master's in Nursing Science, Advanced Practice Nurse Pain Management, University of Virginia Health System, Pain Services, Charlottesville, VA

Abstract

The role of political ideology in health care decisions has become increasingly important in the past decade. As science pushes the boundaries of pain relief and life replication (cloning and stem cells) issues it confronts political, ideological and ethical issues that may have detrimental impacts upon patient care, the relief of pain and cures for diseases. For this study, we examine that portion of political ideology which effects one's beliefs about the origins and preservation of life and its effect upon health care policies, specifically patient pain care and basic research regarding stem cells. We hypothesize that ideological belief systems (based partly upon religious beliefs) can have a detrimental effect upon future patient care. While we are aware that it is difficult to distinguish between political and ideological policy entrepreneurship, we also agree with Colin Campbell (2004) that ideological entrepreneurship involves a depth of feeling that makes compromise difficult. It also involves interests tied to a president's or congressperson's political base which are particularly persistent in their policy preferences.

* EKSTERAJ@JMU.EDU, 540-568-6149

** NBE@VIRGINIA.EDU, 434 924-2261

This study examines a number of variables associated with these two areas. First we discuss the new emerging model of patient health care from the perspective of pain management practitioners. Second, we apply the model to two case studies involving political decisions. One decision is the congressionally based Hyde – Nichols bill with implications for patient pain control and the other is President Bush's decision concerning stem cells and its effect upon basic research and ultimately patient care. Third, we derive some lessons from both experiences that might help manage some of the problems concerning patient pain care and basic research in the stem cell area.

Pain Management and the New Emerging Health Care Paradigm

Individuals in our society and culture learn through various institutions (e.g., media, medicine, family) that "pain" is a deeply individual enterprise and that suffering through the pain is merely an accepted part of life. Society and culture profoundly shape our political, ideological and economic experiences. So too, culture shapes our perception of pain. David B. Morris in his classic work, *The Culture of Pain,* suggests that individual minds and specific cultures shape the experience of pain. He notes, for example the use of sentimental cultural pain in Harriet Beecher Stowe's *Uncle Tom's Cabin* and its effect upon political change. Morris warns that we cannot merely view the concept of pain as medical. Pain is not divisible. It exists within the body but also within the mind. After the terrorist attacks on September 11, 2001, pain specialists saw a dramatic increase in pain complaints from chronic pain sufferers. Peter Staats, chief of pain medicine at John Hopkins University, noted that there is a strong mind-body connection involved in pain and cultural events (Goldstein, 2001). Confining pain to the medical realm and viewing it as a neuro-matter is misguided (Morris, 1991, passim).

Morris prefers to view pain not as something that medical science can conquer or solve like a puzzle but rather as a mystery. This is very often how patients view pain. Thus pain is subjective. For example, acute pain differs from chronic pain. Acute pain can and will disappear but chronic pain befuddles patients and doctors alike. It deeply depends upon the perception of the patient. To mitigate the problem doctors must work closely with patients (Morris, 1991 pp. 9-79). But this notion of partnership contrasts with a doctor centered or driven relationship as predominate in the last century. In the past, patients received and respected their physicians due to their education, knowledge, and their remarkable ability to heal. Traditionally, doctors would decide what the correct treatment was and present it to the patient. This had some undesirable consequences. One such consequence was the paternalistic attitude reflecting the predominant cultural value that women were more emotional and may exaggerate pain complaints. Many doctors were more likely to dismiss women's complaints of pain as hysteria or neurosis. In a study cited by Post et al. (1996 p. 350), a personal and cultural gender bias of doctors and nurses continues to exert a negative influence over medication choices for analgesia and sedation. This balance of power is fundamentally shifting. Today, many health-care recipients demand relationships that respect and protect autonomy, and enhance well-being (Post, et al.1996 p. 349). Patients recognize the limitations of their doctors as well as medical science. They desire more of a partnership when it comes to healing.

If pain is perception, if it is emotional as well as physical, if it befuddles patients and doctors alike, if it requires a partnership between patient and doctor then remediation of pain must involve a broader societal approach. Dealing with pain must involve the patient, the clinicians and a society that comprehends the integration of the mind and body. It must also involve recognition of how the cultural role of pain within a society impacts treatment. There is a political consequence to such cultural pain. For example, cultural pain afflicts the socio-economically disadvantaged. A recent study, (Anderson et. al 2000) demonstrated that physicians are more likely to underestimate pain severity in minorities and women and offer insufficient analgesics for pain associated with cancer.

As Morris notes, "All pain -especially all chronic pain, is an interdependent, inseparable, multidimensional union of the two elemental human forces that the Greeks called psyche (mind) and soma (body)" (Morris, 1991 p. 158). In the treatment of pain a war of sorts broke out between the peripheralist (the dominant) and the centralist (the challenger) paradigm. Theorists from the peripheralist paradigm suggest that pain can be traced to organ or nerve damage. In other words it is organically initiated. The centralist school believes that while some pain may be peripheral, pain is more complex. For example there exists the notion of intractable pain, which serves as a paradox to peripheral notions (Morris, 1991 p. 167). If the centralists are correct the treatment of pain will be more complex and multidisciplinary in nature. It will require greater education, specialization and interdisciplinary efforts aimed at both the body and mind. It will require an extraordinary partnership with patients and a new attitude among physicians and health care providers. It may require societal and cultural choices concerning pain, suffering and dying. These choices will ultimately be economic and political in nature involving issues such as, stem cell research, the right to die, terminal patient pain relief and efforts to remediate pain by physicians, which may conflict with our nation's drug laws. As Rowlingson (2001) notes:

> It's quite fascinating to think about moving from a traditional medical model where doctors directed nurses and nurses delivered the care to patients that had pain. With this perspective you realize that the person with the symptoms is at the end of the treatment chain. The concept of patient-centered care is revolutionary, and it shouldn't be. There will be enormous concerns amongst health care workers in having to work in that kind of environment where the patient has been empowered to be the one in charge.

We stand on the cusp of a paradigmatic change in pain care. However, in this era the old paradigm is very much alive. As this paradigm begins to shift, society will be confronted with anomalies. The organic model has been increasingly unable to provide answers for these anomalies. The old paradigm has fostered many myths, which have been reported by the media and adopted by other institutions as conventional wisdoms. As Daniel B. Carr (1999) notes:

> The popular media convey inaccurate ideas about pain and pain treatment in this country. The result is misperception on the part of the public that lead to the generation and spread of myths about pain. Some of these myths are: Pain should be endured. Pain

> is inevitable: analgesia cannot be provided conveniently. It is wrong if terminal patients die a little sooner because of pain-relieving medication. Good patients endure pain and do not distract physicians from treating their disease. Pain medication leads to addiction. Patients are sometimes in so much pain they desire death, and there is nothing we can do about that.
>
> These statements are all incorrect and, viewed in the light of modern knowledge, are easily refuted. Nevertheless, these myths are prevalent and adversely affecting not only patient's thinking but also medical practice. Education is the means by which these myths will finally be silenced.

While Dr. Carr emphasizes the direct impact from the media, the medical profession cannot escape blame. How many general practice physicians prescribe the lower and "safer" doses of pain killing medication because of their mistaken belief that pain medication leads to addiction (Hansen, 2000 pp. 2317-18)? Misunderstanding of pain, pain treatments and effects of medication lead physicians to fear legal penalties and disciplinary action (Shapiro, 1996 p. 363). Information about what is legally and medically acceptable for the treatment of pain is rarely conveyed to state medical boards (Hyman, 1996 p.338). Some members of the medical profession are precipitating a silent crisis of patient under-medication across this nation.

The myths Carr describes are, in part, the result of the conflict between the old peripheralist paradigms with the new centralist paradigm. The public generally accepts these myths. They have meaning and directly and indirectly affect our economics and politics with respect to the management of pain.

The conflicting nature of these two paradigms must also be placed within the context of the two competing health care paradigms. The old patient care paradigm was essentially paternalistic and is being supplanted by a new patient autonomous paradigm (Post, et al., 1996 pp. 348-49). Paradigms change slowly over time. Already we can observe some progress. Health care regulatory agencies such as Centers for Medicare and Medicaid, (CMS) formerly known as HCFA, and Joint Commission on Accreditation of Healthcare Organizations (JCAHO) are incorporating pain management standards into the evaluation and accreditation process (PM Standards JCAHO: 1999 online). This comes as a response from the public, pain management organizations and pain specialists. However, our economics and politics may be trapped within the old paradigm. Efforts at education will help change attitudes, laws and markets but not without resistance and real cultural pain. These economic impediments work against the alleviation of patient suffering and against the emerging patient centered health care paradigm. One example of the potential impact of political ideology upon societal patient pain care is reflected through the Hyde-Nichols Bill.

The Hyde-Nichols Bill

The now defunct Hyde-Nichols Bill otherwise known as the Lethal Drug Abuse Prevention Act (House Bill 2260, Senate Bill 1272) was proposed in 1998. It essentially amended the 1970 Control Substance Act to prevent physician assisted suicide and had not been designed to influence pain care. The state of Oregon had passed a physician assisted suicide statute known as the Death with Dignity Act. Both Representative Hyde (R-Ill) and

Senator Nichols (R-OK) were opposed to this nascent state movement due to their strong pro-life positions. Their bill would have barred doctors from using federally regulated drugs (schedule II drugs) to aid in patient suicide. Under this proposed act any physician *intentionally* using a controlled substance for the purposes of assisting in a patient's suicide could lose their license. In addition, the act established a board of experts which would advise the Drug Enforcement Agency (DEA) in murky cases where it was not clear if the physician was assisting in patient suicide or aggressively providing pain control (Orentflicker & Caplin, 2000 pp. 255-256).

The language of the bill caused uproar in the health care community. For example, how could one determine the intent of the physician? The line between providing pain control and aiding in patient suicide is very fine. Health care professionals feared the establishment of a board of experts presumably drawn from the health care field to advise the DEA. They believed that the DEA would play a larger role in their field. This coupled with the low levels of knowledge concerning pain relief could prove detrimental to the alleviation of pain nation wide. Both the American Medical Association and the National Hospice Organization opposed the bill (Orentlicher & Caplin, 2000 pp. 255-8).

The bill represented the intersection of political ideology with health care and health care regulation. The bill reflected the pro-life ideological persuasion of both Representative Hyde and Senator Nichols. Legislating according to ideology and not science could have negative consequences for the alleviation of patient suffering. If ideology or religious beliefs and science conflict which path should society choose? We are now witnessing the same type of debate concerning the stem cell issue (Connolly, et. al., 2001; Will, 2001; Cohen, 2001). This issue tends to fit Campbell's (2004) concept of ideological policy entrepreneurship in that it involves a situation where compromise is difficult and significant portions of both Hyde and Nichol's persistent political base (pro-life interests) drive this decision.

Due to the heavy criticism of the bill it was amended and re-named in 1999. The Pain Promotion Relief Act eliminated the board of experts advising the DEA. It also included language that permitted the use of controlled substances for pain management and palliative care even if the use of such a substance may increase the risk of death. There was a section that addressed the education of law enforcement officials about the legitimate use of controlled substances for pain and palliative care. With these changes some interest groups like the American Medical Association originally endorsed the bill. However when other health care groups criticized the bill the AMA rescinded its support (Reuters Health 2000). What were the criticisms of the amended bill?

The amended bill still leaves the line between acceptable palliative care and unacceptable suicide defined by physician intent. It is difficult to determine intent. Physicians would worry that law enforcement officials could interpret intent when no intent existed. The murky line between acceptable care and end of life decisions could also affect physician intent. All of this could force doctors to be conservative and under-treat pain. Physicians utilizing aggressive drug therapies could not only lose their license but also be subject to criminal penalties. This would also constrain doctors from administering adequate pain care. The act could increase a physician's legal liabilities again resulting in under-treatment of pain. The bill would give the DEA a mandate to focus on end of life care. The act would empower the DEA to conduct investigations in an area of health care they know very little about. The act would contravene the principles of federalism by overturning the Oregon statue on physician assisted suicide. Finally, the act would reduce innovation and experimentation in the area of palliative care. Physicians would be reluctant to try experimental drug therapies for fear of prosecution (Orentlicher, 2000 pp. 2933-5).

As a result of these criticisms the amended bill died. The lack of knowledge concerning pain care combined with increased government regulation and monitoring, physician fear and political, ideological motives all combined to kill the bill. In its wake the American Pain Society is presently formulating the National Pain Care Policy Act of 2001. This bill would:

> Create a White House Conference on Pain Care. Create a national Center for Pain Research within the National Institute of Health. Provide funding for pain and palliative care education and training through the Agency for Health Care Research and Quality. Create pain care standards within Medicare and Choice plans to ensure payment for appropriate interdisciplinary pain care within these plans. Require an annual report on Medicare expenditures for pain care services, which will require CMS (formerly HCFA) to treat pain medicine as a physician specialty. (Ashburn, 2001 p. 2).

Many of the provisions of this act were incorporated into the National Pain Care Policy Act of 2003 which was submitted to the United States House of Representatives by Representative Rogers (R. Mich.) on April 29, 2003 (United States House of Representatives, 108th Congress 1st Session, 2003). As of mid fall 2003 no hearings were scheduled. The purpose of the bill was to begin the discussion concerning pain care in the Congress. There is a long legislative road yet to travel. While this budding debate was taking place in the Congress, Bush Administration Attorney General John Ashcroft acted to block Oregon's assisted suicide law by authorizing DEA agents to punish doctors prescribing federally controlled drugs to terminally ill patients (Eggen and Connolly, 2001pp. A1, 3). This was quickly followed by federal judge temporarily blocking the order (Booth, 2001 p. A2) and a United States District Court ruling upholding Oregon's law (Booth, 2002 p. A1). In addition, there were reports in 2003 that federal prosecutors working for the Drug Enforcement Agency were increasing their efforts at prosecuting pain doctors (Kaufman,2003, A1,A5).

One could view these developments as the normal course of political events surrounding public policy issues and to an extent this is true. However, one could also view these events as an example of political ideology driving policy to the detriment of patient pain care management as practiced by physicians across the country. Without an effective patient pain control bill time is wasted and patients suffer. Furthermore the movement toward a patient centered model of health care is impeded. We will draw a few lessons from this experience but first we turn our attention to the stem cell debate.

The Stem Cell Debate

Research on embryonic stem cells represents the holy grail of basic scientific knowledge concerning the future of health care. These cells have the potential to revolutionize the way we look at cures for all sorts of long term illnesses from diabetes to various forms of muscular diseases. The one defining characteristic of these cells is their *pluripotent* nature.

They can differentiate themselves into other types of cells. As such they have enormous potential in research affecting areas of the body such as the liver, brain and skin among others. Due to their adaptability they can literally change the course of research on certain diseases. For example, instead of researching better ways to deliver insulin to a diabetic patient in managing the disease, these cells may actually form new pancreatic cells which would replace the old nonfunctional pancreas. This would represent a quantum leap in research for a whole host of diseases and disorders (http://europa.ev.int/comm/research/ quality of life/stemcells/about.html).

There are ethical problems and dilemmas associated with this type of research. Presently, human embryos are the most promising source of stem cells. Usually these can be provided by in vitro fertilization clinics. The process requires the destruction of the embryo to obtain stem cells. This has touched off an ethical and moral debate concerning the wisdom of conducting such research. Most of the controversy is religious in nature. For many organizations there are strong religious objections to destroying these human embryos. The Catholic Church, for example, considers all embryos potential life. In their view, a ban on federal funds for such controversial research will prevent embryo destruction.

Early in the debate, other sources for these stem cells were touted. For example, a report from the United States Department of Health in 2001 indicated that in addition to embryos, stem cells might be obtained from the blood cells of an umbilical cord or from adult tissue such as bone marrow or from mature adult tissue (United States Department of Health, 2000, Report From the Chief Medical Officer's Expert Group…., Executive Summary, 1: Cheng, et al 2003, 131-42). However, new research has indicated that these possibilities were at best optimistic (Weiss, 2003, A6; Chambers et. al, 2003). New research concluded that early research (see, McKay 2000, 361-64) concerning cells derived from human adult bone marrow was misinterpreted. This is important because it may leave embryonic derived stem cells as the only viable source for promising research at this time. This, in turn, implies that the controversy will continue.

The federal government supplies the largest amount of funding for basic research. Privately-funded stem cell research is one approach scientists are pursuing but these resources are limited. Stem cell research was banned by Congress in 1995. The Congress attached the ban to an appropriation bill funding the National Institute for Health. Since 1995 no public funding has been available for this research (AAAS Policy Brief; Stem Cell Research, September 5, 2003, 2). In 1999, the Department of Health and Human Service indicated that publicly funded research could be conducted on these stem cells as long as the destruction of the embryos for the derivation of these cells was privately funded. They also began to draft regulations for these purposes. In effect, the 1999 guidelines permitted research only on cells derived from embryos which were the remnants from fertility treatments at fertility labs. These were donated with consent for research by patients (AAAS Policy Brief; Stem Cell Research, September 5, 2003, 4-6). The United States Department of Health took a position in 2000 that research on nuclear transplantation of human embryo cells should continue subject to safeguards largely modeled after those established in the United Kingdom (United States Department of Health 2000, 44). President Bush's Secretary of Health and Human Services, Tommy Thompson advocated the use of public funds for this research as did some Republican members of the house and Senate, including the influential

Senators Hatch from Utah and Thurmond from South Carolina (AAAS Policy Brief: Stem Cell Research, 2003 September 5,). In fact, Senator Hatch distinguished “totipotent” cells which form after fertilization from “pluripotent” cells because the latter could never become a human being. Hatch supports research on “pluirpotent” but not “totipotent” cells (Berger, E 2001). During the 2000 presidential campaign, candidate Bush declared his opposition to any federal funding for this type of research. The stage was set for a dramatic presidential address to the nation in August of 2001. President Bush would make a decision affecting the course of basic research in this nation and all potential future patients who might benefit from this research.

The Bush Decision

President George W. Bush faced a conflicted decision. On the one hand most conservatives in Congress and Christian and evangelical groups opposed any funding for “pluripotent” stem cell research. During his campaign, Bush promised he would not extend public funding to this research. On the other hand, Democrats, liberal groups and a handful of Republican congress people and senators along with Secretary of Health and Human Services, Tommy Thompson and some members of the White House staff urged the president to allow public funding (Berger E. 2001). Faced with a politically and ideologically charged environment Bush attempted to split the issue. He allowed federal funding for research using existing stem cell lines only because the life and death decision had already been made in these instances. Bush prohibited public funding on all other stem cell derivatives existing in fertility clinics nationwide. He also created a new Council on Bioethics chaired by Dr. Leon Kass of the University of Chicago for advice on the ethical and moral implications of this type of research.

The address to the nation is instructive of the president’s thinking and influence. The following phrases are taken from the address; “church,” “different faiths,” “same faith,” “potential for life,” “a great deal of thought and prayer,” “moral hazards,” “beginnings of life,” “ life is a sacred gift from our Creator,” “without crossing a fundamental moral line.” Bush openly admitted that, “My position on these issues is shaped by deeply held beliefs. I am a strong supporter of science and technology……. I also believe human life is a sacred gift from our Creator. (The White House Press Release, 2001.). The president went on to encourage stem cell research from placenta and adult human bone marrow sources. The speech demonstrates the faith based influences upon presidential decision making in this area of controversial research and the impact of Christian and evangelical groups upon such decisions in this administration. These groups however were not supportive of the president’s limited decision (http://www.americancatholic.org/News/StemCell/default.asp). Of course, how limited the decision was is itself debatable. Kinsley (2003, A25) notes that President Bush was originally expected to ban research involving future embryos but his decision also banned existing embryos in fertility clinics destined for destruction. This, in part, could explain the muted evangelical criticism.

We argue that this decision also reflects Campbell’s (2004) definition of ideological entrepreneurship. The President admitted the deeply religious beliefs that influenced his decision-making in his speech. While he did not entirely satisfy his Christian evangelical base he did appease them and offer those interests further opportunity to influence these decisions in the future. Their persistence on this issue makes such an outcome likely.

We note that this ideological policy entrepreneurship is occurring not only in the area of health care policy but also in other areas like basic foreign and environmental policy (Campbell, 2004). Indeed, in environmental policy leading researchers were mobilizing against a Bush administration plan which would block new federal regulations unless the science used to justify these regulations is cleared by a peer review process overseen by the executive branch's Office of Management and Budget. Many scientific researchers believe this would unnecessarily politicize the process (Weiss, 2004, A16). All of this comes in the midst of questions concerning President Bush's general orientation to scientific policy. Both the National Research Council and the Union of Concerned Scientists called upon the Bush administration for greater public access to objective scientific information (Gugliotta and Weiss, 2004, A2). There was also a push for a new source for information on the science of terrorism by the Congress. A two and a quarter million dollar grant by the MacArthur Foundation went to the American Association for the Advancement of Science to link policy-makers who need scientific information with the academic community (Kaufman, 2004, A19).

Regarding the stem cell decision, critics immediately attacked the decision on a number of fronts. First, of the stem cell lines that Bush identified, how many were truly viable for research purposes? Stanford University indicated that of the seventy-eight stem cell lines in the world only about one dozen might be in good enough shape for research purposes (PBS Online Newshour Update, December 11, 2002). Second, are these cells genetically diverse enough for research purposes? Third, what is the safety of the existing cells concerning future human implantation? Fourth, the viability of the long run research in this area might be significantly slowed by utilizing only these existing stem cell lines. In fact, The National Academy of Science in September of 2001 concluded that in the long run new embryonic stem cell lines will be necessary (AAAS Policy Brief: Stem Cell Research, 2003). Fifth, many critics predicted that the United States would lose basic research scientists to other nations that have more liberal laws regarding stem cell research. This has not happened as of 2003 largely because the states have stepped in to create research friendly zones for this particular type of research. California has enacted legislation to attract these researchers (Gershon, 2003). Stanford University received a twelve million dollar grant from an anonymous donor to establish an institute that will create new stem cell lines (PBS Online Newshour Update, 2002, December 11). The effect of the Bush ban upon scientists leaving the nation to conduct research elsewhere is largely determinative of how long the ban lasts and the research questions current experiments lead to. Clearly, the jury is still out.

There were other problems with the ban that apparently were not considered in the decision making process. Fins and Schacter (2002: 269-70) argue that the decision is unwise for two reasons. The first involves patent law and the second principles of informed consent. Regarding patent law the authors note:

> ...Ordinarily, upon the expiration of a patent, the patentee may no longer effectively prohibit others from formulating the product. Patent law temporarily limits the monopoly

> so that the invention reaches the public. The president's policy subverts the construct of patent law because it prohibits federally funded researchers from using the disclosed knowledge to create new cell lines even after the patents expire, as such researchers will not have critical access to the ingredient they need to exploit the knowledge. This will be to the detriment of ensuing research and potential utilization in therapeutic product development.

Fins and Schacter further note that the Bush policy will provide an advantage to the initial holders of these patents secured before August 2001. The artificial market constraint imposed by the Bush decision will work against the purpose of our anti trust and patent laws to promote innovation and enhance consumer welfare. Thus the decision will ultimately affect future patients in this area. As Fins and Schacter (271) argue, "To restrict scientific enquiry through restraint of markets upon which biomedical science increasingly depends could deprive a generation of patients of clinical advances…"

The patent argument is becoming central to these debates. In November of 2003 Senator Sam Brownback (R-Kan.), a pro life supporter, added language to a Senate Appropriations bill which would ban patents on human embryos. Brownback followed the lead of Rep. David Joseph Weldon (R-Fla) who, in a bill passed by voice vote in the House, banned patents on human organisms (Weiss, 2003, A4; Weiss, 2003, A19). The Weldon bill has been stalled in the Senate largely due to the efforts of Senator Hatch (R-Utah) (Novak, 2003, A31). These actions touched off a philosophical debate on the role of free markets and nascent human development. It marks the first time Congress has intervened in such issues and potentially pits the Congress against the U.S. Courts.

A second problem involves the principle of informed consent. Are the consent forms which existing donors signed adequate after the Bush decision? Many of these donors were assured that their identity would be lost in a sea of research consent forms. However, with the Bush limitation these donors will be far more concentrated and perhaps conspicuous. Did these donors fully appreciate the economic value of their donations? The authors, Fins and Schacter (273) ask, "What would happen if individuals who had consented to these procedures reconsidered their decision based on the high-profile nature of these sanctioned cell lines and the public persona of their DNA?" How might these circumstances affect future donors in this area? Might it chill basic research? If these questions are answered in the positive they will undoubtedly affect future patients by slowing research.

There are other problems concerning the President's Council on Bioethics. The Council consists of eighteen members but only three scientists and two clinicians. There were charges that the Council has strong conservative, theological and religious leanings and might be skeptical of science and technology in general (Check, 2002, January 24, pp.351-352). The president's panel recently called for a ban on radical reproductive procedures such as creating animal/human hybrids. However, draft reports initially included politically divisive language indicative of a heavy philosophical and scientific debate among the members (Weiss, 2004, A2). Given the nature of the Bush address heavily laden with religious belief and conviction this should not be a surprise. In fact the Bush Administration has come under attack for its misuse of science in a number of policy areas (Weiss, 2003). In addition, the president has recently created new concerns about his objectivity with respect to news in general. The president noted in an interview with FOX news, "The best way to get the news is from objective sources. And the most objective sources I have are people on my staff who tell me what's happening in the world" (Kinsley, 2003).

Objective information is extremely important in this basic area of research. One of the dangers already occurring is the pollution of objective, peer reviewed studies with other reports, news articles and studies of dubious distinction. For example in 2002 the United States Senate Judiciary Committee held hearings on cloning. During this hearing senators and the news media routinely utilized articles and reports of dubious scientific validity. Many of these articles and reports had not been subject to scientific peer review procedures. In fact Stanford biologist Irving Weissman warned the senators that such reports should not be used because they might unduly influence the legislative debate (Check, 2002, February 14 p.722). This is a major problem in a democratic society which relishes first amendment principles. The danger of un-objective reports crowding out scientific reports is ever present and growing. The problem increased when the president joined the debate against therapeutic cloning and argued that, "Research cloning would contradict the most fundamental principle of medical ethics – that no human life should be exploited or extinguished for the benefit of another" (Check, 2002, April 18). If such an information environment is allowed to continue belief will trump science to the detriment of patient care. Irving L. Weissman noted in the *New England Journal of Medicine* (2002, 1578):

> Unfortunately, there are few in Congress or the President's council who can evaluate the scientific and medical issues in order to make an appropriately informed decision. Too often in recent Senate hearings, the views expressed by senators have been based on articles in newspapers and popular magazines rather than reports of the National Academies or articles in peer-reviewed journals. Some journalists are failing the public trust by publicizing findings that have not been published in the scientific literature or independently replicated.

How do these developments in the basic stem cell research area relate back to the newly emerging patient centered health care model? First, scientific technology and inquiry has reduced the amount of time between basic research discoveries and their clinical applications. Basic scientific discoveries are emerging more rapidly today than twenty-five years ago. The human genome project is indicative of such a pace. When non-scientific obstructions are placed upon basic scientific research, the time to discovery and clinical application will be interrupted. Needless suffering will occur because of these obstructions. This sets back the newly emerging paradigm for it reduces the opportunities for patient control and empowerment with respect to a whole host of diseases and disorders. How can patients be empowered in a society that places impediments on basic scientific research, artificially alters the information environment and interferes with the free market results of basic research? Such a society is more in keeping with the old paternalistic health care model rather than the new patient centered one. What are the lessons from the ideological based decisions in the area of pain care management and basic stem cell research?

Conclusions and Recommendations

It would be easy to state that ideological or religious based decision making should be tempered by objective scientific inquiry. Wishing will simply not bring this about. There are some basic suggestions that might help further the cause of objectivity in these most important areas. In the area of pain care the importance of education cannot be overstated. Health care professionals and their associates including insurers, state medical boards, law enforcement agencies and general practice physicians must be better educated in pain control especially as it affects life and death issues. As Dr. Stephen Long noted with respect to pain care:

> If you compare other medical specialties, we get the patients with problems that many primary care doctors do not often see or poorly understand, or haven't been educated to treat properly. Not because they are ill informed but because they do not have any specialty training in pain management. Traditionally, little if any education regarding pain is presented to medical students or while in residency. They rely on what they read or hear historically. So pain specialists become the first line of defense for many of these patients after they have gone from doctor to doctor. We get a lot of referrals from orthopedic and neurosurgeons who are more specific and more geared toward an accurate pain diagnosis. They are asking us (pain care specialists) to intervene and to implement some changes and take care of those patients. As pain care specialists we make many of the initial diagnoses and provide education for not only the patient but the health care provider who referred the patient. We also direct the physical therapist who may not be accustomed to treating a specific condition in a chronic pain patient. We help the pharmacist because the pharmacist may or may not be used to prescribing the doses of medication for the chronic pain indications. There is also a lot of time spent educating the insurance provider as well. The pain care specialist spends an enormous amount of time providing education and that time is not reimbursable.

If Dr. Long is correct about the educational role of the pain care specialist how can we as a society increase the education concerning pain care to empower the patient? First, medical schools can and should make pain care a primary field of study. This will increase the education of doctors who later serve as advisers to drug enforcement organizations, state medical boards and insurance providers. Second, pain control standards should be nationally known and publicized. The JCAHO standards for pain management are beginning to set standards within institutions for this purpose. Third, congressional hearings should be held on the newly proposed National Pain Care Policy Act separate from euthanasia or 'Right to Life' agendas. Fourth, yearly workshops and symposiums on standards and new developments in pain care should be provided to all health care providers. Fifth, the best scientific evidence available should be the foundation for information when legislating public health issues thus avoiding public policy decisions based on alternative or ideological agendas. A process for dissemination of public information based on the scientific evidence can become a citizen's expectation rather than the exception. Throughout the 1990's the federal government, through the Agency for Healthcare Research and Quality (AHRQ) has provided research support and policy guidance, in collaboration with experts from the public and private sectors. Despite a federal (The Healthcare Research and Quality Act of 1999, Part B, TitleIX,

Section 911(a)) mandate that AHRQ identify systems to assess research results, rating the strength of the scientific evidence underlying health care practice, and make systems widely available, this model has been largely ignored in pain care legislation and stem cell policies. (www.infopoems.com/)

What specific suggestions can be gleaned from the stem cell decision? Any congressional or presidential advisory body should have a *balanced* membership between scientific researchers in the area of inquiry and a mix of ethicists, physicians and journalists otherwise such a panel should be suspect. The staff of such an advisory board or panel should employ legal help in a variety of areas including privacy and economic concerns. Information at any congressional hearing or nationwide symposia should be categorized by order of its scientific value. Over the last 15 years the medical community has used a systematic method for evaluating several independent clinical trials and outcomes. This system allows for more objective appraisal of evidence, which may lead to resolution of some present disagreements. In addition, it can reduce the probability of false negative results and help speed up the process to the introduction of effective treatments. (Egger and Smith.1997). The science should be presented based on systematic research-based evidence rather than on expert opinions or clinical experience thereby minimizing bias and ideology. A hierarchy should be established with scientifically, peer reviewed research receiving the highest level or category of information. Participants, journalists, members of congress and the general public should be made aware of the levels of information categorization.

These suggestions, while basic in nature, can go a long way in helping to temper the impact of ideology upon public policy decision making. We can never eliminate ideology nor should we. However, in the twenty-first century the pace of scientific discovery and medical advances should not be impeded by fear and misinformation. Ideology must be balanced by scientific knowledge to satisfy the emerging patient centered paradigm in health care. The future of patient care demands no less.

References

American Association for the Advancement of Science (AAAS). Online. (2003, September 5). AAAS Policy brief: Stem cell research. *Center for Science, Technology, and Congress* http://www.aaas.org/spp/cstc/briefs/stemcells/

Ashburn, M. A. 2001. "National Pain Care Policy Act of 2001." *American Pain Society Bulletin.* 11: 2.

_____ . 1996. "Effort to Establish a National Pain Research Center Progresses." *American Pain Society Bulletin.* 6: 8-9.

Ashburn, M. A. and P.S. Staats. 1999. "Management of Chronic Pain." *Lancet* 353: 1865-9.

Berger, E. (2001, July 11). Research avenue adds fuel to stem cell controversy. *Cnn.com http://cnn.com/2001/HEALTH/07/11/stem.cell.fact*

Booth, W. 2002. "Oregon's Assisted Suicides Upheld." *The Washington Post.* 18 April. A1.

_____ . 2001. "Judge Blocks Sanctions Over Assisted Suicide." *The Washington Post.* 9 Nov. A2.

Campbell, C. (2004) Unrestrained ideological entrepreneurship in the Bush II advisory system: An examination of the response to 9/11 and the decision to seek regime change in Iraq. In C. Campbell & B.A. Rockman (Eds.). The George W. Bush Presidency: Appraisals and Prospects. (pp74-75). Washington, DC: CQ Press.

Carr, D. B. (1999). Pain in the Media. (Address delivered at *Pain In The Public Eye: 18th Annual Scientific Meeting of the American Pain Society.* Fort Lauderdale, Florida, October 22, 1999.

Chambers, I., Colby, D., Robertson, M., Nichols, J., Lee, S., Tweedie, S., Smith, A. (2003, May 30). Functional Expression Cloning of Nanog, a pluripotency Sustaining Factor in Embryonic Stem Cells. *Cell.*Vol. 113: 643-655.

Check, E. (2002, January 31). Senators square up over cloning. (News) *Nature.* Volume 415, p. 461.

_____ . (2002, January 24). Panels' conflicting views cloud legal future of human cloning. (News) *Nature.* Volume 415, pp. 351-352.

_____ . (2002, February 14). Cloning agenda 'skewed' by media frenzy.. (News) *Nature.* Volume 415, p. 722.

_____ . (2002, March 07). Call for cloning ban splits UN. (News) *Nature.* Volume 416, p. 3.

_____ . (2002, April, 18). Bush rallies opposition to cloning. (News) *Nature.* Volume 416, p. 669.

Cheng, L., Hammond, H., Ye, Z., Zhan, X. & Gautam, D. (2003). Human adult marrow cells support prolonged expansion of human embryonic stem cell culture. *Stem Cells 21,* Volume 2: 134.

CNN.com. (2001,August 10, 2001). Bush to allow limited stem cell funding. (http://cnn.com/2001/ALLPOLITICS/08/09/stem.cell.bush)

Cohen, R. (2001). An Ethical Travesty. *The Washington Post.* 14 August. Op-Ed.

Connolly, C., Gillis, J. & Weiss, R. 2001. "Viability of Stem Cell Plan Doubted." *The Washington Post.* 20 August. A1, 4.

Dennis, C. (2002, March, 07). Stem-cell reverse angers Australian biologists. (News) *Nature.* Volume 416, p. 4.

Department of Health, Office of the Chief Medical Officer. (2000, June). A report from the Chief Medical Officer's expert group reviewing the potential of developments in stem cell research and cell nuclear replacement to benefit human health.

Eggen, D. & Connolly, C. (2001). Ashcroft Ruling Blocks Ore. Assisted Suicide Law. " *The Washington Post.* 7 Nov. A1, 3.

Egger, M. & Smith, G. D. (1997). Meta-analysis: Potential and promise. *BMJ.* No. 7119, vol315.

Europa European Union Online. (2000) About stem cells – Seven important questions on stem cells. (http://europa.ev.int/comm/research/qualitylife/stemcells/about.htm/

Finns, J.J & Schacher, M. (2002) Patently Controversial: Markets, morals, and the president's proposal for embryonic stem cell research. *Kennedy Institute of Ethics Journal.* Vol. 12, 3,:265-278.

Goldstein, A. (2001, October 1). Terrorism tied to jump in pain problems. *The Washington Post..* Section A, pp. 1&6, Col 1.

Gershon, D. (2003, April 24). Stem-cell research. *Naturejobs 422,* pp. 928-929.

Gugliotta, G. & Weiss R. (2004, February 19). President's scientific policy questioned. *Washington Post,* Section A p2.

Hansen, G. (2000). Assessment and Management of Pain: Letters. *The Journal of the American Medical Association.* 284: 2317-2318.

http://www.americancatholic.org/news/StemCell/default.asp

Hyman, C. S. (1996). Pain management and disciplinary action: How medical boards can remove barriers to effective treatment. *Journal of Law, Medicine & Ethics.* 24: 338-43.

InfoPOEM, Inc. http://www.InfoPOEMS.com

Joint Commission on Accreditation of Healthcare Organization. August, 1999. Pain assessment and management standards. [Online]. Available: *http://www.jcaho.org/standards/pm*

Kaufman, M. (2003, Dec. 12). Worried Pain Doctors Decry Prosecutions. *Washington Post.* Section A p.1,4.

_____ . (2004, February 9). Linking Lawmakers, Scientific Knowledge. *Washington Post,* Section A, p.19.

Kinsley, M. (2003, October 17). Bush's Filtered News. *Washington Post,* Section A, p. 29.

_____ . (2003, November 31). Kabuki and Stem Cells. *Washington Post,* Section A, p25.

Knight, J. (2002, February 14), American Red Cross turns it's back on stem-cell grant. (News). *Nature.* Volume 415: pp. 719.

Long, Stephen. (Director of Pain Management Center) (2001, October 16) *Steve Long Interview* Richmond, Virginia.

Max, M. B. (1990) Improving Outcomes of Analgesic Treatment: Is Education Enough?" *Annals of Internal Medicine,* 113. 885-888.

McKay, R. (2000, July 27). Stem cells – hype or hope. *Nature, 406,* pp.361-364

Morris, D. B. (1991). *The Culture of Pain.* Berkeley: University of California Press.

Novack, R. D. (2003, December 15). New Jersey's End Run On Cloning. *Washington Post.* A31.

PBS Online NewsHour Update. (2002, December 11, 4:45pm EST) *Stanford University to create new stem cell lines.*

Orentlicher, D. (2000). Legislation and End of Life Care. *Journal of the American Medical Association.* 283:2933-5.

Orentlicher, D., Caplin, A. (2000). The Pain Relief Promotion Act of 1999: A serious threat to palliative care. *Journal of the American Medical Association.* 283:255-258.

Philipkoski K. (2001, Aug. 9, 1:40 pm). Bush OK's stem cell funding. *Wired News.* (http://search.wired.com/wnews/default.asp?query=Bush%20OKs%20Stem%20Cell%20 Funding)

Post, L. F., Blustein, J., Gordon, E. & Dubler, N. (1996). Pain: Ethics, culture, and informed consent to relief. *Journal of Law, Medicine & Ethics.* 24: 348-59.

Reuters Health. (2000, Jan.17). Medical ethicists claim Pain Relief Promotion Act of 1999 is bad nedicine. *Medscape.com* http://primarycare.medscape.com/reuters/2000

Rowlingson. J. (Professor of Anesthesiology and Director of Pain Management Services at the University of Virginia), (2001, March 14) *John Rowlingson Interview.* University of Virginia. Charlottesville VA.

Weissman, I.L. (2002). Stem Cells – Scientific, medical, and political issues. *New England Journal of Medicine, 20,* Vol. 346: 1576-1579.

Weiss, R. (2003, August 8). Bush misuses science, report says: Democrats say data are distorted to boost conservative policies. *Washington Post,* Section A, p. 15.

_____ . (2003, October 13). Stem cell research is questioned: Bone marrow's potential overstated, study shows. *Washington Post,* Section A, p. 6.

_____ . (2003, November 17). Funding Bill gets Clause on Embryo Patents: ban Inserted into appropriations measure Stirs Political, Philosophical Debate. *Washington Post.* A4.

_____ . (2003, November 25). Hill Negotiators Agree to Bar Patents for Human Organisms. *Washington Post.* A19.

_____ . (2004, January 15). Peer review plan draws Criticism. *Washington Post,* Section A p19.

_____ . (2004, January 16). Bioethics Panel Calls for Ban on Radical Reproductive Procedures. *Washington Post,* Section A p. 2.

The White House, Release from the Office of the Press Secretary. (2001, August 9, 8:01pm). Remarks by the president on stem-cell research. *http://whitehouse.gov/news/releases /2001/08.*

Index

A

B

C

D

E

F

G

H

I

J

K

L

M

N

O

Q

R

S

T